GENERAL ENERGETICS

GENERAL ENERGETICS

Energy in the Biosphere and Civilization

VACLAV SMIL

University of Manitoba
Winnipeg, Manitoba
Canada

A WILEY INTERSCIENCE PUBLICATION

JOHN WILEY & SONS

New York · **Chichester** · **Brisbane** · **Toronto** · **Singapore**

Library of Congress Cataloging in Publication Data:
Smil, Vaclav.
    General energetics: energy in the biosphere and civilization/
Vaclav Smil.
        p.   cm. — (Environmental science and technology)
    Includes bibliographical references and index.
    ISBN 0-471-62905-7
    1. Bioenergetics.   2. Energy budget (Geophysics)   I. Title.
II. Series.
QH510.S64   1991
574.5'22 — dc20   90-42106
                                             CIP
Printed in the United States of America

10 9 8 7 6 5 4 3 2 1

SERIES PREFACE
Environmental Science and Technology

The Environmental Science and Technology Series of Monographs, Textbooks, and Advances is devoted to the study of the quality of the environment and to the technology of its conservation. Environmental science therefore relates to the chemical, physical, and biological changes in the environment through contamination or modification, to the physical nature and biological behavior of air, water, soil, food, and waste as they are affected by man's agricultural, industrial, and social activities, and to the application of science and technology to the control and improvement of environmental quality.

The deterioration of environmental quality, which began when man first collected into villages and utilized fire, has existed as a serious problem under the ever-increasing impacts of exponentially increasing population and of industrializing society. Environmental contamination of air, water, soil, and food has become a threat to the continued existence of many plant and animal communities of the ecosystem and may ultimately threaten the very survival of the human race.

It seems clear that if we are to preserve for future generations some semblance of the biological order of the world of the past and hope to improve on the deteriorating standards of urban public health, environmental science and technology must quickly come to play a dominant role in designing our social and industrial structure for tomorrow. Scientifically rigorous criteria of environmental quality must be developed. Based in part on these criteria, realistic standards must be established and our technological progress must be tailored to meet them. It is obvious that civilization will continue to require increasing amounts of fuel, transportation, industrial chemicals, fertilizers, pesticides, and countless other products; and that it will continue to produce waste products of all descriptions. What is urgently needed is a total systems approach to modern civilization through which the pooled talents of scientists and engineers, in cooperation with social scientists and the medical profession, can be focused on the development of order and equilibrium in the presently disparate segments of the human environment. Most of the skills and tools that are needed are already in existence. We surely have a right to hope a technology that has created such manifold environment problems is also capable of solving them. It is our hope that this Series in Environmental Sciences and Technology will not only serve to make this challenge more explicit to the established professionals, but that is also will help to stimulate the student toward the career opportunities in this vital area.

Robert L. Metcalf
Werner Stumm

PREFACE

Energy is the only life and is
from the Body;
and Reason is the bound or outward
circumference of Energy.

Energy is Eternal Delight.
—William Blake
The Marriage of Heaven and Hell

Thus does the Devil define energy as he corrects what he feels to be one of the principal errors of sacred codes:

That Energy, call'd Evil, is alone from the Body;
& that Reason, call'd Good, is alone from the Soul.

This book is preoccupied with more pragmatic and less contentious propositions, but it too owes its existence to fascination with energy's elusive qualities.

The allure of energy—its unifying, underlying, omnipresent, multi-faceted, indeed transcendental nature—has been reflected for millennia in the feelings of awe and reverence connected with fire, light, heat, growth, and motion, and for several centuries in intensifying scientific and engineering efforts seeking to unlock the fundamental laws of nature and to expand civilization's mastery of energy flows. As a result, there exists a profusion of revealing writings on many aspects of energetics, most notably on the fundamentals of thermo-dynamics and on the principles of bioenergetics. In contrast, compre-hensive coverage of energetics on the organismic and systemic level—embracing the enormously heterogeneous biospheric energy storages and flows, as well as civilization's energy sources, conversions, and uses—has been surprisingly rare. The explosion of interest in energy studies since the early 1970s has changed little in this regard, as even the broader, integrative works have a clear contextual ("energy crisis") bias, focusing overwhelmingly on supply of fossil fuels, generation of electricity, and alternative ways of satisfying future demand.

Few volumes have surpassed this traditional scope to include, in varying proportions, appraisals of ecosystemic and preindustrial energetics as well as attempts to establish stronger links between physical and social facets of the inquiry; yet even these syntheses overlook large segments of energetics, ranging from an appreciation of geoenergetics and complexities of heterotrophic existence to puzzles of human nutritional requirements and interference in grand biospheric cycles.

Consequently, it seemed to me that there is a place for a book that would embrace all essential energy storages, flows, and conversions in a unified and systematic manner, a book that would combine an encyclopedic sweep and richness of detail with an evolutionary overview and analytical crispness and would not shy away from grand synthesizing generalizations or from acknowledging the inability to offer such statements—a book whose titular adjective would refer both to a comprehensive scope and to a unified, synthesizing coverage.

As good a reason as that may be to write a book, I do not feel that filling a gap in the existing scientific coverage is enough to justify the project. The strongest commonality of the wide range of my energetic interests during the past twenty-five years has been an incorrigible fascination with unruly and fuzzy realities in preference to abstract models and dubious generalizations. Complexities and peculiarities of the real world and counterintuitive outcomes of many of its processes have seemed to me always more appealing than the indubitable elegance and allure of theoretical models.

This preference requires a from-the-ground-up approach where gradual appreciation of details and cumulative acquisition of the widest possible scope of information precede any attempts at generalization. And such effort, I believe, is an irreplaceable foundation for rational decision making required to manage the increasingly complex affairs of modern civilization.

The late 1960s and the 1970s marked the emergence of big interdisciplinary and international concerns requiring such an approach: environmental pollution, energy supply, food security, quality of life, economic cooperation. During the 1980s the scope of these concerns rapidly progressed to the global level—and planetary management is becoming inevitable if we want to prosper as a species during the next century and beyond. An integrated approach is imperative to appreciate both the unities and the peculiarities of the complex whole: no sensible global management can be based on increasingly specialized but almost irretrievably splintered knowledge.

Advances of modern civilization can be directly related to higher availability of useful energy, but no true understanding of energetics can start or end with fossil fuels and electricity. Evolution of the biosphere and the scope of human action are circumscribed by grand planetary radiation fluxes; virtually all heterotrophic life is an energetic extension of photosynthesizing plants; recent advances of our species should be seen as the latest achievements of a long-lasting quest for mastering larger, and more concentrated, flows of energy; and the effects of this accelerating quest have not only remolded our economies and our social organization; they are also degrading air, waters, land, and biota on local and regional scales and they may be altering the global environment in a number of undesirable—even perilous—ways.

One of the key approaches to understanding this grand unity made up of countless peculiarities is to focus on its actual prime movers and to study the variety of energy flows governing the physical make-up of this planet, its life forms, and its civilizations. This yin–yang book attempts to do just that. Its aims are broad interdisciplinary coverage—and the richness of detail; clear and specific analyses—and syntheses rooted in the presented information; commonsense generalizations—and repeated attention to their qualifications and limits.

The means are a systematic and evolutionary coverage, but one whose boundaries and flavors are also clearly influenced by personal experiences, preferences, and fascinations. This book takes just one of many possible routes to a better understanding of planetary and civilizational energetics; it cannot be devoid of lapses and errors.

And if it falls far short of its ambitious mark, there is, as always, consolation in the wisdom of an ancient sage. Lao Zi, noting that it is the void space which makes bowls and houses useful, wrote:

> So advantage is had
> From whatever is there;
> But usefulness rises
> From whatever is not.

VACLAV SMIL

Winnipeg, Manitoba, Canada
August 1990

CONTENTS

Contents   **xiii**

1

THE UNIVERSAL LINK: ENERGETICS AND ENERGY

And the things best to know are first principles and causes. For through them and from them all other things may be known but not they through the things covered by them.... But these things, the most universal, are perhaps the most difficult things for men to grasp, for they are farthest removed from the senses.

—Aristotle
Metaphysics

Ari. vs. induction ?

Aristotelian preference for the knowledge of what is most universal finds the largest realm of inquiry in energetics, the study of transformations creating the inanimate universe and sustaining life on Earth. Nothing needs to be excluded from this realm: any process can be rewardingly analyzed in terms of its underlying energy conversions; any object, as well as any bit of information, can be valued for its incorporated energy content and for its potential contribution to future energy transformations.

But underlying energetic commonalities are only a part of the fascination of energy studies. A no less exciting endeavor is to uncover, explain, and compare countless specific expressions of these commonalities throughout the enormous heterogeneity of the world we live in; long-lasting fascination with this unified heterogeneity led me to write this book.

Before proceeding with systematic segmental analyses, reductions to common denominators, and synthesizing evaluations it is essential to appreciate the milestones of the study of energetics and to soundly understand basic variables and generalizing approaches used in these inquiries.

1.1 EVOLUTION OF ENERGETICS

Studies of energy phenomena have always straddled many disciplines: sometimes in isolation, often in parallel, later with profitable transfers and fusing of ideas—and since the closing decades of the nineteenth century with increasingly frequent attempts at finding general patterns and offering grand syntheses (Mach 1896; Stallo 1900; Brody 1945; Cardwell 1971; Martinez-Alier 1987). Many outstanding creators of classical and modern science were students of energetics, but this fundamental yet diffuse field of inquiry—spanning physical, life, and social sciences—has never acquired the clear identity of chemistry or anthropology.

The first isolated brushstrokes of this vast canvas were put down long before those decades of astonishing intellectual ferment at the close of eighteenth century. Etymology of the word *energy*—from the Greek *en* (in) and *ergon* (work)—reveals as much as it hides. There is no record of it before Aristotle (Düring 1966), who gave it a primarily kinetic meaning: "The word 'actuality' (*energeia*) which we connect with 'complete reality' (*entelechia*), has, strictly speaking, been extended from movements to other things; for actuality in the strict sense is identified with movement" (*Metaphysics*, Theta 3).

The word and its cognates filled a much larger conceptual niche for Greeks than for modern scientific usage (Lindsay 1974). In *Ethics* *energeia* stands in opposition to mere disposition, *hexis*; in *Rhetoric* it carries the vigor of the style. The verb *energein* meant to be in action, implying ceaseless motion, work, production, change. The classical concept of *energeia* was a philosophical generalization, an intuition embracing the totality of transitory processes, the shift from the potential to the actual. The perception was clearly holistic and qualitative.

Roman civilization, Islam, dynastic China, and medieval Europe advanced many ingenious solutions to everyday energetic challenges, but systematic harvest of the underlying understanding was poor, and even many founders of modern science did little to extend it. To Galileo heat was an illusion of the senses, an outcome of mental alchemies; Bacon thought that heat could not generate motion or motion heat; and Newtonian mass, momentum, and force had little pressure relevance for men experimenting with steam engines or pondering the energetic basis of living organisms. But these manual and mental experiments laid the foundations of energetics: scientific understanding has been perfected on the basis of particularistic definitions and quantitative assessments. Only later came the profusion of laws, conjectures, theories, and expectations as the burgeoning nineteenth-century science pursued its multifaceted inquiries.

Practical roots of this knowledge are evident in James Watt's (1736–1819) contributions. His steam engine revolutionized industrial production while his invention of a simple indicator, a recording steam gauge, opened the way for detailed studies of engine cycles, which contributed immeasurably to the emergence of thermodynamics during the following century. But Sadi Carnot (1796–1832) most likely never heard of Watt's indicator; his was a purely abstract approach and this explains why his perceptive contributions were fully appreciated only decades later.

Carnot attacked the ways of producing kinetic energy from heat, a conversion "considered independently of any mechanism or any particular agent" in order to set down the principles applicable "to all imaginable heat engines, whatever the working substance and whatever the method by which it is operated" (Carnot 1824). French science made other critical early contributions. Antoine Lavoisier's (1743–1794) suggestion of the equivalence between heat output of animals and men and their feed and food intake was the necessary foundation for the subsequent studies of heterotrophic metabolism.

The writings of Dumas and Boussingault (1844) were immediately

influential, and their explanations of grand biospheric cycles underpinning the planetary bioenergetics were essentially correct. At the
same time Justus Liebig (1803–1873) was the first scientist to ascribe
the generation of carbon dioxide and water to food oxidation (Liebig
1843), offering a basically correct view of heterotrophic metabolism.
He also introduced the powerful concept of limiting nutrients applicable equally well to autotrophs and heterotrophs. And during the
1850s Edward Smith (1857) collected about 1200 samples of expired
air, clearly establishing that the extent of respiration must be
regarded as a function of one variable, the ingested food, and one
constant, a distinct individual metabolic rate.

During the 1840s advances in understanding human energy needs,
conversions, and expenditures were closely associated with the
genesis and progress of general energetic principles in the work of
Julius Robert Mayer (1814–1878), a German physician who made the
famous observation of brighter venous blood in persons living in the
tropics, which led him to a correct assessment of the balance between
food intake and oxygen consumption. Mayer saw muscles as heat
engines energized by oxidation of the food and offered calculations
proving the sufficiency of food's chemical energy to supply the
mechanical energy necessary for work as well as to maintain the
homeothermy.

All of this led him to estimate that mammals can work with
efficiency of about 20% and, much more important, to establish the
mechanical equivalent of heat and thus to formulate the law of
conservation of energy (Mayer 1851), commonly known as the first
law of thermodynamics. Independent, and more accurate quantification of the equivalence of work and heat came from an English
physicist, James Prescott Joule (1818–1889), whose first publication
put the conversion rate at 838 foot-pounds and a later revision at 772
foot-pounds (Joule 1850), a difference of less than 1% from the
actual value. The third independent formulation of the law was the
work of a German physiologist, Hermann von Helmholtz (1821–
1894), who, like Mayer, wrote about forces rather than energies
(Helmholtz 1847).

Soon afterward William Thomson (Lord Kelvin, 1824–1907) identified the Sun as the principal source of kinetic energies available to
man and wrote about "a universal tendency in nature to the dissipation of mechanical energy" (Thomson 1853). Rudolf Clausius (1822–
1888), one of Germany's most influential physicists, sharpened this
understanding by showing that heat energy at low temperature is the
outcome of these dissipations, by expressing the process in quantita-

tive terms, and by naming the transformational content entropy, a term derived from the Greek *trope* for transformation. As the energy content of the universe is fixed, but its distribution is uneven, its conversions seek uniform distribution and the entropy of the universe tends to maximum (Clausius 1867).

This second law of thermodynamics—the universal tendency toward atropy, heat death, and disorder—became perhaps the most influential and most frequently misunderstood cosmic generalization. Only at absolute zero (–272°C) is the entropy nil: this third law completes the set of thermodynamic fundamentals. Josiah Willard Gibbs (1839–1903), an American physicist, applied thermodynamic concepts to chemistry and introduced the important notion of free energy. This energy actually available for doing work is determined by subtracting the product of temperature and entropy change from the total energy entering a chemical reaction.

The second law exercised a powerful influence on scientists thinking about energetic foundations of civilization. Edward Sacher (1834–1903), a little known Austrian science teacher, viewed economies as systems for winning the greatest possible amount of energy from nature and tried to correlate stages of cultural progress with per capita access to fuels (Sacher 1881). Contrast of rising fuel demands and inexorable thermodynamic losses led to anxious calls for energy conservation. Wilhelm Ostwald (1853–1932), a Nobel laureate in chemistry, formulated his energetic imperative, an admonition to waste no energy but to value it—"Vergeude keine Energie, verwerte Sie!" (Ostwald 1909)—as mankind makes the inevitable transition to a permanent economy based on solar radiation.

Another Nobel laureate in chemistry, Frederick Soddy (1877–1956), inquired into "how long the natural resources of energy of the globe will hold out," was the first to make the frequently quoted distinction between utilizing natural energy flows and fossil fuels—"The one is like spending the interest on a legacy, and the other is like spending the legacy itself"—and anticipated "a period of reflection in which awkward interviews between civilization and its banker are in prospect" (Soddy 1912).

The twentieth century brought a fundamental extension of the first law with Albert Einstein's follow-up of his famous relativity paper (1905). Soon after its publication Einstein, writing to a friend, realized "that the principle of relativity in conjunction with Maxwell's fundamental equations requires that the mass of a body is a direct measure of its energy content—that light transfers mass" (cited in Miller 1981). During the next two years Einstein formalized this "amusing and attractive thought" in a series of papers firmly establishing the equivalence of mass and energy.

In the last of these papers he described a system behaving like a material point with mass

$$M_0 = \mu + E_0/c^2$$

and noted that this "result is of extraordinary importance because a physical system's inertial mass and energy content appear to be the same thing. An inertial mass μ is equivalent with an energy content μc^2" (Einstein 1907). The equation requires that any mass emitting energy must be diminished, a fact of no practical importance in chemical reactions.

For example, combustion of 1 kg of hard coal (requiring 3 kg of O_2 and releasing about 30 MJ of energy) will diminish the mass of the two reactants by about 10^{-10}, a reduction too small to measure. In contrast, in nuclear reactions the reduction is obvious. Einstein was aware of this difference, noting that the practical demonstration of the law was difficult but writing that "for radioactive decay the quantity of free energy becomes enormous." Fissioning of 1 kg of U^{235} releases about 8.2 TJ of energy—about 2.7×10^5 more than the same mass of coal—and diminishes the uranium mass by 1 g (loss of 10^{-3} of the initial mass).

The last decades of the nineteenth and the first half of the twentieth century also brought great gains in bioenergetics. Experiments performed by von Pettenkofer, Voit, Lusk, and Atwater determined fairly accurate energy balances of living organisms so that Max Rubner (1902) was able to offer a systematic account of human metabolism including all the essentials of modern understanding. Max Kleiber uncovered the allometric relationship between the basal metabolic rate and the body mass of heterotrophs (Kleiber 1932), a beginning of fascinating studies of energetic scaling. And Samuel Brody published his monumental synthesis of bioenergetics (Brody 1945).

An important theoretical step forward came when Alfred Lotka, American mathematician and biologist (1880–1949), formulated a law of maximum energy. For biota it is not the highest conversion efficiency but the greatest flux of useful energy that is maximum power output, which is most important for growth, reproduction, and species radiation. Consequently, living organisms do not convert energy with the highest supportable efficiencies—but rather at rates optimized for the maximum power output (Lotka 1925).

During the 1930s came rapid advances in understanding nuclear energy including the discovery of the neutron, the first laboratory demonstration of fission, and the first correct explanation of energy processes in stars (Bethe 1939). At the same time a fundamental change started to affect the scientific method. After centuries of progressive compartmentalization came a gradual formulation of general system approaches based on recognition of underlying commonalities and interactive complexities. Russian ecologist Vladimir Ivanovich Vernadsky, with his powerful idea of the biosphere (Vernadsky 1929), and German biologist Ludwig von Bertalanffy, with his systematic theoretical look at biology (Bertalanffy 1932–1942), were the pioneers of the approach in life sciences. And Tansley's (1935) ecosystem enriched bioscience by one of its most important unifying concepts.

Just before the end of the Second World War Schrödinger (1944) attempted to solve the dilemma of living systems which are creating and maintaining exquisite order from disordered elements dominated by carbon. He explained this puzzle by introducing the idea of "non-equilibrium thermodynamics": organisms create the improbable state of order by importing free energy from outside, and processing it to generate a lower entropy state within. They thrive on negentropy. These ideas, often embedded in the newly flourishing general system studies, opened up the still lively debate about the behavior of open systems ranging from cells to civilizations, and about the thermodynamic fundamentals of bioenergetics (Prigogine 1947; Bertalanffy 1968; Morowitz 1968; Brooks and Wiley 1986; Weber et al. 1988).

But outside of bioenergetics there has been relatively little interest in general energy studies. The best explanation of this neglect lies in the increasing abundance and decreasing real cost of principal commercial energies between 1945 and 1973; in those circumstances there was little need to heed Ostwald's energetic imperative or to do complex energy studies. Works by Ubbelohde (1954), Cottrell (1955), Thirring (1958), Hubbert (1962), and Odum (1971) were the exceptions. Hubbert's innovative analysis of the cycle of mineral production became influential when it correctly predicted the peak of crude oil production in the contiguous United States. Odum's book contributed an integrative ecoenergetic approach to environment, power, and society.

The two energy "crises" (1973–1974 and 1979–1981) gave rise to an unprecedented interest in energy affairs. Yet most of this attention was still very particularistic, too beholden to the perceptions of the day, preoccupied with immediate (and improperly interpreted) problems, and hence highly error-prone in its ignorance of broader settings and implications.

But our understanding of other fundamental segments of energetics raced ahead with new syntheses ranging from delineation of historic patterns of energetic innovation (Needham 1954–; White 1978) to accounts of primary biospheric productivity (Lieth and Whittaker 1975). This multifaceted interest continued in the 1980s with publication of surveys and syntheses ranging from geoenergetics (Verhoogen 1980) to improvements in conversion efficiency (Gibbons and Blair 1989), and from animal scaling (Schmidt-Nielsen 1984) to new transportation fuels (Sperling 1988).

This enormous heterogeneity of energy studies is fascinating but it makes synthesis difficult. There is a need for defining and imposing a manageable set of key variables which are applicable across the vast field of general energetics. Only such systematic quantification can lead to meaningful attempts at grand generalizations, which almost always must be modified by qualitative caveats. Appreciation of basic concepts and levels of energy and power and of complexities inherent in conversion and efficiency calculations including the key variables (Appendix) are essential before embarking on systematic inquiries in energetics.

1.2 APPROACHES TO UNDERSTANDING

Primary analytical tools of this book belong to two essential classes of measurements which are applicable to all phenomena: no matter if the processes are animate or inanimate, natural or anthropogenic, they can be profitably studied in terms of energy and power densities and intensities. These rates relate energy and power (including time) to space and mass; all fundamental existential variables can be thus embraced in a few simple yet powerful measures.

Energy density (J/m^2) reveals the concentration of resources, a critical determinant of extraction or harnessing methods and costs and of associated infrastructural needs. When applied to natural ecosystems the measure is a fine surrogate for species diversity, in heavily managed ecosystems it informs about harvest possibilities, in urban and industrial areas it expresses the levels of intensification either in terms of habitation densities or as energy incorporated in structures and infrastructures—and in all instances it reminds us of inexorable limits to future growth.

In this sense *power density* (W/m^2), with its built-in time reference, is an even more valuable measure; in many ways it is the most important variable in energetics. This measure is of fundamental physical importance in determining the maximum performance of individual energy converters in all cases where the transformation of energy can be envisaged as proceeding within a certain volume with one form of energy supplied into this volume across its surface and the transformed energy leaving that surface.

The flux of energy through the working surface of such converters is the product of the velocity of propagation of a disturbance (be it a kinetic or a heat flow) and its energy density (measured in J/m^3), $\frac{J}{m^3}$ which is limited by the physical properties of the working medium. The resulting product, sometimes called the Umov–Poynting vector, was favored by Peter Kapitza—Rutherford's student and an eminent Russian theoretical physicist—as both fundamental and a convenient characteristic of energy transformers (Kapitza 1976).

Where combustion is involved the vector is the product of gas pressure, the square root of temperature, and a constant dependent on the molecular composition of the gas. Reliance on the vector also illustrates the often irrelevant nature of high conversion efficiencies. Fuel cells may be directly transforming chemical energy into electricity with high efficiencies, but low diffusion rates in electrolytes limit the power density to some 200 W/m^2 of the electrode, too low ever to become a centralized base-load supplier for modern high-energy society.

The second use of power densities is for the rates of energy flow per square meter of surface area (rather than per square meter of the working surface of a converter). This broader measure is perhaps the most critical parameter determining both the structure and the operating modes of energy conversion systems (Häfele and Sassin 1977; Smil 1984). There will be systematic comparisons of power densities for all important natural and anthropogenic energy conversions.

Power intensity (W/g) is used with autotrophic and heterotrophic metabolism as well as with inanimate energy transformations; when looking at the performance of prime movers the measure is reversed (g/W) to emphasize the importance of lightweight conversion devices in transportation. *Energy intensity* (J/g or MJ/kg) or *energy cost* of anthropogenic conversions informs about energy invested in extraction, production, and distribution of goods and in provision of services, and about effective substitutions. *Energy intensity of energy* (J/J or simply a fraction of gross energy content) appraises the net energy gain of fuels and electricity; differences in *energy intensities of economies* (J/constant monies) illustrate the variability of national performances. And *metabolic intensities* (g/J) call attention to material requirements of energy fluxes and to their environmental impacts.

Using these general measures, and augmenting them by more specific quantifications, this book proceeds in an evolutionary sequence. Chapter 2 examines terrestrial energy fluxes, starting with solar radiation and concluding with the energetics of geomorphic processes. This sets the stage for reviewing both the bioenergetic fundamentals and various specific and ecosystemic peculiarities of plant and animal kingdoms. Chapters on primary productivity and heterotrophic conversions are followed by a closer look at human energetics, stressing surprising uncertainties of our knowledge.

From this point on the book concentrates on the energetic limits and exploits of our species. Brief notes on the energetics of the simplest human societies are followed by analyses of traditional farming and preindustrial complexification based on animate power and biomass fuels. Energetics of modern civilization is covered by focusing first on its resource foundation and key conversion techniques—that is, by appraising stores and combustion of fossil fuels, generation of electricity, and development of inanimate prime movers. Then the perspective widens to embrace general patterns and trends of energy use in modern societies and to survey numerous environmental implications and fascinating socioeconomic complexities accompanying this high-energy way of life.

The book closes with a juxtaposition: grand energetic generalizations (energy as the measure of all things) concerning planetary flows, maintenance and evolution of life, and advancement of civilization are contrasted with inadequacies inherent in approaches which—fundamental and universal as they are—are still insufficient to be the measures of all things.

2

PLANETARY ENERGETICS: ATMOSPHERE, HYDROSPHERE, LITHOSPHERE

Wherefore again and again the earth deserves the name of mother which she has gotten, since of herself she created the human race ...
—Lucretius
De rerum natura

Compared to other terrestrial planets—Mercury, Venus, and Mars—Earth is unmistakably unique. Its atmosphere has an improbable composition: instead of the expected 98% of carbon dioxide there is the dominance of molecular nitrogen and oxygen and presence of trace gases (CH_4, N_2O, NH_3) whose concentrations violate the rules of equilibrium chemistry to an infinite extent (Lovelock 1979). Its average surface temperature is not the expected 500 K (227°C) and the surface pressure is not a few megapascals. During the past 3 billion years its surface temperature has remained between 0 and 20°C and the mean surface pressure is just 101.325 kPa (1 atm). These conditions ensure the continuous presence of liquid water and the evolution of complex life.

This uniqueness does not end at the surface as the internal differences are no less profound (Head and Solomon 1981). The Earth's lithosphere is rather rapidly recycled into the underlying mantle as the incessant motions of giant plates open up and close oceans, form great mountain belts, and generate earthquakes and volcanic eruptions. The uniqueness of these tectonic arrangements results from the existence of an internal heat "engine" in the mantle and from the lateral rigidity of the lithospheric crust. Plates with horizontal dimensions of megameters are pushed away from mid-ocean ridges as distinct units to be subducted at deep-sea trenches.

The Earth's unexpected atmosphere is almost certainly a creation of planetary life whose rise and complexification have been energized by solar radiation. Although only a very small part of the huge solar flux is converted by photosynthesis into phytomass that sustains all heterotrophic species, solar radiation absorbed by atmosphere and terrestrial surfaces plays no less essential a role in maintaining environments suitable for perpetuation and evolution of life.

Without surface temperatures high enough to keep water liquid (at least for a part of the year), to evaporate it in large quantities in order to set in motion large-scale redistribution of heat and moisture, and to allow for rapid rates of biochemical reactions there would be no excitation of chlorophyll, no rains, no absorption of nutrients through roots, no decomposition of dead organic matter. Without the tectonic processes driven by terrestrial heat there would be no re-creation of continents and oceans, and hence a vastly reduced differentiation of landforms and seafloor, which would have a profound effect on the evolution of life.

Consequently, the following accounts of planetary energy balances and of numerous thermal and kinetic energy flows in the atmosphere, hydrosphere, and lithosphere set the grand stage for all of the subsequent inquiries into the energetics of life and civilization.

2.1 SOLAR RADIATION

Not everything on Earth owes its existence to solar radiation: it does not determine the rate of geotectonic processes nor does it energize chemoautotrophic bacteria, notably those in the deep ocean darkness near hot vents where they oxidize hydrogen sulfide and support larger organisms, including white clams, crabs, and giant tube worms (Macdonald and Luyendyk 1981). Otherwise the whole pyramid of life rests on photosynthetic conversion of solar radiation to phytomass while the features of the planetary surface have been molded by waters and winds driven by solar energy.

The source of the radiation is a star with magnitude of 4.48, spectral type G, mass of 1.991×10^{33} g, radius of the visible disk 696 Mm, average density 1.41 g/cm^3, surface temperature about 5800 K—just one of similar millions among the 100 billion stars of our galaxy (Friedman 1986). About 5 billion years old, it still has some 5 billion years to go before transforming itself into a red giant whose expanding photosphere will obliterate any evidence of this planet and its temporary civilizational achievements. The Sun's energy is produced within the innermost quarter of its radius, enclosing a mere 1.6% of its volume. There, with temperature at about 15 MK and density of 158 g/cm^3, thermonuclear fusion of hydrogen into helium consumes 4.3 Mt of matter every second, releasing 3.89×10^{26} J. Assuming isotropic radiation, 64 MW are passing through every square meter of the photosphere.

When this radiation reaches the Earth's orbit its approximate power density will be the quotient of total luminosity (3.89×10^{26} W) and the area of a sphere with the orbital radius (average 150 Gm), that is, about 1370 W/m^2. This flux is known as the solar constant, a misnomer as it has gradually increased with the aging of the star, and it also has short-term fluctuations. These were accurately measured only after the launching of the Solar Maximum Mission (SMM) satellite in February 1980. Linear regressions of daily means for the first five years (Fröhlich 1987) showed maximum periodical dips on the order of 0.2% and revealed a decrease of 0.018%/year but the long-term pattern is almost certainly one of oscillation.

There is a paradoxical contrast between the remarkable constancy

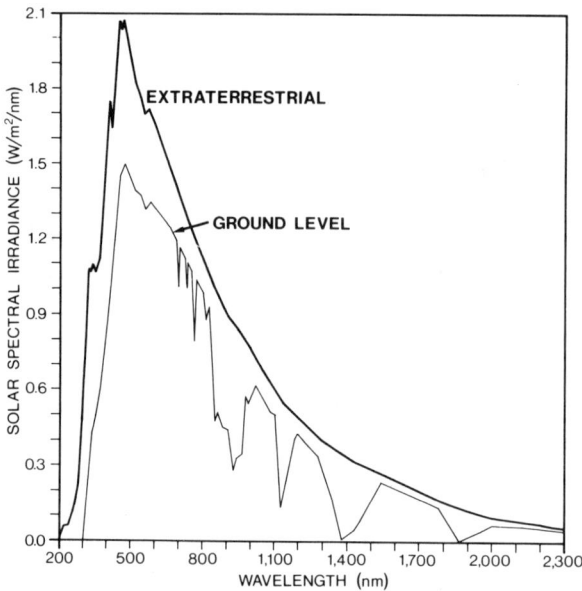

Figure 2.1 Extraterrestrial spectrum of solar radiation closely resembling that of a black body radiating at 6000 K, and the toothy-looking spectrum of solar radiation received at the ground plotted from data in Thekaekara (1977).

of climate over the 3 billion years of life complexification and the roughly 30% increase of solar luminosity (Crowley 1983). Can it be that the Earth's climate is an almost intransitive system (Lorenz 1976) capable of amazing adjustments to maintain quasi stability over very long time spans? Such a reality would be in accord with Lovelock's (1979) Gaia theory.

For all practical terrestrial purposes we may safely return to the total irradiance as a constant. By 1987 its value was almost exactly 1367 W/m^2. The spectrum of this flux extends from 10^{-8} to more than 10^3 cm, and its curve closely resembles that of a black body radiating near 6000 K (Fig. 2.1). The wavelength of the maximum emission is almost exactly 500 nm, near the lower limit of green light (491–575 nm). Division of this spectrum into ultraviolet, visible, and infrared is based on its effects on biota. Ultraviolet (up to 400 nm) carries only a bit less than 9% of the total flux; its highest frequencies are lethal to most organisms (but thresholds vary enormously), the lower ones are germicidal and skin-burning.

Visible light (from 400 nm of the deepest violet to 700 nm of the darkest red) carries about 38% of solar energy. Our eyes are most sensitive to green (491–575 nm) and yellow (575–585 nm), and maximum visibility is at 556 nm (3.58×10^{-19} J/photon). Infrared radiation carries 53% of solar energy and most of its heat comes from wavelengths shorter than 2 μm carrying less than 1×10^{-20} J/photon.

Ellipticity of the Earth's orbit accounts for seasonal fluctuations of ±3.3% in the total radiation, and the daily means of irradiance incident on a horizontal surface at the top of the atmosphere are also a function of the latitude. Although the disk with the Earth's diameter (12.74 Mm) intercepts only a negligible fraction of total solar output (4.5×10^{-10}), the rate of this intercept, 174.26 PW, and its annual aggregate, 5.495×10^{24} J, are enormous: in 1990 global consumption of fossil fuels was nearly 10 TW, or 0.005% of the solar irradiance. Total resources of fossil fuels are perhaps as large as 1.7×10^{23} J—but that would be no more energy than in the solar intercept in fewer than 11 days!

Even without any atmospheric interference the average insolation per square meter of the planet's spherical surface would be only one-fourth of the intercept value for the disk of the same radius, or about 340 W/m^2. And because the Earth's clouds and surfaces reflect about 30% of incoming radiation, the total actually absorbed by the atmosphere and by the ground is no greater than 240 W/m^2, or some 122 PW (3.85×10^{24} J); this flux is the actual energizer of the Earth. The incoming radiation is considerably weakened and altered during its passage through the atmosphere: the spectrum of the solar radiation reaching the biosphere differs substantially from the extraterrestrial curve as the absorption by a variety of atmospheric molecules and particles produces a generally depleted and toothy-looking profile (Fig. 2.1).

2.2 RADIATION FLUXES

Incoming radiation undergoes its first major atmospheric modification in the stratosphere, whose ozone (O_3) is a strong UV absorber. Troposphere and oceans are decisive in determining the global radiation balance but only a thin layer (up to 1 m) of the lithosphere is involved in daily temperature variations, the annual ones extend mostly to less than 20 m, and even the ice ages affect just the topmost 1–2 km, the merest pellicle compared to the average thickness of the crust, whose thermal regime is governed from below.

Modern knowledge of planetary heat budgets has been assessed by Budyko et al. (1963) and Kessler (1985), advances in global radiation balance observations from satellites are reviewed in Gautier and Fieux (1984) and Houghton (1984), and a wealth of detail concerning solar energy flows in the biosphere has been gathered and evaluated by Gates (1980), Miller (1981), and Rosenberg et al. (1983). Visible light passes largely unaffected through the clean troposphere but water and CO_2 as well as CH_4, N_2O, and, again, O_3 absorb much of the infrared flux; the total absorption profile shows two peaks in the near infrared and three extended blocks between 1.5 and 2, 2.5 and 3.5, and 4 and 8 μm.

Absorption can eliminate 11–23% of incoming radiation, and scattering can return 1–11% and send 5–15% of the beam downward as diffuse sunlight, leaving 56–83% in the direct beam. Total mid-day peak irradiances under cloudless sky in middle latitudes would then be 970–1203 W/m^2. Cumulonimbus clouds reflect 90% and absorb nearly all the rest of incident radiation; wispy cirrus may let 90% of the light through.

A convenient expression of the prevalence of cloudiness is the percentage of possible sunshine, ranging seasonally from less than 10% in the world's cloudiest regions (northern Pacific between the the Aleutians and Kamchatka, the Atlantic between Newfoundland and Greenland) to around 90% in subtropical deserts (peaks in Chile's Atacama, southern Egypt, and northern Sudan). Many surfaces reflect as efficiently as thick clouds. The albedo (portion of the incident radiation reflected to space) of fresh snow is between 80 and 95% (42 and 70% for older layers), hence the snow-covered parts of the Northern Hemisphere affect greatly the planetary energy balance. Albedos of calm seas are just 7–8%, forests reflect 10–20%, crops 18–30%. Planetary mean fluctuates between 30 and 32%, with clear late summer minima. Deforestation has been the largest contribution to about a 1.5% increase of planetary albedo compared to 6000 years ago.

The most notable characteristic of the outgoing radiation is its shifted wavelength: whereas the Sun radiates as a black body of 5800 K with the peak flux at 500 nm, the curve of terrestrial radiation corresponds roughly to that of a 300 K (27°C) warm sphere with the peak flux at 10 μm, far into the IR zone. Various atmospheric gases absorb the outgoing radiation and maintain the biosphere sufficiently warm for life. But since the advent of large-scale fossil fuel combustion there has been a concern about constant buildup of atmospheric carbon dioxide resulting in tropospheric warming. Generation of other greenhouse gases increases the risks of rapid climatic change.

Graphic representation of the global radiation balance (Fig. 2.2) charts the multiplicity of flows with single approximate estimates given first as percentages of Q_0, the mean incident solar radiation at

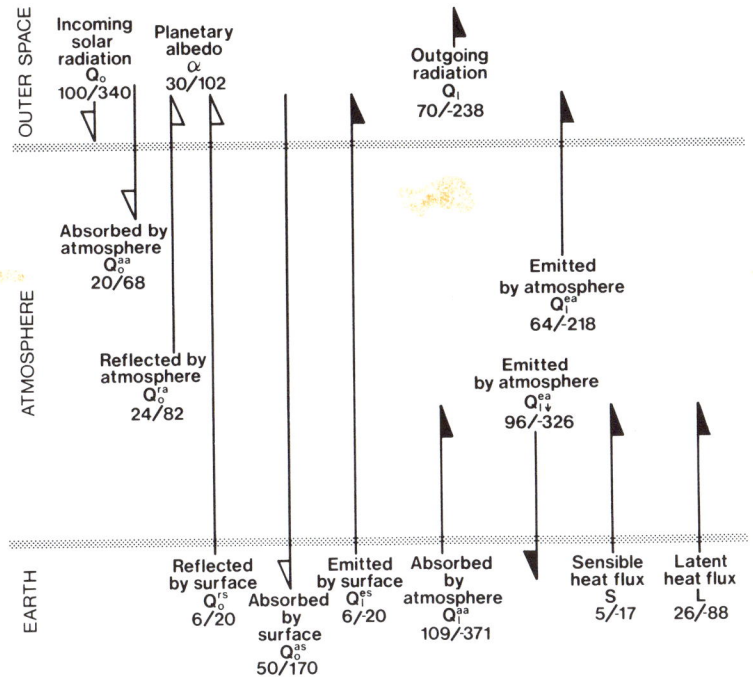

Figure 2.2 Global radiation flows expressed as the percentages of incoming solar energy (the first number) and in watts per square meter. Left-barbed arrows mark the incoming radiation, the right-barbed arrows the outgoing; shortwave fluxes have open arrows, reradiated longwave flows are filled. All values are approximate averages.

the top of the atmosphere, and then as averages. Flows of long waves from the surface to the atmosphere and of the absorbed energy reradiated by the atmosphere to the surface exceed 100% of the incoming total because internal stores of energy are circulated between the two reservoirs in addition to the solar throughput.

The most surprising outcome of satellite radiation measurements has been the fact that each hemisphere is independently in radiation equilibrium with outer space (Slingo 1982). During some months net planetary radiation does not equal to zero for the entire Earth. Total insolation is primarily the function of cloudiness: the highest values are in the cloudless high-pressure belts of subtropical deserts (Fig. 2.3). A fairly regular poleward decline is unmistakable. There is also a little appreciated absence of major differences in total insolation between the humid tropics and temperate latitudes, but the annual march of these similar totals is quite different.

Tropics have little seasonal or daily variation: Manaus, Brazil

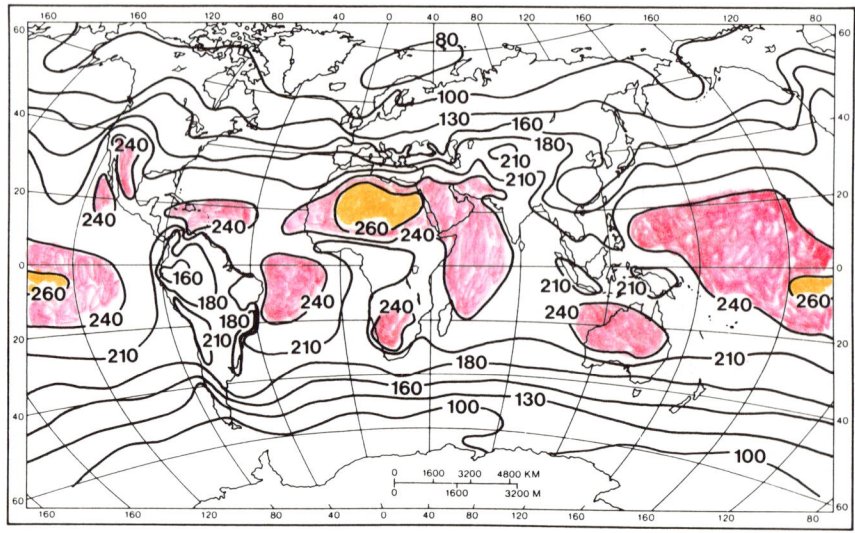

Figure 2.3 Total annual insolation in watts per square meter. (From Budyko et al. 1963).

(3°S), has a daily average of 195 W/m^2, with the highest values in August just 16% above and the lowest in April 13% below the mean. Ames, Iowa (42°N), averages 167 W/m^2, just 15% less than Manaus, but its December flux is 41% below and its July insolation 55% above the mean. Implications for agriculture are obvious: seasonal plants maturing in 2–6 months will, when provided with adequate moisture and nutrients, perform no worse, and those requiring long days will do much better, in the temperate regions.

The diurnal pulse of total radiation ranges from zero up to 800–1200 W/m^2 during summer in mountains and on high plateaus and up to 600 W/m^2 in humid, cloudy lowlands. Radiative impoverishment of the tropics is seen even in the peak midday fluxes: Jakarta's 550–580 W/m^2 is no different from the summer fluxes in Edmonton or Yakutsk. Local cloudiness can also result in large differences within short distances. Global average of the flux is just short of 170 W/m^2 for oceans and about 180 W/m^2 on land, a flow totaling about 88 PW. This surface heating warms soils, bodies of heterotrophs, and all of our shelters: every building is solar heated.

The annual global average of the net radiation flux is just over 100 W/m^2. Maxima over the mid-ocean waters exceed 150 W/m^2 and the values are positive over the entire ice-free ocean, going to as little as 25 W/m^2 at the boundary of floating ice. In contrast, even the annual sums are negative in the Antarctic (about –5 W/m^2) and are around zero in the central Arctic. Latent heat flux is determined by the net radiation at the surface and over the land areas by the availability of water. With water's extraordinarily high heat of vaporization—about 2.45 kJ/g, although the rate is temperature-dependent and the exact formula is $2475 - 2.26T$ (in J/g)—evaporation of 1 mm/day requires 28–29 W/m^2. This much is an annual mean even in high-latitude seas, up to 70°N in the Atlantic and to about 60°S in Antarctic waters.

The highest latent heat flux, close to 200 W/m^2, is associated with warm waters of the Kuroshio and Gulf streams. Global mean of the flux is close to 90 W/m^2, which means that energy expenditure on the water cycle proceeds at a rate of about 45 PW, corresponding to daily evaporation of around 3 mm, that is, an annual total of 1.1 m. Only 13% of this flux originates from the continents. Conversion of radiation into latent heat is essential for the maintenance of all eco-systems; the process is "pumping" water for photosynthesis and recycling it between soils and atmosphere. Forests can increase this flux by an order of magnitude. Latent heat flow in deciduous woods may be just 15 W/m^2 after leaves fall and up to 150 W/m^2 with full foliage.

Sensible heat flux is relatively unimportant as long as there is sufficient water for evaporation. Peaks of its annual means, around 80 W/m^2, coincide with the areas of high aridity while most of the oceans have yearly means below 10 W/m^2. About 60% of the global average of 17 W/m^2 derives from the land and some 45% of radiation absorbed annually by the continents goes into this turbulent exchange with the atmosphere. In dry inland locations the daytime means can be up to 200 W/m^2 and the maxima can reach up to 400–500 W/m^2. Relatively humid middle latitudes have peaks around 150 W/m^2 and averages of 50–60 W/m^2.

Only the central areas of large cities will go much above these levels, especially where the absorbed radiation and the latent heat in moist air inside high-rise buildings are removed by air conditioning into the surrounding space. There the sensible heat fluxes may rival those above the deserts, reaching over 300 W/m^2 in the early after-noon. Sensible heat flux is usually of minor importance in all well-watered ecosystems, but it is a critical defense against overheating as plants translocate heat from the soil to leaves and from exposed leaves to the shaded ones to maintain tissue temperature within optimum range.

Finally, the flux, whose assessment seldom appears in standard treatments of planetary heat balances, which consider just the net surface emission of longwave radiation: the immensely important incoming longwave radiation, a downward flow emitted between 4 and 50 μm by the triatomic atmospheric molecules. Whereas the stratospheric ozone contributes 15–20 W/m^2, carbon dioxide adds 70–75 W/m^2 and variable water vapor concentrations emit 150–300 W/m^2. This flux is just an internal subcycle, a temporary delay in the outward flow of reradiated heat, but it is a critical determinant of tropospheric temperatures and the largest source of energy in nearly every ecosystem (Miller 1981).

The annual global mean of the flux is around 320 W/m^2, with midlatitude continents receiving around 300 W/m^2 and cloudy equatorial regions up to 400 W/m^2. Diurnal ranges are most commonly between 20 and 50 W/m^2 and the flow loses relatively little (25–30%) of its importance even in winter. Momentary variations can be large: a passing radiating cloud can boost the flux by 25 W/m^2. Anthropogenic changes in atmospheric levels of various "greenhouse" gases are the major sources of its increase.

Criticality of water in the planetary energy cycle is obvious. In the atmosphere it is the principal absorber and radiator of incoming and outgoing radiation; in soils and plants it is the carrier of latent heat; and, above all, in the ocean it is the planet's largest reservoir of warmth—everywhere it is an unsurpassable thermal "flywheel." That these feats can be accomplished with little temperature change results from the most peculiar attributes of the water molecule.

Water's specific heat (4.185 J/g·°C) is 2.5–3.3 times higher than that of common land surfaces (soils, rocks) and its heat capacity (specific heat per volume) of 4.185 J/cm^3·°C is approximately twice as high as that of clays and about six times that of many dry soils. Its absorptivity in long wavelengths and its low viscosity make it an ideal heat carrier and its anomalously high heat of vaporization enables relatively high rates of energy flows as latent heat. And water's heat of fusion, 334 J/g, is larger than that of similarly structured compounds.

Oceans dominate the planetary energy balance. About 80% of all radiation intercepted by the Earth (total of 173.5 PW) enters the atmosphere above oceans. With 50% reaching the surface and with average oceanic albedo of 6%, oceans thus receive about 65 PW, nearly twice as much energy as is absorbed by the whole atmosphere and four times as much as the continents. As they also absorb about two-thirds of the downward longwave atmospheric radiation (roughly 110 PW), their global annual heating rate totals some 175 PW.

Virtually all of the absorption occurs in the top 100 m and the high specific heat of water restricts the range of possible temperature amplitudes. Downwelling carries the absorbed energy into greater depths and the equatorial waters and the southern seas between 30 and 50°S are the main zones of heat storage. Principal vertical transport to the surface is upwelling. Energy release from the ocean to the atmosphere is dominated by thermal radiation (about 80%) and by evaporation as latent heat (about 16%). The global pattern of this energy release shows the highest fluxes generated by strong northward ocean currents. Attempts to measure and deduce heat transport at major meridional transocean sections show a large degree of uncertainty. The only agreements concern the heat transport in the Atlantic (northward along the entire length) and its magnitude at 24°N (1–1.2 PW).

Curiously enough, the northern Pacific does not appear to carry a larger flux and the flow may be actually equatorward (Anderson 1983). In the southern Pacific about 0.2 PW is transferred poleward across 24°S. The western Pacific may be a considerable heat source for the southern Atlantic, as the South Indian Ocean, most likely with twice as high a rate, is for the Pacific (Ellett 1982). Unexceptionable generalizations would be to point out that the oceanic circulation of heat involves interbasin fluxes of 0.1–1 PW and that the zonally averaged meridional component of the mean oceanic energy flux is comparable to that of the atmospheric circulation.

Although these grand processes are essential for maintaining the habitable environment, the magnitude of energy transfers generated by planetary heat flows is most directly appreciated by experiencing air and water motion on smaller scales, ranging from tropical cyclones to floods. These fluxes—be they beneficial or catastrophic—have molded civilization's farming potential and defined many of its existential limits.

2.3 AIR AND WATER IN MOTION

A small fraction of solar radiation is absorbed by the atmosphere, continents, and oceans and is converted into a variety of kinetic phenomena. Some of these fluxes are highly effective in redistributing heat but they have other indispensable biospheric roles, above all cycling water, pollinating plants, and denuding the continents. For all civilizations these fluxes have represented risks of material damage and catastrophic loss of life—as well as a promise of potentially inexhaustible sources of energy.

There is no way to ascertain directly the portion of planetary heating which has to power the atmospheric motion in order to offset energy dissipated in turbulence (aloft) and friction (at the surface). The best estimate puts the share at about 2% (Lorenz 1976), or roughly 7 W/m^2, and 3.5 PW for the whole planet—but other estimates are as low as 1.2 PW. Assuming average density of 1.2 kg/m^3 and mean speed of about 10 m/s, kinetic energy of 1 m^3 of moving air equals about 60 J and for the total atmosphere (5.1×10^{18} m^3) it adds up to roughly 300 EJ, a total larger than the annual world energy consumption in the mid-1980s. But it is the first total, the solar recharge, which is the absolute limit on utilization of wind energy.

We do not know what share of this flux can be extracted without any significant changes of planetary circulation. In any case, only the winds moving in the lowest few hundred meters above the surface can be practically intercepted. On the global scale about 35% of wind energy is dissipated within 1 km of the surface (about 1.22 PW, or nearly 2.5 W/m^2) and Gustavson (1979) opted for 10% of this near-surface dissipation as an upper limit of practical wind energy utilization.

This total of about 122 TW (0.24 W/m^2) is very unevenly distributed, and large temporal and spatial variations complicate efficient utilization. Successive annual wind speed averages can differ by 30%, and wind contours plotted on the basis of standard observations are a poor guide to precise location of wind machines as displacements of less than 100 m can result in 20% higher speeds, and, as the power goes up with the cube of wind speed, in 70% higher outputs. Merriam (1977) aptly likens the search for the best wind sites to mineral prospecting: only a detailed search can pinpoint optimum locations. The fastest steady winds, carriers of the most concentrated energy, are too powerful for electricity generation, as are the rapidly moving cyclonic systems, which are also too irregular and too short-lived.

Kinetic energy of common thunderstorms sweeping typically 0.5–1.5 $\times$ 10^8 m^2 with winds of 15–25 m/s during 8–15 minutes will be between 30 and 300 TJ, and their total power will range from 75 to 600 GW. Simply prorated, their power density would be 1.5–4 kW/m^2, but most thunderstorms release their kinetic energies aloft, the impact in the lowermost 100 m above the ground is equal to just between 30 and 100 W/m^2, and the vertical surfaces have to withstand power densities up to 20 kW/m^2. Kinetic energy of hurricanes, whose 30–50 m/s winds (maxima up to 90 m/s) can affect up to 1×10^6 km^2 for many hours, is surprisingly small when prorated over the whole impact area (mostly just 200–500 W/m^2), but the places passed by the eye of the cyclone often are subject to brief spells of energy releases as intense as 0.5–1 MW/m² of vertical surface.

The impact of tornadoes is frequently even worse. Based on the characteristics of 10,826 tornadoes in the United States (Schaeffer et al. 1980), we know the means of their paths' width (126.7 m), length (9.6 km), speed (60 m/s), and duration (160 seconds). Total energy in the average 100 m tall tornado funnel is about 275 GJ, its power is 1.7 GW, and its power density is about 1.4 kW/m². Wind cross-sectional densities average about 135 kW/m² and maximum wind gusts release up to 1.35 MW/m² against the structures. The largest tornadoes have power up to about 100 GW and their path can be as wide as 1.5–2 km.

While the high kinetic power densities of cyclones cause much damage, total energies in destructive winds are only a tiny fraction of huge cyclonic latent heat content. Thunderstorms leaving behind just 1 cm of rain in 20–40 minutes over 1–2 × 10⁸ m² release 2.5–5 PJ (1–4 TW) of heat, 10–100 times the kinetic energy total. Most of this energy goes into heating the atmosphere. Latent heat transferred by hurricanes is usually between 10 and 50 EJ (120–580 TW). An average year sees 80–100 tropical cyclones with winds faster than 20–25 m/s; about half of these become hurricanes (Anthes 1982). Only a few tenths of 1% of the planetary latent heat flux are discharged by these cyclones.

The Earth's most pronounced seasonal landward water-borne transfer of solar energy collected over the tropical oceans is the Asian monsoon (Webster 1981). Its course affects lives of about 2 billion people in an area totaling nearly 10 Gm². Continental precipitation averages about 10 Tm³ during six months, a latent heat transfer amounting to almost 1.5 × 10²⁴ J at a rate of 1.5 PW and land power density of some 150 W/m². Compared to its latent heat content, kinetic energy of precipitation is quite small. Even a fairly heavy rainfall of 2 cm/h with raindrops falling with terminal velocity of 6 m/s will have impact energy of just 360 J/m², that is, a power density of 100 mW/m²—but its effect may be large as it breaks up topsoil unprotected by vegetation.

Although the resulting runoff may look far more damaging than the pounding of raindrops, their kinetic energy, and hence their erosive power, is far stronger. Terminal raindrop velocity is primarily a function of drop diameter (2 m/s for diameters of 0.5 mm, about 9 m/s for the largest sizes around 5 mm) although strong driving winds, boosting the drop velocity by the reciprocal of the cosine of the rain's angle of inclination off the vertical, can make a substantial difference (Wischmeier and Smith 1978). Even ordinary velocities of 6 m/s convey kinetic energy of 180 J/m^2 for every centimeter of rain; should the runoff equal one-third of all precipitation, its kinetic energy, with average speed of 1 m/s, would be just 1.65 J/m^2, a rate two orders of magnitude smaller than the rain's kinetic energy density.

Hailstones are much more damaging. Even the largest raindrops (diameter 5 mm) hitting with velocity of about 9 m/s have just 2.6 mJ of kinetic energy; in contrast, hailstones with diameters of 2 cm have kinetic energy about 75 times higher (0.2 J). Kinetic energy of an intense hailstorm adds up to about 500 J/m^2 (10–20 W/m^2), 100–200 times the rate of heavy rainfall. At the point of impact kinetic power releases are easily equivalent to 10^3–10^4 W/m^2, enough for heavy damage.

Before leaving the atmosphere, a few notes on lightning are appropriate. Exact origins and means of cloud electrification remain elusive (Williams 1988). Most of the high-current strokes originate in cumulonimbus cloud and their power goes up with the fifth power of the cloud size. Energy available in an average cumulonimbus for dissipation in lightning is on the order of 4 GJ; single strokes discharge 20–500 MJ or, with paths of 2–5 km, between 10 and 100 kJ/m. The bulk of this energy is dissipated within 10 microseconds, resulting in an enormous effective power of 1–10 GW/m (Hill 1979; Uman 1982).

Once the precipitation reaches elevated ground the gradual release of its potential energy is usually a subdued affair. Meltwater from 1 m of snow accumulated during winter in a high mountain valley would, when dropping on the average 1 km to reach the main valley channel within one month, reduce its potential energy at a rate of mere 380 mW/m^2. Total potential energy of high annual precipitation (2 m) in maritime mountains (3 km above the sea level) would be just short of 60 MJ/m^2, a large total renewed annually. But the importance of these potential energies is much greater than the power densities can indicate.

In Miller's (1981) accurate characterization this "quiet and unrelenting" flux "gives a common direction to ecosystem functions." Moreover, since it is in a form amenable to conversions into such useful work as wearing away rocks and moving and removing the products of weathering, it is a key agent of geomorphic processes. Sudden releases of water's potential energy occur only in falling avalanches and after breaching of elevated lakes. A slab of dry snow measuring just 10 × 10 × 20 m, weighing about 500 t and hanging

500 m above the lower slopes of a mountain valley, contains potential energy equal to nearly 2.5 GJ; its unimpeded, focused, narrow-path discharge can easily result in tragedy. Kinetic energy of such a falling mass, assuming average speed of 30 m/s, would be 225 MJ, resulting in vertical power densities around 110 kW/m^2, a rate directly comparable to the vertical wind density of tornadoes.

The biggest known catastrophic water release was the discharge of glacial Lake Missoula in Washington (Baker 1981). Maximum discharge was up to 21×10^6 m^3/s (for comparison, global discharge of rivers is now 1.2×10^6 m^3/s) with speeds averaging 20 m/s. Total kinetic energy of this unparalleled breach was about 400 PJ, its power most likely 1–2.5 TW, and its vertical kinetic power density up to 13.5 MW/m^2, 10 times the impact of the fastest tornado gusts. These enormous flows were capable of total reshaping of about 8000 km^2, moving boulders up to 10 m in diameter and creating such impressive features as the Columbia River's Grand Coulee.

The more orderly stream flows are indispensable for transporting huge loads of sediments, accreting nutrient-rich alluvial plains, and also providing mankind's first practical inanimate source of mechanical energy and inexpensive electricity. On the debit side are the destructive energies of great floods. The best estimate of annual global river runoff is 38,230 km^3, with one-third of the total coming from Asia and a bit over one-quarter from Latin America (Berner and Berner 1987). With the mean continental elevation of 840 m (Ridley 1979), the potential energy of this runoff is about 314 EJ and the theoretical global hydrogeneration capacity then would be almost exactly 10 TW.

Such a total is obviously unattainable. The principal limitations are the considerable seasonal variability of flows, physical impossibility of complete utilization owing to a multitude of competing uses, unsuitability of many locales for dam construction, and the engineering impossibility of converting the flux with perfect efficiency at full capacity. Global estimates of potential hydro energy capacity are prepared by assembling national and regional appraisals, which estimate the capabilities of suitable sites and assume average stream flows tapped with a 50% utilization factor.

Comprehensive compilation of these estimates puts the annual global potential at about 35 EJ of electricity—92% in projects larger than 1 MW, the rest in small stations—or roughly 11% of the theoretical potential (World Energy Conference 1986). Continental distribution of this potential is very uneven, with Asia accounting for 40% and Latin America 20% of the total; China has about 15%, the

Soviet Union 12%. About 60–80% of total capacity eventually may be developed; the European share is already above 60%, the African share below 10%.

Waterfalls are, of course, the sites of the highest concentrated releases of potential energy of flowing water. Average height and discharge estimates (Czaya 1981) translate into the maximum of 16.25 GW for the Inga Falls on the Zaire; 5.2 GW for the Sete Quedas on the Paraná and 4.9 GW for the spectacular 72-m tall Iguassú in South America; Niagara's potential power totals 3.4 GW Stream velocities are unevenly distributed throughout a channel but the flows are typically around 0.5 m/s in wide lowland rivers and 2–3 m/s in floods; none are known to exceed 9 m/s even over rapids.

Cross-sectional kinetic power density of a flooding stream then can be up to about 13 kW/m^2, less than in many thunderstorms and an order of magnitude below tornado or avalanche impacts—but considerably longer durations of typical floods and their ability to weaken the foundation soils translate into structural damages much larger than the power density value would indicate (threshold for structural damage is 18 kW/m^2).

Stream competence, the maximum movable weight of individual bedload pieces, varies with the sixth power of water velocity; stream capacity, the ability of moving total bedload, goes up with the cube of the velocity (Goudie 1984). Average shares of the three components are difficult to estimate on a global scale, but a ratio of 4:5:1 may be a good approximation. There are many published estimates of the global sediment load carried by rivers, all uncertain. The total is at least 15 billion and possibly 30 billion t/y (Judson 1968; Selby 1982); the highest rate would be equivalent to at least 260 PJ of lost potential energy, or less than 0.1% of the total loss in the global runoff.

Kinetic energies of ocean waters include the currents redistributing the absorbed heat, wind-generated waves, seismic and volcanic waves (tsunamis), and tides. Aggregate energies of these flows are large but average power densities are low. Total power of ocean currents has been estimated at 100 GW (Isaacs and Schmidt 1980), that is, 0.3 mW/m^2 of ice-free ocean surface. Only several major currents concentrating about half of the global flux could be considered as potential sites for commercial conversions. For example, the Florida Current between Bimini and Miami rates about 20 GW, or 1.6–2.2 kW/m^2.

Wind-generated waves abound on the ocean. Their total power was put at as much as 90 PW and the renewal rate at no less than 1

TW. My estimate, using average height of 1.5 m and period of 6 seconds, is about 40 PW for the whole ice-free ocean; this would be about 110 W/m² of the ocean surface. Linear power density is much more relevant for potential energy utilization than the estimates of areal totals. As the power goes up with the square of the wave height, a 4 m high wave with a 6 second period will have nearly twice the power of a 3 m wave (45.9 vs. 25.8 kW/m).

The world's highest reported waves, encountered in the Pacific by the USS *Ramapo* on February 7, 1933, had the height of at least 33 m and the period of 14.8 seconds (Bascom 1959), containing enormous power of 7.7 MW/m. In contrast, wavelets in largely placid inland seas may carry no more than 1 W/m. Measurements of linear power densities in such stormy waters as the North Atlantic show rates between 25 in the Irish Sea and 91 kW/m west of Scotland (Ross 1979). Development of practical conversion techniques will be very difficult.

When the waves break on the shores their kinetic energies are rapidly dissipated. The kinetic energy of a 2 m wave is about 5 kJ for every square meter of beach, or in the neighborhood of 1 kW/m², a tremendous amount of work when considering the incessant nature of the pounding. Seismic sea waves, tsunamis, generated by underwater earthquakes or volcanic eruptions, travel thousands of kilometers at speeds between 550 and 720 km/h with little loss of power. Virtually invisible at sea, they rise tens of meters in shallow waters. Big seismic waves may be hitting coastlines with as much as 200–500 MW/m² of vertical surface and generate onshore horizontal impact power densities on the order of 10^1–10^2 MW/m², far surpassing powers of cyclones and avalanches.

Waves generated by gravitational power of the moon and the sun have the great advantage of predictability as there are two tidal cycles roughly every 24 hours. There is also considerable regularity of tidal ranges, with variations ranging from differences among the two successive cycles to monthly and annual fluctuations. Average day lengthening of 1.5 ms/century caused by almost linear decrease of the Earth's rotation since the Paleozoic Era represents an input of about 3 TW into the tidal friction. About 90% of this flux is dissipated in the ocean. This is just 7.5 mW/m², but as most of this dissipation takes place in shallow seas, regional rates are much higher.

The maximum possible amount of energy to be extracted in one tidal cycle goes up with the square of the tidal range. Only combinations of high tidal amplitudes (at least 5 m) and coastal features

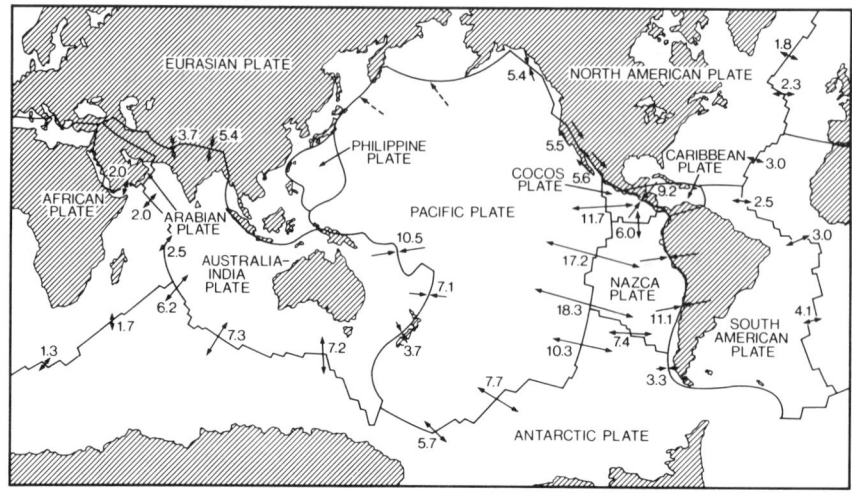

Figure 2.4 Tectonic plates tiling the planetary surface. Arrows show the principal direction of spreading; numbers are the annual rates of relative spreading in cm.

allowing for natural, effective impoundment can provide sites for commercial conversions of tidal energy. The Earth's highest tidal ranges are in Nova Scotia's Bay of Fundy (6.47–11.71 m mean, 7.50–13.30 m spring tides), in Alaska's Cook Inlet (average 7.5 m), in southern Argentina (5.9 m), along the coast of Normandy (5.0–8.4 m), and in the White Sea bays (up to 6.8 m). Detailed evaluation of 28 best potential sites adds up to total theoretical power of about 360 GW (Merriam 1978), or some 13% of the planetary tidal power.

This closes the survey of atmospheric and hydrospheric energy fluxes. The remainder of this chapter is concerned with tectonic and geomorphic phenomena. In these cases, the spectacular energy releases (volcanic eruptions or massive landslides), not unlike the atmospheric energy flows, account for a very tiny fraction of terrestrial heat transformation but they may have profound existential effects—while the slow but gargantuan plate motions have been one of the principal determinants of life's evolution.

2.4 LITHOSPHERE

The demise of vertical geology in the 1960s revolutionized our understanding of the Earth's crust (Shea 1985). Large oceanic plates created at the mid-ocean ridges (Macdonald and Fix 1990) spread with relative speeds of 2–18 cm/y (Fig. 2.4). An average of 3 cm/y along some 60,000 km of ridges winding around the planet would create annually about 1.8 km^2 of new ocean floor; the formation of the current floor of about 310 million km^2 thus would have taken only about 175 million years. Movements of the rigid crust must be driven by vast convection currents in the Earth's mantle. Their prime mover is the release of planetary heat, a flow so massive that explaining the crustal dynamics as effects of the Earth acting as a heat engine requires a surprisingly small efficiency.

No more than 0.1% of the planetary heat release is needed to form the ocean floors, to shape the continents, and to power earthquakes and volcanoes. Unlike the movements deep in the mantle, the Earth's heat releases are easily measurable, although the planet's earliest thermal history remains speculative (Sleep and Langan 1981). Global spreading rates during the Archean period were at least 4 and up to 25 times greater than the recent mean. The plates were most likely smaller, thinner, easier to fracture, and more rapidly recycled.

A substantial share of the Earth's heat flow comes from basal cooling, the rest from radioactive decay of crustal elements. Experimental measurements of iron temperatures at immense pressures simulating the conditions within the Earth's core came up with figures 2000–3000 K higher than the previous estimates (Williams et al. 1987). These results imply that a very high share of heat in the mantle originates in the core. Still, the radioactive decay must be a major constituent of the total flux. Principal heat-producing crustal radioactive isotopes are U^{235}, U^{238}, Th^{232}, and K^{40}. The greatest uncertainty concerns K^{40}, whose decay generates 36×10^{-13} W/g (compared to 97×10^{-9} W/g for uranium) but whose crustal concentrations have been put as low as 170 and as high as 800 ppm (Verhoogen 1980). The lower estimate produces an annual flux of 24 TW, the higher, 38 TW.

Numerous measurements provide detailed regional and age subdivisions (Davies 1980; Sclater et al. 1980). In the oceans where thicker sediments prevent major losses owing to hydrothermal circulation the heat flow decays uniformly from rates in excess of 250 mW/m^2 for floors younger than 4 My to 46 mW/m^2 for those older than 120 My, reaching an equilibrium value of 38 mW/m^2 after 200 My. On continents the youngest regions average 77 mW/m^2, the oldest shields (over 800 My) just 44 mW/m^2, with the nonradiogenic share declining to a constant rate of 21–25 mW/m^2 after 200–400 My.

Sclater et al. (1980) calculated the total heat loss through the oceans at 30.4 TW, through the continental shelves at 2.8 TW, and through the continents at 8.8 TW for a grand total of 42 TW. With

24–38 TW, radioactivity alone could supply at least 55 and up to 90% of this flux. Other recent estimates have been similar, assigning about 70% of the total flux to deep oceans and 30% to continents and their shelves and ranging from 39.2 to 43 TW, or 77 to 84 mW/m^2 of the Earth surface. Roughly a quarter of the global heat flow (10 TW) is transported by hydrothermal circulation in young oceanic crust.

The mean heat loss density of oceans is 95±10 mW/m^2 and that of continental crust is 55 ± 5 mW/m^2; the global average of about 80 mW/m^2 is minuscule in comparison with the mean insolation of 160 W/m^2. Creation of new lithosphere in oceanic ridges releases about 26 TW, accounting for about 90% of the heat loss through oceans. Nearly a third of this loss occurs in the South Pacific, where the spreading rates are fastest. The whole Pacific accounts for half and the Atlantic and Indian oceans each for one-fifth. Radiogenic heat loss in the ocean is small, less than 5% of the total flow. Africa and North America have the highest total heat loss (each about 16% of the continental sum), and Australasia has the top mean heat flow density at about 64 mW/m^2.

Highly concentrated maxima are the result of hydrothermal venting along the mid-ocean ridges. Exploratory dives recorded waters up to 350°C ejected from relatively small openings at rates of 25–330 MW (10^6–10^7 W/m^2), levels equaled or surpassed only by volcanic explosions. By comparing these enormous fluxes to the total oceanic heat loss, it is clear that such intensive ventings must be short-lived. Appraisals of the Earth's heat as the source of useful energy differ by up to six orders of magnitude. Total heat content of the outer 10 km of the heat crust is at least 1.25×10^{28} J and that of geothermal reservoirs utilizable for electricity generation is at 4.1×10^{22} J, when the depth is reduced to 3 km at no more than 83 EJ.

Creation, collision, and subduction of plates are accompanied by frequent earthquakes: about 95% of all tremors occur at the edges of plates and 90% are recorded in the circum-Pacific Belt, where the fastest moving oceanic plates are colliding with, or sliding past, the continental plates (Kanamori and Boschi 1983). The remainder of the quakes is associated mostly with volcanic hot spots. During this century earthquakes have caused more deaths than volcanic eruptions, floods, and cyclones combined—but energetically they are only a minor component of tectonic processes.

The magnitude of earthquakes, defined empirically by Richter (1935), was first correlated with total energy releases by Gutenberg and Richter (1942); since then the relationship has been repeatedly modified. All of the proposed conversions have the form $\log_{10} E =$

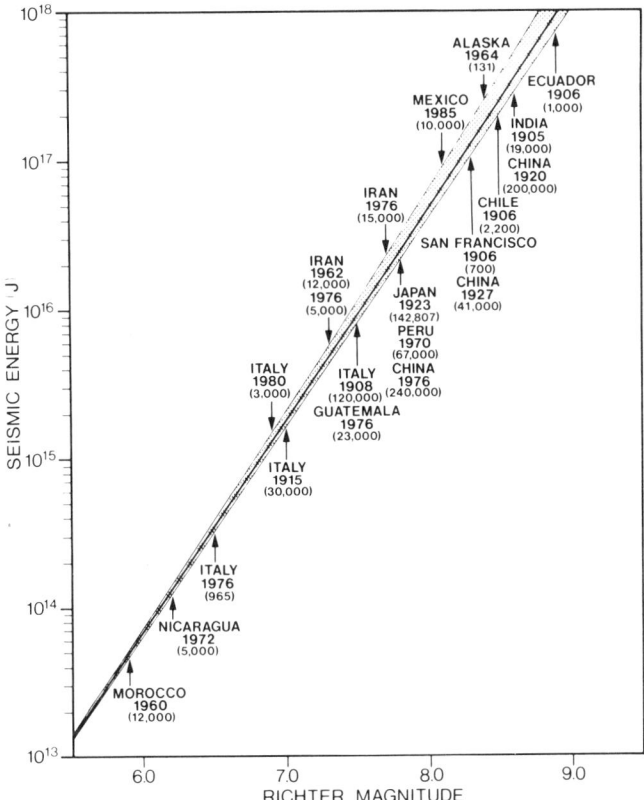

Figure 2.5 Seismic energies of earthquakes calculated from magnitude lists in Gutenberg and Richter (1942) and Kanamori and Boschi (1983). Mexico's September 1985 quake had Richter magnitude of 8.1. Casualty numbers are in parentheses.

$a + bR$, where E is the energy released as seismic waves (in ergs), R is the Richter magnitude, and a and b are empirical coefficients ranging, respectively, from 6.1 to 13.5 and from 1.2 to 2.0. Richter's earliest conversions used 11.3 and 1.8, Scheidegger (1982) uses 12.32 and 1.42, and Okamoto (1984) prefers 11.8 and 1.5 based on the Japanese experience.

The largest recorded earthquakes have magnitudes between 8 and 8.9 and hence their corresponding energy releases would be between at least 48 PJ and 1.41 EJ. Seismic energies of major twentieth-century earthquakes are plotted in Fig. 2.5. Casualties have little correlation with earthquake magnitudes; population densities and the resistance of buildings are far more important factors. Between 1900 and 1980 the frequency of earthquakes with magnitudes larger than 7.0 averaged 10 a year, corresponding to the seismic energy release of about 450 PJ. This is no more than 0.03% of the Earth's total annual heat loss. Average annual release of seismic energy from all recorded earthquakes may be about 300 GW; the addition of strain energy accumulated in irreversible deformations and of friction-generated heat along the faults may raise the total to about 1 TW (Verhoogen 1980), no more than 2.5% of the global heat flow.

Brevity of the quakes makes for extraordinarily high power ratings and densities. A magnitude 8 quake lasting just 30 seconds would rate about 1.6 PW and its power density, in terms of damaged area, could be as high as 80 kW/m^2 (with uniform prorating over the area with radius of 80 km subjected to strong ground shaking). A magnitude 6 tremor of the same duration with shocks felt throughout 100 km^2 would have an overall surface power density of about 23 kW/m^2.

Volcanic eruptions are an even more pronounced example of the disparity between the surprisingly low aggregate annual energy release on global scale and large and spectacularly displayed local or regional power. A very liberal assumption of 30 Pm^3 of lava ejected since the Cambrian would imply the total cooling and crystallization loss of some 1.7×10^{26} J (1.7 kJ/g), compared to the ordinary conductive heat loss of at least 5×10^{29} J during the same period. The overwhelming concentration of active volcanoes along the subduction edges of tectonic plates leaves no doubt about their genesis as a part of grand horizontal lithospheric processes driven by ocean floor spreading.

An obvious measure of the intensity of volcanic eruptions is the volume of ejecta. Some intensity scales put the 1815 Tambora eruption ejecting at least 150 km^3 of solids at the top of historic volcanic events (although more recent estimates are as low as 30 km^3) followed by Krakatoa in 1883 with 18 km^3. But there are many eruptions releasing huge quanta of energy without spectacular ejections into the atmosphere. Kinetic energy of the ejected material is only one of many energy transforms ranging from the change of potential energy in the height of the magmatic column to energy fractioning or deforming the surrounding crust. Invariably, thermal energy is dominant, at least 10 and up to 1000 times greater than the other releases.

The best monitored volcanic eruption, that of Mount St. Helens, Washington, on May 18, 1980, provides an excellent illustration of the dominance of thermal energies, which accounted for 96% of the total flux of 1.73 EJ (Decker and Decker 1981). Nearly half of these energies was invested in the ash cloud reaching above 20 km. By far the largest total energy releases during modern eruptions were those of Laki (Iceland, 1783, 85 EJ) and Tambora (Sumbawa, Indonesia, 1815, 84 EJ). The only historic eruption surpassing the Laki event was the creation of Santorini caldera in the Aegean Sea about 1500 B.C.: its total energy release of about 100 EJ lifted about 70 km³ of fragmentary ejecta. This is dwarfed by the creation of a 2000 km³ caldera during the Toba (Northern Sumatra) eruption 70,000 years ago. Such giant destructions do not happen more frequently than once in a million years (Francis 1983).

During the past two centuries there were only eight eruptions releasing more than 1 EJ and resulting in an average of 1.25 EJ/y. This averaging is done here strictly for power calculations as these eruptions are much less frequent than large earthquakes; at roughly 40 GW they would equal merely 0.1% of global heat loss. Even if smaller eruptions and constant venting of hot gases in active, non-erupting volcanoes raised the total to 200 GW [Elder's (1976) estimate of total annual rate of volcanic heat loss], this would still be no more than 0.5% of global thermal flux. Verhoogen (1980) puts the current maximum volume of continental lava outpouring at less than 1 km³/y and estimates the highly uncertain rate of oceanic eruptions at close to 4 km³/y. The global total of less than 5 km³/y would then correspond to a heat loss of about 800 GW, still no more than a small percentage of planetary heat flow.

But power ratings of individual explosions are huge. The main eruption of Mount St. Helens (1.7 EJ) lasted 9 hours, giving an average rating of 52 TW. In comparison with earthquakes, an eruption releasing 80 EJ in 10 hours will have power rating of over 2 PW, roughly equal to a magnitude 8 earthquake lasting 30 seconds. Similarly, more common eruptions releasing just around 500 PJ will have power ratings equivalent to magnitude 7 earthquakes, around 15 GW.

Modern volcanological literature has favored comparisons of eruptions with explosions of nuclear bombs. Mount St. Helens' 1.7 EJ would be equal to energy released by about 32,000 Hiroshima bombs. Such comparisons ignore different ways and durations of energy release. In nuclear explosions the fluxes are instantaneous: blinding light is released within milliseconds, infrared radiation causing death and thermal burns is emitted in vast amounts 0.2–3 seconds after the eruption, and the blast wave, containing about 50% of all latent energy, travels at initial velocities of 10^2 m/s (Committee for Compilation, 1981).

In contrast, volcanic eruptions may last hours, days, even months and, as is the case of Hawaii volcanoes, may consist overwhelmingly

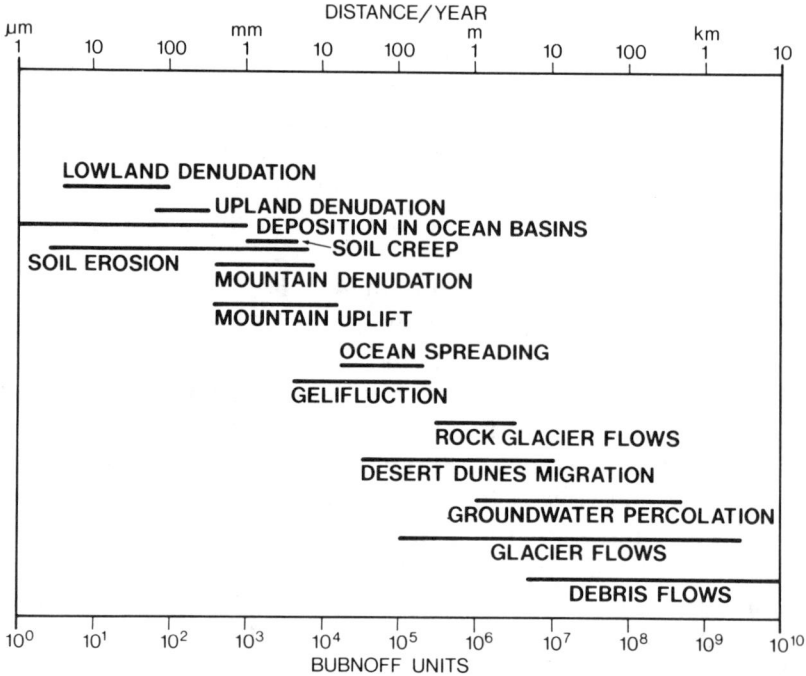

Figure 2.6 Rates of important geomorphic processes. Data from Caine (1976) and Selby (1985).

of slow upwellings of hot lava with few explosive effects and large energy releases dispersed along fracture lines or over molten surfaces of lava pools. Hot lava usually spreads only over several square kilometers, ballistic projectiles fall on up to 10 km², and tephra deposits (airborne fragments) can affect 10^2–10^6 km². Consequently, it is impossible to calculate typical impact power densities of volcanic eruptions. Their ejection power densities can be enormous: with tephra clouds rising up to 80 km and with initial ejection velocities up to 500 m/s (Kittleman 1979) they can be on the order of 10^8–10^9 W/m², unequaled by any other concentrations of terrestrial power.

The geomorphic change is overwhelmingly a matter of gradual processes working at tiny rates. Tectonic uplift and soil formation are in this category, as is most of the large-scale erosional denudation. Average rates of such gradual processes are expressed conveniently in Bubnoff units, with 1 B = 1 m/My (Selby 1985), a denudation equivalent of 1 μm/y or 1 m³/km²·year. Figure 2.6 summarizes the rates of all important slow geomorphic processes. Considerable ranges for most processes are not surprising: differences in relief, effective precipitation, and groundcover combine to give rates ranging up to over three orders of magnitude. Mean denudation rates in the Himalayas can be sustained at nearly 10 kB, in many lowlands they are no more than 10 B; on smaller scales, soil erosion in climax forests may be mere 4–5 B, in badly damaged croplands nearly 2 kB.

As the erosion reduces the potential energy of the landscape, the annual global rate of denudation—say, 80 B, or with the average density of 2.5 g/cm^3 about 200 t/km^2·y—would be equivalent to a change of roughly 240 PJ or about 7.5 GW, 50 μW/m^2 of continental surfaces. If recent erosion rates have been roughly equal to the increase in potential energy of continents, then the sum of 7.5 GW also represents the power going into the uplift of terrestrial formations (just 0.02% of the global heat flux).

Actual field measurements of various erosion rates have been infrequent (Rapp 1960; Caine 1976). Their results indicate that most geomorphic processes proceed at power densities ranging from 10^{-7} to 10^{-6} W/m^2. Maximum reported rates per square meter were as follows: surface wash 200 nW, snow avalanche debris and rockfalls 500 nW, soil creep and solifluction 2 μW, solute transport 20 μW, and earth slides and mudflows 50 μW. Maximum combined power of these sediment movements would be 10^{-5} W/m^2, the same order of magnitude as the global estimate for 80 B denudation, and a minuscule fraction of potential energy loss in runoff (easily 10^{-2} W/m^2) or geothermal flux (also 10^{-2} W/m^2). Romans knew it well: "*gutta cavat lapidem non vis sed saepe cadendo.*"

In closing, just a few calculations of the planet's encounters with extraterrestrial bodies are noted. Collisions with meteorites are frequent but as these solids follow circumsolar paths in the same direction as the Earth the collision speeds are no higher than 15 km/s and the effects are localized. Still, the impact power densities are enormous, as shown by the Arizona meteorite that created the Barringer Crater about 25,000 years ago. Weighing about 60 Gg and traveling at about 11 km/s, this meteorite had kinetic energy of about 12 PJ, impact power of 700 PW, and impact power density of around 600 GW/m^2. But this is a negligible effect compared to a head-on collision with a comet.

Speed of at least 50 km/s and mass of 500 Pg would result in enormous kinetic energy of about 6×10^{23} J. Even if 90% of it were dissipated in the atmosphere, the impact energy of 6×10^{22} J would be equivalent to explosion of 150,000 hydrogen bombs of 100 Mt each (here the instant energy release makes the comparison appropriate). Global consequences of such a collision would involve a profound climatic change. There has been much interest in tracing the end of the Mesozoic era to such an event but there are alternative scenarios for the Cretaceous extinctions (Stanley 1987). None of the unknown number of such collisions during the past 3.5 billion years was powerful enough to derail the evolution of life resting on the capture of solar radiation by photosynthesizing organisms—that is, on complexification and diffusion of bacterial and plant life.

3

PHOTOSYNTHESIS: ENERGETICS OF PRIMARY PRODUCTION

The Sun appears to be poured down, and in all directions indeed it is diffused, yet it is not effused. For this diffusion is extension.

—Marcus Aurelius
Meditations

And what an extension! All complex life on this planet is but a transmutation of that light—and photosynthesis is the agent of this marvel. Absorption of sunlight and the subsequent sequence of photochemical and thermochemical reactions in the chloroplasts of green plants are the most important energy conversions on the Earth. The plants provide (directly or after being first eaten by the animals) all our foods; their immediate harvests (as wood and crop residues) or the extraction of their fossilized remains (as coals and hydrocarbons) supply all our fuels—all the richness of heterotrophic life and all the intricacies of human civilizations are thus energized by photosynthesis.

For all of its internal complexity and external plant variety the process has one energizer, the absorption of photons by a small group of excitable pigments; the synthetic sequence always shares the core of an intricate multistep reductive pentose phosphate cycle; its rates are determined by the same limiting factors in all terrestrial environments; and the bulk of the newly formed phytomass is surprisingly uniform in its energy content (Govindjee 1982; Edwards and Walker 1983; Foyer 1984). These facts make it easier to survey the bioenergetic essentials of photosynthesis.

3.1 PROCESS ESSENTIALS

The only way to reduce atmospheric carbon dioxide is through the multistep reductive pentose phosphate (RPP) cycle whose sequence was detailed for the first time in the 1950s by Melvin Calvin and his colleagues (Bassham and Calvin 1957). The RPP pathway is made up of 13 enzyme-catalyzed reactions and it has three key sequences, carboxylation, reduction, and regeneration (Fig. 3.1). The RPP cycle must be preceded by the generation of adenosine triphosphate (ATP) and nicotineamide adenine dinucleotide phospate (NADPH), the two compounds energizing the CO_2 fixation (which can proceed in the dark) as well as all of the biosynthetic reactions in the light.

Every photosynthetic sequence starts with light absorption by pigments (chlorophylls, carotenoids, or phycobilins) in thylakoid membranes of chloroplasts. Chlorophylls, present in all autotrophs, have absorption peaks falling in two distinct, narrow bands, between 420 and 450 nm and between 630 and 690 nm; photosynthesis is energized by a combination of blue and red light. Blue light is actually no more effective than red because the higher excited state of chlorophyll decays rapidly (in just 10^{-13} second) to the level of red-driven excita-

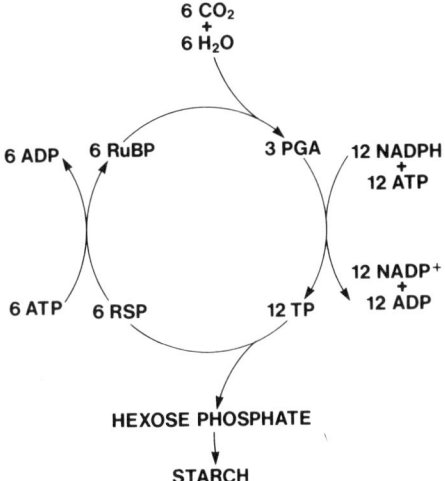

Figure 3.1 Simplified form of the reductive pentose phosphate metabolic pathway—or Calvin or C_3 cycle—in chloroplasts. (Danks, S.M., et al., 1983.)

tion, making the red state the common starting point of the photosynthetic process. Photosynthetically active radiation (PAR, between 400 and 700 nm) is thus only about one-half of total insolation but light is uncommonly a limiting factor of photosynthesis.

Energy efficiency of the actual carbon assimilation is very high. To reduce one molecule of CO_2 requires three molecules of ATP and two molecules of NADPH in the RPP pathway. Free energies of the two compounds are, respectively, about −29 and −216 kJ/mol. Reacting compounds contribute 519 kJ while the difference between the broken (in H_2O and CO_2) and newly formed (in sugars) bonds is about 465 kJ/mol. Theoretical efficiency of the process is almost exactly 90%, a great contrast with the much lower efficiency of the whole photosynthetic sequence.

The minimum quantum requirement for the synthesis of three ATP molecules needed for the reduction of each molecule of CO_2 depends on the H^+/ATP ratio. With $2H^+$/ATP the minimum quantum requirements would be six but synthesis of two molecules of NADPH would raise this to eight quanta for each molecule of assimilated CO_2. Energy content of a quantum depends on the light frequency, varying inversely with the wavelength. Assuming the mean PAR wavelength at 550 nm the energy content of an average-sized quantum would be 3.61×10^{-19} J (Planck's constant multiplied by the light frequency). One einstein (1 mol, or Avogadro's number, 6.02×10^{23}) of green photons would have an energy of 217 kJ and 8 einsteins would supply 1.736 MJ of radiant energy; the maximum theoretical efficiency of photosynthesis then would be almost 27% (465 kJ/1736 kJ).

In reality only the light absorbed by chlorophyll can drive photosynthesis; as the average share of PAR in the direct light is only about 43%, the theoretical efficiency maximum is reduced to about 12%. A portion of the useful light is reflected by the plants and a portion is transmitted through canopy leaves. Losses caused by reflection and transmission of green light total at least 10% (up to 25% has been suggested as a reasonable average) and the photosynthetic efficiency maximum would be around 11%.

This is an absolute, immovable top—at least until evolution develops an entirely new mechanism of photosynthesis (Good and Bell 1980). An ideal leaf exposed at a 90° angle to direct sunlight (with density 7 W/dm^2) would then reduce about 250 mg $CO_2/dm^2 \cdot h$ and synthesize 170 mg/dm^2 of new phytomass, converting the radiant energy to chemical bonds with a power density of 0.8 W/dm^2 (8 mW/m^2). With an average leaf weight of 2.85 g/dm^2 the metabolic intensity would be about 280 mW/g. This leaf tracking the bright light for 10 hours would produce about 1.7 g of new photosynthate. A field of such ideal leaves would fix daily 1.7 t/ha of new pytomass and where the growth could continue for the whole year 620 t/ha would be added.

Actual short-term increments of new phytomass are at best 50% and large-scale averages are merely 10% and all the way down to 2% of the best theoretical performance. Reasons for this large disparity are the costs of respiration and the inevitable losses accompanying rapid rates of photosynthetic reactions. To conserve as much light as possible during the limited hours of intensive insolation photosynthetic rates must be quite fast, but this rapidity results in two kinds of large inefficiencies. First, unless the plant's enzymes can keep up with the radiation flux coming in the excited pigments, the absorbed energy will be reradiated as heat; utilization must be immediate as the chlorophyll molecules cannot store the sunlight.

No less important is the fact that the rapid photosynthetic rates maximizing plant growth and improving the chances of survival entail large irreversible losses, since evolutionary success was paid with conversion inefficiencies. A generally applicable modal estimate of these losses is impossible; a minimum of 20–25% would be common, lowering the best photosynthetic efficiency to 8.2–8.8%. Carbon dissipated in metabolic processes and in the maintenance of the photosynthetic system and its supporting structures ranges from less than

20% in such immature fast-growing plants as high-yielding crops to virtually 100% in aged forest trees. With respiratory losses at 40% the peak plant growth efficiencies would be around 5%.

By far the most widespread limiting factor in carbon fixation is the availability of water. Photosynthesis is impossible without an extremely lopsided tradeoff between CO_2 and H_2O. Water vapor pressure is much higher inside the plants than outside and the difference of these two pressures is two orders of magnitude higher than the difference between external and internal CO_2 levels. Even if the relative air humidity averaged 50% (hence the difference of water vapor pressures would be 2.13 kPa) and the internal CO_2 level was near 0 ppm (pressure difference, with 340 ppm of ambient CO_2, at 34 Pa), 1 mol of CO_2 could not be gained for less than 98 lost moles of H_2O (quotient of the two pressure differences multiplied by 1.56, the ratio of H_2O and CO_2 mobilities). The lowest actual transpiration loss would be 400–500 mol of H_2O for each mole of CO_2 fixed.

Plants photosynthesizing with such a high water utilization efficiency initially follow a fixation sequence different from that described by the Calvin cycle (Fig. 3.1). Instead of reducing CO_2 with ribulose 1,5 bisphosphate carboxylase (RuBP) and producing phosphoglyceric acid (PGA) containing three carbons, they use the enzyme phosphoenol pyruvate carboxylase (PEP) in their mesophyll cells to form oxaloacetate, a four-carbon acid. This acid is reduced to malate (another four-carbon acid), transported into chloroplasts of the bundle sheath cells where CO_2 is regenerated, and only then used in the Calvin cycle. Pyruvate released during the CO_2 regeneration goes back to mesophyll cells to be regenerated into PEP (Fig. 3.2).

Species using this cycle have been labeled C_4 plants as opposed to C_3, or Calvin cycle only, plants. The two types also differ anatomically: there is no significant differentiation in mesophyll and bundle sheath cells in C_3 plants, whereas C_4 species have an arrangement where vascular conducting tissue is surrounded by a bundle sheath of large, thick-walled cells containing chloroplasts. PEP carboxylase has a greater affinity for CO_2 than RuBP carboxylase, resulting in higher concentrations in the bundle sheath cells and hence a more efficient functioning of the Calvin cycle. Moreover, as the RuBP, which can also use oxygen instead of CO_2 and thus produce rather than reduce CO_2, is only in the bundle sheath cells of C_4 species where O_2 concentrations are low, the opportunities for such an oxidation (photorespiration) are practically eliminated, further raising the photosynthetic conversion efficiency.

In C_3 plants only the reduction of atmospheric O_2 to 2% or greatly

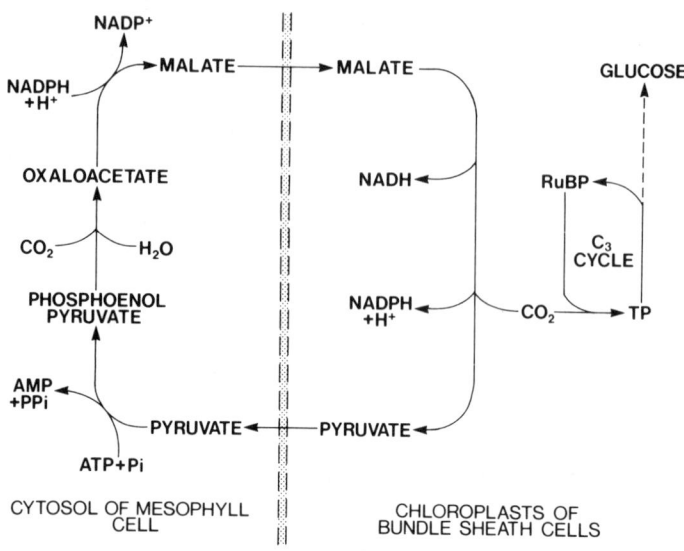

Figure 3.2 Basic outline of the C_4, or Hatch–Slack, photosynthetic cycle. (Danks, S.M., et al., 1983.)

elevated ambient CO_2 levels would eliminate large photorespiration losses. There is also no light saturation in C_4 species, whereas C_3 plants saturate at irradiances around 300 W/m^2; optimum temperature for net photosynthesis is 15–25°C in C_3 plants but 30–45°C in C_4 varieties, and water requirements of around 500 mol per mole of CO_2 fixed are only about one-half of common C_3 needs (900–1200 mol, but up to 4000 mol may be required by some species). C_4 plants thus thrive in warmer and drier environments. Store and Teeri (1978) found a combination of summer evaporation and dryness ratio to be the best predictor of C_4 distribution, a clear confirmation of the pathway as an adaptation to aridity.

Naturally, a C_4 pathway needs more energy than the Calvin cycle alone: additional ATP is required to energize the regeneration of pyruvate to PEP but overall net conversion efficiencies are much higher. Maximum daily growth rates of C_3 and C_4 species are, respectively, 34–39 and 50–54 g/m^2, a 40% difference, which is especially significant in food production (Monteith 1978). Daily maxima averaged over the whole growing season show still greater difference: with 22 g/m^2 C_4 plants are about 70% ahead of C_3 species fixing 13 g/m^2.

Crassulacean acid metabolism (CAM) is the other important modification evolved by succulent plants growing largely in arid and semiarid environments to minimize H_2O losses. This is done by absorbing CO_2 during the night and, as in C_4 species, converting it initially into C_4 acids. During the day with stomata closed sunlight energizes decarboxylation of these acids and C refixation into carbohydrates via the RPP cycle. These processes, unlike those in C_4 species, are not spatially separated: they take place at different times in the same cells. All CAM plants are succulents with considerable internal water stores and many species can totally suspend any CO_2–H_2O environmental exchange for weeks or even months.

Transpiration ratios of CAM species are as low 50, with typical values between 70 and 150, losses of an order of magnitude lower than in C_3 plants. CAM metabolism has worldwide distribution with Crassulaceae being the most important family in the northern hemisphere and Cactaceae in the Americas. Throughout the tropics there are numerous epiphytic Orchidaceae and Bromeliaceae, some of them able to shift between C_3 and CAM in response to environmental stresses.

Numerous intricacies of photosynthetic energetics remain unknown but certainly one of the most surprising weaknesses in our knowledge of life is our patchy understanding of phytomass stores and productivities. Our lack of satisfactory appraisal of photosynthesis on planetary and ecosystemic scales is more troubling than remaining gaps in our biochemical understanding.

3.2 PHYTOMASS STORES

At a time when we are gathering information from the surfaces of other planets we are still surprisingly ignorant about the distribution and total quantity of the Earth's phytomass. Of course, its variability (over 2×10^5 species) poses enormous challenges for any large-scale appraisals. Phytomass densities are largely determined by the presence of enclosed air spaces and minerals, with the extremes ranging from just 0.14 g/cm^3 for floats of aquatic macrophyta to over 1.2 g/cm^3 for heavily silicified diatoms. Water shares may be as low as 5% for mature seeds, as high as 95% for young shoots. The only way to ensure comparative uniformity is to express all masses in absolutely dry terms. Their densities are then 0.2–0.8 g/cm^3, with organic matter accounting for around 95% of most species but with inorganic substances making up as much as 50 or even 70% of the dry mass in many aquatic plants.

Four carbon compounds are the principal constituents of phytomass: monosaccharides (40% C, 15.5 kJ/g), disaccharides (42% C,

16.5 kJ/g), and polysaccharides (44% C, 17.5 kJ/g), which form the bulk (66–72% in softwoods, 74–80% in hardwoods) of wood, and lignin (63% C, 26.4 kJ/g, 19–30% of wood). Larger shares of energy-dense lipids (39 kJ/g) are usually only in seeds. Standard phytomass-to-carbon conversion value is 0.45, or about 17.5 kJ/g, and ecosystemic means range from 17.1 for tropical rain forests to 20.1 kJ/g for boreal forests (Lieth 1975).

There are fundamental uncertainties regarding both the areas of major ecosystems and their representative storages. Poor forest statistics, rapid deforestation in the tropics, and the lack of a uniform agreement on the classification of forested land are especially frustrating (Smil 1983). The best inventories indicate that the world's closed forest (canopies covering at least 20% of the ground) occupied no more than about 25×10^6 km^2 in the mid-1980s, with all natural forests within the tropics still over 10×10^6 km^2. To avoid the appearance of unwarranted accuracy I used heavily rounded values to estimate the global standing phytomass at a bit over 1×10^{12} t, prorating to stores of 160 MJ for each square meter of continental surface (Table 3.1).

In energy terms, forests store 90% of the total, grasslands 5%, and cultivated lands 0.5%. Tropical rain forest contains most of the world's species (Golley and Medina 1975; UNESCO 1978; Ayensu 1980). A typical Central Amazonian site may contain nearly 95,000 plants belonging to more than 600 species and stratified in six distinct layers over each hectare of land (Klinge et al. 1975). Emergent and canopy trees, altogether just some 50 species dominated by Leguminosae and Euphorbiaceae, store at least 80% of the standing phytomass.

Almost 67% of aerial phytomass is in the stemwood of dicot species, mere 2.5% in leaves. Total phytomass amounts to 450–500 t/ha (about 25% in roots) and there are nearly 300 dry t/ha of dead organic matter, nearly 90% in the soil. The forest's diversity means that a single tree species will store at most 5% of all phytomass. The most important reason for this high diversity is to increase spacing among the adult trees of the same species and thus to avoid high mortality of juveniles caused by concentrated heterotrophic attack (Janzen 1970). Clark and Clark (1984) found the seedling survival positively correlated with distance to adult and negatively with local conspecific seedling density.

Seasonality reduces heterotrophic pressure on temperate and boreal forests. Only a handful of species dominates in mixed forests; in many temperate rain woods and in boreal ecosystems a single

Table 3.1 Areas, Total Phytomass, and Average Storage of Major Ecosystems[a]

Ecosystem	Area (10^6 km^2)	Phytomass Storage (t/ha) Range	Mean	Total (10^9 t)	Energy Storage (MJ/m^2)
Forests	35	60–900	250	850	460
Tropical	10	150–600	300	300	510
Temperate	10	60–900	250	250	490
Boreal	15	60–450	200	300	400
Woodlands	15	20–200	75	110	140
Grasslands	30	10–200	20	60	35
Tropical	20	10–200	20	40	30
Temperate	10	10–200	20	20	35
Wetlands	5	1–200	75	40	140
Deserts	20	0–20	5	10	10
Farmlands	15	1–80	5	5	5
Cities	5	1–100	10	5	20
Tundras	10	1–40	5	5	10
Total	135	0–900	85	1085	160

[a] Based on data in Whittaker and Likens (1975), Olson (1982), and Smil (1983). All values are rounded to the nearest 5.

species may store the bulk of the site's phytomass (Persson 1980; Reichle 1981). Just one or two species dominate the world's highest accumulations of phytomass, the old-growth coniferous forests of western North America (Edmonds 1982). These ecosystems shelter the oldest living plant, a 4600 year old bristlecone pine, as well as the largest living tree (indeed, the planet's most massive living creature, although most of its phytomass is dead wood): giant sequoia (*Sequoiadendron giganteum*), growing over 100 m tall and surviving for over 3600 years.

Phytomass accumulations in these forests are enormous. A century-old forest dominated by Sitka spruce or noble fir can store about 900 t/ha, older stands of Douglas fir and noble fir may have up to 1700 t/ha, and the maximum levels for coastal redwoods are at 3500 t/ha (about 68 TJ/ha)—with all of these values *excluding* roots! Even the richest tropical rain forests will not store more than a quarter of this huge total; among the temperate deciduous ecosystems the richest cove forests of the Great Smoky Mountains store no more than 600 t/ha. As for the partitioning of phytomass, woody aboveground tissues in temperate and boreal forests account typically for 70–80% of total storage, roots for 15–25%. Needles of coniferous trees commonly store 4–8% of all phytomass, but some species store only around 1.5%, a share closer to the usual share of leaves of deciduous trees.

Tree phytomass of commercial interest, the so-called merchantable bole, includes, in its American definition, only stems with diameter over 12.5 cm at breast height above 30 cm stump and up to 10 cm diameter at the top. Exclusion of smaller stems, stumps, branches, and tops leaves about half of the temperate tree phytomass as timber. The FAO has been estimating worldwide mean growing stocks of timber at 90–103 m^3/ha for coniferous trees and 85–96 m^3/ha for broadleaves. In the tropics, Lanly's (1982) averages for the two kinds of trees in productive closed forests are 174 and 183 m^3/ha. Depending on the species, conversions of 1 m^3 freshly felled wood to oven-dry equivalents (in kg) requires multiplication by 370–420 for common conifers and 430–570 for temperate deciduous trees.

In spite of the enormous diversity of habitats and species there are unmistakable commonalities determined by energetic imperatives. In ecosystems with environmentally limited photosynthetic rates investments in annual renewal of leaves would far surpass the costs of their prolonged retention. Where leaf growth is energy-expensive, their life spans must lengthen to amortize the higher cost; thus evergreens are dominant in all subxeric and boreal habitats (Chabot and Hicks 1982). Their leaves have high weight per unit area (6.3–15 mg/cm^2 in dry weight, compared to the deciduous range of 2.9–7.8 mg/cm^2) as their photosynthetic tissues are diluted with structural supports and protective coatings.

The self-thinning rule, describing plant mortality owing to competition in crowded and aged stands, is another important general energetic imperative. Total tree mass (P) in a stand can be increasing independently of the stem density (p) until an intensive competition sets in, lowers the density, and limits the maximum weight (w_{max}) of surviving trees. This is known as the -1.5 rule where $w_{max} = k \cdot p^{-1.5}$ (k being a specific constant). Since w equals P/p, an alternative formula proposed by Westoby (1984) is $P = k \cdot p^{-0.5}$ (Fig. 3.3). The rule has three fascinating features: time can be ignored because mortality depends only on phytomass accumulation; this makes the thinning rate slower when conditions for growth are worse; and the location of the thinning line varies remarkably little for different species and growing conditions. The rule fits not only trees but also

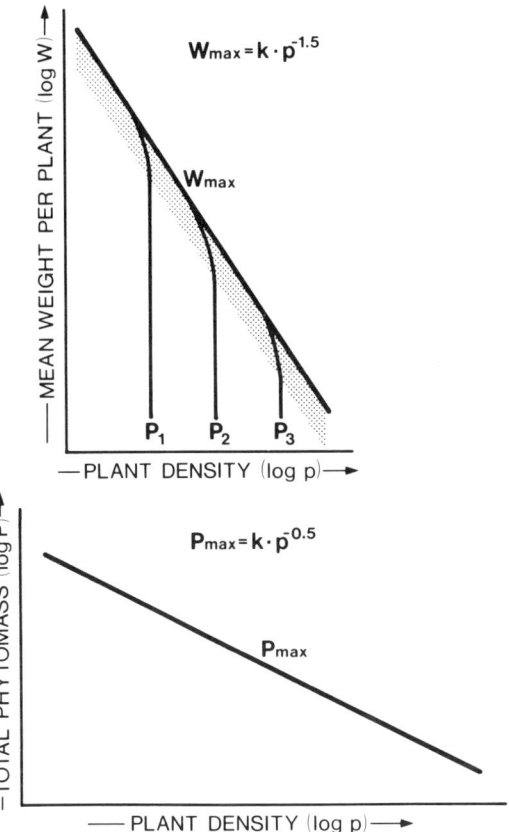

Figure 3.3 Two graphic presentations of the self-thinning rule using the symbols defined in the text.

shrubs, herbs, ferns, and mosses, spanning masses of more than 10 orders of magnitude.

This means that a monospecific even-aged stand with 100,000 stems per hectare can have plants averaging no more than 0.2–0.4 kg and adding up to the maximum of a bit over 40 t, whereas 100 trees per hectare would average over 6 t/tree and have a mass of more than 600 t/ha. Economies of scale are obvious: every tenfold concentration of photosynthetic conversion brings roughly a thirty-two-fold increase in the average plant size. Advancing concentration of phytomass results in development of long-lasting large-sized structures requiring durable construction (dense wood) and effective protection (barks or defensive chemicals) against heterotrophs. Growth rates will be much slower than in short-lived perennials or annuals, but the longevity of structures will be as much as 10^4 times greater.

Grasslands store an order of magnitude less phytomass per unit area than forests. Except for the tall tropical grasses, they have 2–5 times more standing phytomass underground than in the canopy shoots. Recorded extremes of live-shoot phytomass range from 76 g/m^2 for a dry formation in India to 2088 g/m^2 for humid Panamanian growth (Coupland 1979). Greatest live-shoot phytomass is associated with subhumid tropical and maritime climates or with temperate habitats with natural irrigation. Underground phytomass ranges from about 50 to nearly 5000 g/m^2, with the lowest accumulation in the tropics (median value about 700 g/m^2) and the highest values in temperate grasslands (median about 1700 g/m^2), an expected inverse correlation with temperature, the key determinant of decay rates.

A wealth of agricultural statistics may seem to make appraisals of crop phytomass quite easy but these figures rarely refer to the total aboveground stores and they inform about the underground phytomass just in the case of root crops. In traditional settings near-complete harvesting of aboveground phytomass is still common as crop residues (mostly cereal straws) are gathered for feed, fuel, and raw material in households and for rural manufacture; throughout Europe and North America crop residues are largely recycled. Estimates of residue production are commonly expressed as multipliers of crop harvests (dry-matter residue yields in relation to field weight of harvested crop).

As residue yields are determined by so many factors (above all cultivars, weather, soils, fertilization, irrigation, height of cutting), no single values can serve reliably either in generalized estimates or in localized appraisals. The only solid generalization is about the declining straw/grain ratio in modern short-stalked, high-yielding grain varieties. Consequently, residue multipliers for wheat, rice, and barley are now largely between 1 and 1.5 whereas in the traditional varieties they were between 1.5 and 2.3. Hybrid corns have straw/grain ratios between 0.9 and 1.2, traditional varieties commonly up to 2.5. For legumes the multipliers for vines are 0.5–1.0, for tubers 0.2–0.3, for sugarcane tops 0.2, for cotton stalks and leaves about three times the harvest of lint and seed.

For cereals the global means of total above-ground (grain and residue) phytomass are just around 5 t/ha; the best national levels are about twice as high. Averages prorated for the whole growing season or the whole year will be considerably lower. For example, for Canadian wheatfields the mean phytomass will be just over 1 t/ha. Addition of roots would raise these totals 15–33%. The FAO's worldwide crop production statistics make it relatively easy to estimate the global peak field phytomass. Mid-1980s dry-weight harvests of cereals, legumes, tubers, vegetables, oil, sugar, and fiber crops added up to about 2.3 Gt and residues increased the total to 5.5 Gt. This prorates to just over 6 MJ/m^2, storage two orders of magnitude smaller than in forest ecosystems.

Photosynthetic organisms in the ocean range from monocellular phytoplankton and benthic algae to large seaweeds, symbiotic corals, and vascular plants such as seagrasses. The smallest single-cell autotrophs in the ocean—ultrananoplankton of bacteria and blue-green algae—have diameter less than 2 μm; nanoplankton, 2–20 μm, includes diatoms, coccolithophores, and silicoflagellates; and diatoms and dinoflagellates are the most common kinds of microplankton (20–200 μm in diameter). Phytoplankton densities are highest near the coast and decline oceanward: there may be 10^3 microphytoplankton cells per milliliter on the continental shelf but only 10^0–10^1 cells/mL in the open ocean. Aggregated and colonial phytoplankton is characteristic of nutrient-poor tropical waters (Sargasso Sea).

Benthic autotrophs are mostly mixtures of algae and vascular plants, rocky coastal zones often support dense stands of macroalgae, and the reef-building corals live symbiotically with photosynthesizing dinoflagellates. Standing stock of intertidal phytomass is lowest in the tropics, highest on the boreal shores. Average density of algal beds and reefs is most likely no greater than 2 kg/m^2. Estuarine waters may have the same order of phytomass densities while the means for upwelling regions are two orders of magnitude and for open ocean three orders of magnitude smaller. Published global estimates fit mostly between 170 and 200 Mt, but Whittaker and Likens (1975) put the total as high as 3.9 Gt. In any case, the total oceanic storage is a tiny fraction of the terrestrial storage, which adds up to at least 250 times larger mass.

3.3 PRODUCTIVITY OF ECOSYSTEMS AND PLANTS

Primary productivity, the rate of synthesis of new living matter in dry weight, is quantified on three levels of declining totals but increasing utility. The gross primary productivity (GPP) includes all of the newly fixed phytomass before it is reduced by autotrophic respiration (R_A) to the net primary productivity (NPP). NPP has become the most frequently used rate in modern bioenergetic studies. But this fundamental measure is also elusive and it represents just a theoretical concept, ignoring heterotrophic respiration (R_H), that is, all

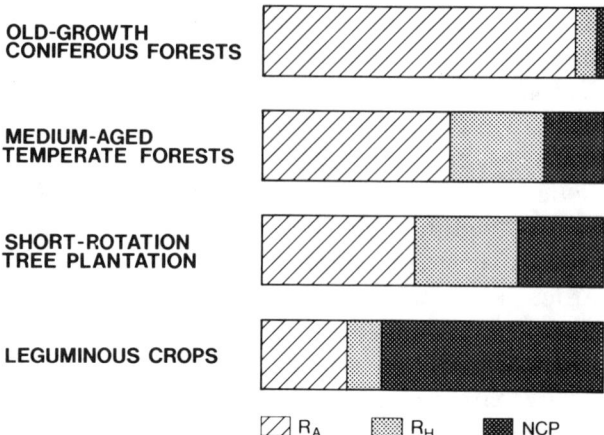

Figure 3.4 Older ecosystems channel more of their productivity into maintenance. Net community productivity (NCP) of mature forests may be nil or it may equal just a small percentage of the gross rate. The other extreme is represented by leguminous crops whose harvests may account for some two-thirds of gross production.

herbivore and microbial consumption. A rate that will inform best about the phytomass increments actually added to the existing stores and available for harvesting is the net ecosystem (or community) productivity (NEP or NCP), the difference between the GPP and the total respiration.

As the total respiration rises with the maturity and complexity of ecosystems the NEP of the most vigorously photosynthesizing communities is quite low (Fig. 3.4). In spite of its prodigious photosynthesis an old-growth Douglas fir forest of the Pacific Northwest stores annually only 2% of its gross fixation—while a grassland whose GPP is an order of magnitude lower stores nearly 25% of its primary synthesis. NPP remains the scientific measure of choice but only NEP, the yield, can be harvested. In many cases the usable yield may be only a part of the aboveground phytomass (in most cereals where straw is recycled), in others virtually the whole aerial part (sugarcane, trees for pulpwood or fuel chips), or the whole plant (algae, cassava roots for food, its stems and leaves for fuel and feed) may be taken.

As with the measuring of standing phytomass, productivity assessments are most difficult with forests (Lieth and Whittaker 1975; National Academy of Sciences 1975; Landsberg 1986). Highly accurate gas exchange studies are practical only on a very small scale $(10^0–10^1 \text{ m}^2)$; fairly reliable complete destructive harvesting is limited by logistics and cost to 10^3 m^2. Good approximations of aboveground

NPP can be based on short-term measurements of fixation rates. Actual net photosynthetic maxima in single leaves can go up to 85 mg $CO_2/dm^2 \cdot h$ for corn (C_4) and 62 mg for rice (C_3). The highest reported values for wild species (around 70 mg) show little difference between C_4 *Spartina alterniflora* and C_3 cattails *Typha latifolia*. As a group, of course, C_4 species have higher single leaf maxima than the typical C_3 range of 15–35 mg $CO_2/dm^2 \cdot h$.

When converted to mass photosynthetic rates these fixations would correspond to maxima of 50–65 $g/m^2 \cdot day$. Indeed, the reported daily growth peaks are 50–55 g/m^2 for C_4 crops (corn, millet, and sorghum) and forage grasses (Napiergrass, *Pennisetum purpureum*) and over 60 g for marsh plants. Typical daily means sustainable for several weeks of the most rapid growth are just around 20 g/m^2 for C_4 plants and 5–15 mg (average 13) for C_3 species. As most of the grain and oil crops mature in three to four months these typical growth rates would mean total phytomass productivities of about 5–25 t/ha.

For woody species maxima in optimum environments are 5–15 mg $CO_2/dm^2 \cdot h$ for conifers, 8–20 for subtropical and tropical evergreens, and 10–20 for temperate deciduous trees. With an average leaf area index of 6 these extremes would translate to daily maxima of 5–20 g/m^2. Actual maximum annual net productivities of forests are seldom above 40 t/ha and mostly between 5 and 20 t/ha, indicating typical daily leaf fixations of 1.5–5.5 g/m^2 for year-round growth or up to 10 g/m^2 for seasonal increments. Tree plantation productivities differ little from those of natural communities. Tropical forests should do better, with annual NPP up to 40 t/ha.

Forage grasses growing year-round under optimal conditions, as well as natural wetlands, would produce anywhere between 20 and 85 t/ha. In contrast, poor temperate and tropical grasslands may produce only 5–10 t/ha while their least stressed counterparts may have NPP ranging form 20 to 35 t/ha. In energy terms the extremes of 5 and 85 t/ha correspond to annual fixations of 8–140 MJ/m^2, giving a power density range of 0.25–4.4 W/m^2. This translates to maximum conversion intensities around 3 mW/g. In forests, where photosynthesizing foliage is only a very small fraction of total standing phytomass, conversion intensities range from 0.3 to 1.5 mW/g of leaf or needle mass but only 25–50 $\mu W/g$ when the total tree phytomass is included.

Estimates of global NPP present uncommon difficulties: too few sites have been studied to make confident spatial extrapolations. Fortunately, NPP has some strong correlations with environmental variables on scales ranging from small vegetation patches to global

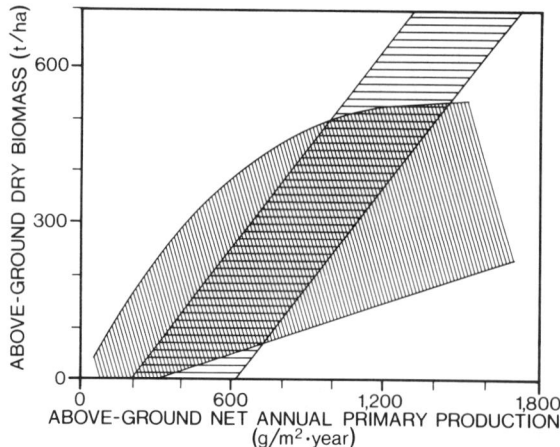

Figure 3.5 Relationships between primary productivity and aboveground phytomass generalized by Lieth (1975) and O'Neill and De Angelis (1981). The later data set (slanted shading) shows a much less linear trend.

patterns. For terrestrial ecosystems the key variables determining plant distribution are the annual temperature and precipitation means. But analysis of the most extensive data set on woodland productivity shows a surprisingly weak correlation between annual NPP and the aboveground standing phytomass for the middle of the sample, between NPP of 600–1200 $g/m^2 \cdot y$ and phytomass of 100–200 t/ha: ecosystemic complexities are not easily fitted into neat linear relationships (Fig. 3.5).

The earliest published estimate of global NPP put its continental total at 36.2 Gt (Schroeder 1919). Estimates during the 1960s ranged between 16 and 182 Gt for the continents and 30 and 72 Gt for the oceans (Bazilevich et al. 1971). Since then the results of the International Biological Programme (IBP) and accumulation of other productivity studies produced estimates scattered loosely around 100 Gt/y, with extremes ranging between 89 (Golley 1972) and 133 (Ajtay et al. 1979) Gt. Using the ecosystem areas and average energy contents listed in Table 3.1 and multiplying them by heavily rounded average productivities results in a total of about 2000 EJ of global NPP, a flux of nearly 65 TW, prorating to 0.5 W/m^2 of land (Table 3.2).

A large part of this production is being continuously discarded in plant litter. Meentemeyer et al. (1982) used a climate-dependent model to estimate the worldwide annual leaf fall at 35.1 Gt and the total litter production at 54.8 Gt. Studies in various ecosystems

Table 3.2 Total Production and Average Productivity of Major Ecosystems[a]

Ecosystem	Average NPP (t/ha)	Total Production (Gt)	Total Energy Conversion (EJ/y)	NPP Flux (W/m^2)
Forests	13	45	830	0.8
Tropical	20	20	340	1.1
Temperate	10	10	195	0.6
Boreal	10	15	300	0.6
Woodlands	10	15	285	0.6
Grasslands	10	30	495	0.5
Tropical	10	20	330	0.5
Temperate	10	10	165	0.5
Wetlands	15	8	140	0.9
Farmlands	5	7	120	0.3
Cities	5	3	50	0.3
Deserts	1	2	35	0.1
Tundras	1	1	15	0.0
Total	9	114	1975	0.5

[a] Calculated from data in Whittaker and Likens (1975), Olson (1982), and Smil (1983). All values are rounded to the nearest 1 or 0.1.

showed ranges of 5–15 t/ha (equivalent to 9–26 MJ/m^2, or 0.3–0.9 W/m^2) in tropical forests (11 t/ha may be a good mean) and mostly 4–8 t/ha (8–16 MJ/m^2) in temperate and boreal biomes where 4.5–5 t/ha may be a typical loss. These rates represent 45–55% of NPP, and as the annual litter fall in grasslands is similar, the weighted average of 50% may be a good approximation. Leaves are the bulk of the tree litter everywhere, between 55 and 75% in temperate forests, less in tropical forests.

Global NPP estimate may be also used for extrapolating the GPP total by assuming average respiration of 70% in forests and 60% in other ecosystems; the Earth's GPP would then be about 335 Gt, an equivalent of 5900 EJ or a bit over 1.4 W/m^2; the real value may easily be up to 15% higher or lower. Some 40% of NPP originates in forests, 25% in grasslands, and 6% in crop fields.

Reliable moisture and high thermal regimes of tropical rain forests translate into average NPP of around 30 MJ/m^2 (1.1 W/m^2), the flux nearly twice as high as the mean of temperate forests; reported maxima reach about 55 MJ/m^2 (1.7 W/m^2). Age-dependent differences in photosynthetic rates (averages of 14 mg CO_2/dm^2·h for early successional trees, 7 mg CO_2 for canopies, and 3 mg CO_2 for under-story plants) are very similar to the typical assimilation of temperate communities.

But there are some major distinctions as far as the intensity of nutrient cycling is concerned. Unlike the rain forests of the temperate zone, which are growing from relatively nutrient-rich mineral sub-strates covered with thick mats of accumulated litter, nutrients neces-sary for photosynthesis reside in the phytomass itself and the fertility of the forest is a matter of their constant rapid recycling with mini-mum losses. Nutrient-conserving adaptations include rapid direct absorption, extensive mycorrhizal symbioses, absence of denitrifying bacteria, leaves scavenging nutrients but resistant to rainfall leaching and heterotroph attack, and quick secondary regrowth in clearings (Jordan and Herrera 1981).

In spite of litter fall rates at least twice as high as in extratropical forests there is no accumulation, since the tropical decomposition constants (quotient of new litter added each year and the accumu-lated surface litter) have values between 3 and 5, while those in temperate deciduous forests are around 0.5, in boreal forests 0.13, and in dry, warm coniferous growth of California as low as 0.01 (Olson 1963). Perhaps the most notable dynamic difference between the tropical rain forest and the three principal biomes of higher latitudes is in the efficiency of nutrient utilization. Boreal coniferous forest can turn 1 kg of nitrogen into more than 250 kg of phytomass, temperate conifers can synthesize on the average about 180 kg, but deciduous forests do much less well (only about 100 kg/kg N) but are still ahead of the tropical rain forest where 1 kg of nitrogen produces only 80–90 kg of new phytomass.

Efficient use of nutrients, moderately high productivity, high efficiency of stemwood production (as a share of total insolation easily twice as high as in the tropical forests), and the high proportion of phytomass in the stem make the temperate forests, and especially their coniferous stands, most suitable ecosystems for man-agement. Coniferous forests have commonly accumulated 30–50 t/ha of litter, an order of magnitude more than the tropical growth, and a single application of fertilizers in these stands may be effective for 10–25 years. In managed forests the harvestable yields, rather than NPP, express the annual productivity. Whereas annual NPP of a mixed temperate forest is most frequently around 1 kg/m^2, NEP will be around 0.5 kg/m^2 and a forester's merchantable bole increment may be no more than 150–200 g/m^2 (most likely, he would express it volumetrically as 2–2.5 m^3/ha).

The FAO's global averages of gross wood increment in commercial forests have been 1.8 m³/ha for coniferous forests and 2.5 m³/ha for broadleaves. For tropical forests, Lanly (1982) used values of 1–2 m³/ha for broadleaves and 1.4 m³/ha for conifers, with actually allowable cuts between 0.5 and 1 m³/ha. The U.S. mean for the country's commercial forests is 2.7 m³/ha, whereas even the most productive forests in China have net wood growth of just 1.5 m³/ha (Smil 1984). In energy terms, typical sustainable worldwide wood harvests range from 4 GJ/ha in many tropical stands to over 20 GJ/ha in northern forests (dry weight equivalents are about 525 kg/m³ for hardwoods and 440 kg/m³ for softwoods).

The IBP summary of natural grassland productivities lists extremes of 702 g/m² in semiarid tropics to 3470 g/m² in subhumid tropical climate (Coupland 1979). Aboveground production accounted for 82–3396 g/m² and in only about a third of cases was it higher than the underground additions. Only in the highly productive tropical grasslands does the share of new production accumulating in shoots exceed 50% of the total. Dependence of grassland productivity on water availability is most evident in dry tropical locations. Curiously, the IBP data offer no clear rainfall–productivity link, or indeed any other easily discernible environmental associations.

With large parts of urban areas in temperate latitudes given to lawns it is interesting to note their very high productivity (Falk 1980): temperate lawns are highly productive grasslands whose NPP, ranging mostly between 1 and 1.17 kg/m², is comparable to the fixation of their natural counterparts. Compared to temperate lawns, average global cropfield productivity is low (Carlson 1980; Lowrance et al. 1984), Annual harvest (NEP) of 5.5 Gt of crops and their residues (or 6 Gt with the roots) prorates to just 400 g (6.8 MJ)/m². Performances of high-yielding crops are much higher.

Wheat or corn yielding 8 t of grain per hectare would have the whole-plant NPP of 15–18 t/ha, equal to that of a good lawn. Those C_4 crops that can fix CO_2 year-round do as well as the best natural grasses and when irrigated and fertilized do better than any other plants. Even the worldwide average of sugar cane NPP is about 30 t/ha and the highest recorded fixation in Java amounted to 94 dry t/ha. These record productivities—up to 160 MJ/m²—put sugarcane well ahead of other crops even after making adjustments for the length of growing period (365 days for the cane, 90–150 days for cereal crops, legumes, and tubers). In terms of solar conversion efficiencies these productivities translate to 0.10–0.70% for C_3 crops and up to 1.5–2.5% for C_4 cane.

High yields of modern cultivars have resulted above all from selection for a higher proportion of the harvested organ, mostly seeds.

Traditional varieties produce no less phytomass than the modern cultivars but the partitioning of their photosynthates is much less desirable. Changes of harvest index, the ratio of grain to the total aboveground production, trace this important shift best (Donald and Hamblin 1976).

Around 1900 the harvest index of unimproved cereals was just between 0.25 and 0.35 as the bulk of their phytomass was stored in long stalks and numerous leaves. Modern high-yielding cultivars have heavy ears born by short stems carrying fewer and narrower leaves. Winter wheats now have typical harvest indices of 0.40–0.42, corn 0.47–0.50, barley 0.51–0.57, and rice around 0.5. These advances brought higher yields without any increases in the rate of the photosynthetic process itself (Evans 1980).

Marine productivity is determined above all by the availability of light and nutrients (Morris 1980; Parsons et al. 1984; Valiela 1984). The depth of the photic zone ranges from just a few meters to 200 m. Photosynthetic rates have characteristic subsurface peaks corresponding to optimum light intensities of different taxa. Blue light penetrates farthest in the open ocean while even in clean waters little red light will be available for pigments below 10 m. Profiles of macronutrient concentrations nearly always show a sharp decline near the surface and a remarkable stability below 100 m, a result of dominant stratification of the ocean. Not surprisingly, areas of the highest NPP coincide with the zones of nutrient enrichment, be it by continental runoff in near-shore waters or by intensive coastal upwelling.

These eutrophic waters average 10^4–10^5 cells per liter and their standing phytomass is 100–1500 mg/m^3 (maxima up to 20,000). In contrast, the oligotrophic regions of the central Pacific or Atlantic Ocean have as few as 100 cells per liter and phytomass below 1 mg/m^3 (Kinne 1983). The best available mapping of NPP in the world ocean clearly shows these disparities between eutrophic and oligotrophic waters: the North Atlantic with extensive continental shelves and upwelling zones off Africa, the Pacific with shelves off Asia and upwelling zones along the coasts of the Americas, and the Northern Indian Ocean have the highest annual production.

The global magnitude of oceanic primary production was greatly overestimated by most of the earliest accounts. Only the 1960s brought lower values and greater consensus. The simplest calculation based on an average of 250 mg C/m^2·day and 361×10^6 km^2 of the ocean gives an annual fixation of about 33 Gt C. The Pacific Ocean produces about a third of this total but the Atlantic is more productive per unit area. An aggregate around 30 Gt C would mean annual fixation of 66 (60–75) Gt, an equivalent of nearly 60% of the total terrestrial NPP and a sum raising the aggregate planetary NPP to some 180 Gt.

Prorated over the world ocean this fixation would average just 180 $g/m^2 \cdot y$, a flux of around 100 mW/m^2, and a value one order of magnitude lower than the productivity of the richest forests and about 20% of average grassland NPP. Tiny nanoplankton commonly accounts for more than 80–90% of the ocean's NPP. The short life span, one to five days, means that annual phytomass turnover rates in the most nutrient-enriched shelf waters are 300–400 while even in the nutrient-poor open ocean they will be around 40–50.

Rapid turnover, low productivity, and low density of standing oceanic phytomass limit any harvesting of phytoplankton. The bulk of the ocean's NPP, coming from fluctuating masses of short-lived diatoms and various flagellates, will always remain beyond any rational consideration for harvesting as a source of energy, food, or fuel. Ocean macrophyta would do somewhat better: among brown algae giant kelp can photosynthesize at rates up to 8 mg $CO_2/dm^2 \cdot h$ (Lobban and Wynne 1981).

The most efficient converters on an ecosystemic scale are wetlands receiving natural nutrient subsidies. Both the tropical and temperate marshes convert commonly 1.5% of insolation and the best sites do at least twice as well. Typical grassland conversions range from 0.13% (a desert grassland) to 1.2% (a mountain formation); Alpine grasslands have efficiencies between 0.05 and 0.13%, Arctic grasses below 0.09%. The best conversions for temperate forests are around 1.5%, for tropical rain forests about 1%, in rich mature coniferous and deciduous forest just 0.4–0.9%, and only around 0.3% for more stressed growths

With annual global NPP averaging only 15 MJ/m^2 and with mean radiation reaching the Earth's surface at 5 GJ/m^2 the mean global efficiency of photosynthesis would be 0.3%. With oceanic NPP averaging 180 g, or roughly 3 MJ/m^2, the global mean of phytoplanktonic fixation efficiency would be 0.06%. Detailed long-term studies measured efficiencies from less than 0.092% in the oligotrophic Sargasso Sea to 0.9% in Nova Scotia's coastal waters and in excess of 5% in Eniwetok Atoll (Parsons et al. 1984). Thus both the terrestrial and marine efficiencies of NPP can differ by two orders of magnitude— and these differences are reflected, but far from perfectly mirrored, by the richness of heterotrophic life, whose energetics is surveyed in the next chapter.

4

HETEROTROPHIC CONVERSIONS: CONSUMER ENERGETICS

For the main factor in the nature of an animal is much more the final cause than the necessary material.... If any person thinks the examination ... of the animal kingdom an unworthy task, he must hold in like scorn the study of man.

—Aristotle
Parts of Animals

Heterotrophic life, precisely because of its total dependence on photosynthesis, evolved countless adaptations to cope with environmental challenges, diffused into nearly every conceivable niche, and eventually resulted in the emergence of global civilization. Photosynthesizing organisms take care of themselves by reducing atmospheric carbon dioxide and producing a vast variety of complex organic compounds. They are autotrophic. (But, as already noted, not all autotrophs are photosynthesizing: chemoautotrophic bacteria derive the needed energies from the oxidation of reduced gases as they provide critical links in cycling nitrogen, carbon, and sulfur.)

Heterotrophs use the wealth of substrates produced by autotrophs to energize their growth, maintenance, and activity: they are incapable of *ab initio* synthesis of complex compounds. They have two basic alternatives in performing their task, anaerobic fermentation or aerobic respiration. The first is limited to simple prokaryotic cells belonging to large phyla of fermenting bacteria (best appreciated in lactic fermentations), methanogenic bacteria, and *Ascomycota*, a fungal phylum containing yeasts responsible for alcoholic fermentation. Aerobic respiration is the way of life for 8 of 16 phyla of *Monera* (bacteria), ranging from myxobacteria to nitrogen fixers, for nearly all fungi, and for the whole kingdom of *Animalia*.

This prevalent mode of heterotrophic existence is examined in the following sections, overwhelmingly by focusing on the life of higher animals. Evolutionary steps to its origin are not difficult to postulate; details and timings are elusive. Free oxygen is the precondition of such existence but the preplanetary matter was clearly anoxygenic and the Earth's secondary atmosphere had only a limited amount of the gas formed by photolysis of water vapor by UV radiation.

Then the rise of photosynthesizing organisms increased the partial pressure of oxygen to the point where some prokaryotes could use aerobic respiration to generate energy in the form of ATP more efficiently than by fermentation. This shift happened repeatedly and independently among many bacterial species.

Energetic advantages of oxidation over anaerobic fermentation are clear. Lactic acid fermentation liberates 195 kJ for each molecule of glucose, alcoholic fermentation yields 232 kJ—while a complete oxidation of that sugar releases 2.8 MJ, a twelvefold to fourteenfold gain. This is the basis for enormous diversification of heterotrophs and for their extensive radiations.

The best way to deal with this diversity on the organismic level is to work with allometric relationships (Pedley 1977; McMahon and Bonner 1983; Peters 1983; Schmidt-Nielsen 1984). A large number of variables determining animal metabolism are scaled exponentially relative to size of the whole body or its parts. These attributes fit the general equation

$$y = ax^b$$

where a is a constant multiplier needed to express the result in particular units, x is the variable of size (most often body weight), and b is an appropriate scaling exponent.

As with the coverage of photosynthesis, I initially concentrate on the essentials of heterotrophic metabolism. This is followed by a look at energy cost of major reproductive and growth strategies and locomotion modes. The chapter closes with an assessment of heterotrophic biomasses and large-scale productivities.

4.1 ENERGY METABOLISM

Polymeric carbohydrates stored in plant tissues must be broken down to their constituent monosaccharides, with glucose being the dominant nutrient. Triglycerides (lipids) are hydrolyzed into glycerol and fatty acids. Amino acids are the third large class of energy sources, but only if carbohydrates and lipids are in short supply; otherwise they go into protein synthesis in consumers' bodies. Energy released by oxidation of these nutrients is partially conserved in ATP. As in photosynthesizing cells, ATP is the principal energy carrier, the key link between cellular catabolism (degradation of nutrient substrates) and anabolism (biosynthesis), locomotion, and active transport of metabolites.

ATP is the universal cellular energy currency owing to its intermediate free energy of hydrolysis (–31 kJ/mol) among the organic phosphates formed during the respiration of glucose, the process of unrivaled importance in energizing heterotrophic bodies. Because of this it can readily donate a phosphate group to produce glucose-6-phosphate (its free energy is just −13.8 kJ/mol) at the outset of glycolysis—and it can be easily formed from a much more exergonic 1,3-diphosphoglyceric acid (free energy −49.3 kJ/mol) later in that multistep process. Biochcmistry of thcsc enzymatically catalyzed reactions is well understood (de Duve 1984).

The maximum energy gain is 38 mol of ATP for each mole of glucose broken down in prokaryotic cells, an overall free energy change of about −2.8 MJ. With −31 kJ/mol available from each ATP transformation to ADP the overall efficiency of the whole sequence would be about 42%. In eukaryotic cells the net ATP gain is a bit smaller, but as the free energy of the compound may be up to −50 kJ/mol in living mammalian cells (a value of −31 kJ/mol is valid only for unimolar concentrations, neutral pH, and 25°C), the overall efficiency may be over 60%.

Respiration of fatty acids yields the maximum of 44 ATP per mole but as the oxidized compounds have higher energy contents than glucose (around 3.4 MJ/mol), the peak efficiencies will be about 60%. And although only two molecules of ATP are gained during the breakdown of glucose to lactic acid, the overall free energy change of this transformation is just −197 kJ and the process, at about 31%, is surprisingly efficient. But in vertebrates it can be sustained only for limited periods of time.

Intensity of ATP generation is stunning (Broda 1978). A 60 kg man consuming daily about 12 MJ of food in carbohydrates (i.e., about 700 g) would make and use no less than 70 kg of ATP (assuming production of 36 molecules of ATP for every digested hexose molecule), more than his total weight. This rate, roughly 3 g ATP for every dry gram of body, is minuscule compared to intensities achieved by respiring bacteria. *Azotobacter* breaking down carbohydrates while fixing large amounts of dinitrogen produces 7000 g of ATP for each gram of its dry mass!

Heterotrophs vastly surpass the Sun in power output per unit of mass. While solar luminosity is immense (3.9×10^{26} W), so is the star's mass of 1.99×10^{33} g. Consequently, the Sun's power intensity averages just about 200 nW/g but the daily metabolism of schoolchildren proceeds at a rate of 3 mW/g of total body weight (15,000 times the power intensity of the Sun) and respiring *Azotobacter* reaches up to 100 W/g, or 500 million times the Sun's rate. ATP-driven energy conversions in heterotrophs are no less awesome in their intensity than the Sun's performance is in its overall magnitude.

Kleiber (1932) collected basal metabolic rate (BMR) data for animals ranging from rats to steers and showed that they depend on total body weight (w in kg) as a function of $3.52w^{0.74}$. Hundreds of species were subsequently added to the original data set but the exponents of recalculated equations always came out close to 0.75. Eventually Kleiber (1961) recommended a rounded expression of $70w^{0.75}$ (in kcal/day) or $3.4w^{0.75}$ (in W) and the straight log-log mouse-to-elephant line became one of the most important generalizations in bioenergetics (Fig. 4.1).

Data for marsupial mammals ($2.33w^{0.737}$) support the approxima-

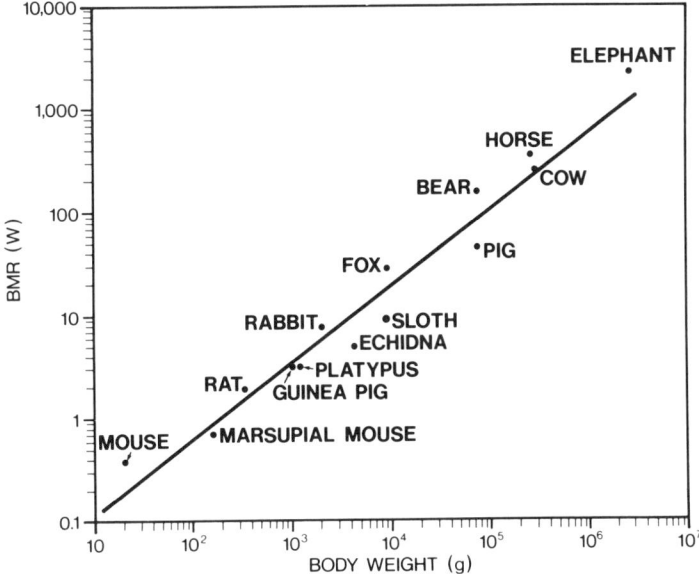

Figure 4.1 Plot of the best available mammalian basal metabolic rates (Eisenberg 1981) shows that both the mouse and the elephant are significantly above the predicted mouse-to-elephant line following Kleiber's $w^{0.75}$ law.

tion of the 0.75 line, but the constant is only about 70% of the eutherian line. BMR of larger nonpasserine birds is also very close to the 0.75 line, but passerine birds have metabolic rates 30–70% above that level. Slopes for various invertebrate groups vary from less than 0.67 to more than 1.0, but most of them are close enough to the 0.75 line to conclude that that slope is representative for all ectotherms (Schmidt-Nielsen 1984). But ectotherm BMRs are only 2.5–5% of equally massive homeotherms.

The best available theoretical explanation of the 0.75 slope is McMahon's (1973) consideration of mechanical requirements of animal bodies. Elastic criteria require a proportional relationship between the cube of the critical breaking length and the square of the diameter (d) of a loaded animal limb or spine. As the weight of these loaded members is a fraction of total body mass (w), their diameter will be proportional to $w^{3/8}$. Power output of muscles depends only on their cross-sectional area (proportional to d^2) and thus the maximum power output is related to $(w^{3/8})^2$ or to $w^{0.75}$. If applicable to any particular muscle, the scaling should rule the whole organism and BMRs should be a function of $w^{0.75}$.

Animal metabolism has declining specific rates (total BMR divided

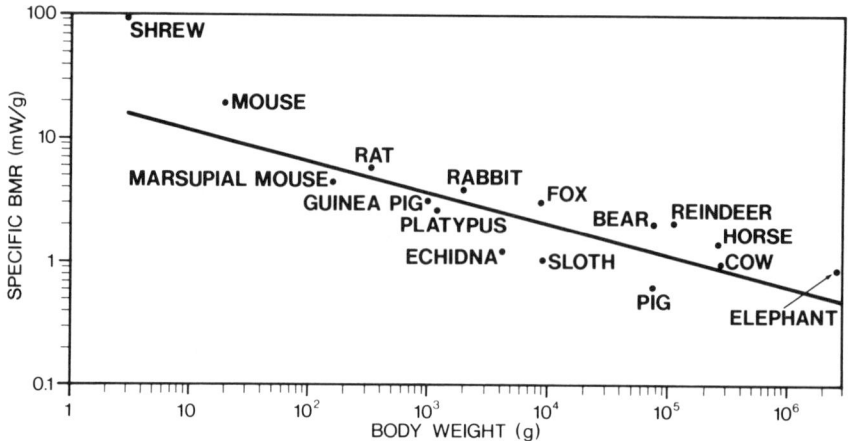

Figure 4.2 Departures from the allometric decline of specific BMR show the precarious position of a shrew—as well as a pig's unusually low rate, which is essential in making the animal an efficient converter of feed. Calculated from data in Eisenberg (1981).

by body mass). A simplified relationship (in W, weight in kg) is $3.4w^{-0.25}$. A 10 g chickadee will need 10.8 mW/g, a 100 kg calf just 1.08 mW/g, indicating that a 10,000-fold jump in body mass reduces the specific metabolism by 90% (Fig. 4.2). This reality limits the size of the smallest warm-blooded animals: creatures lighter than shrews and hummingbirds would have to feed incessantly to compensate for rapid heat losses.

There are two grand strategies of coping with nonoptimal temperatures: ectothermy and endothermy (poikilothermy and homeothermy) with some transitory adaptations. All ectotherms—arthropods, fishes, amphibians, and reptiles—have relatively very low BMRs and poor body insulation. Their thermoregulation is behavioral; that is, they seek optimal microenvironments.

In terrestrial ectotherms body size is a critical determinant of sustainable body temperatures—and hence of their behavior and niches (Stevenson 1985). Large ectotherms with high heat capacity warm up slowly, but their maximum body temperature ranges are narrow (just 3.5–5.5°C in giant tortoises and only 2°C would be expected for an ectothermic 3000 kg dinosaur), they are able to maintain a larger gradient between body and ambient temperature, and their considerable thermal inertia allows for longer periods of daily activity. In contrast, small insects unable to raise their temperature above the ambient level can readily control it by moving around. But many flying insects can be endothermic during the short periods preceding their takeoffs when they warm up their flight muscles by shivering.

Portable environments of endotherms are highly uniform, 36–40°C for most mammals and 38–42°C for birds. These temperatures are already approaching the level of protein decay. But why are not the homeotherms regulating at lower levels, perhaps just around 20°C? Regulation at such levels would call for lower metabolic rates but for high rates of evaporative cooling: with surface body temperatures below the ambient level there would be no conduction or convection heat loss.

These high evaporative heat losses would pose excessive risks of desiccation and would restrict the radiation of homeotherms in arid climates. Maximization of evaporative cooling would also require sparse insulation, which would restrict the diffusion in cold environments. And lower temperatures would reduce the efficiency of most body systems, making such a low-level homeothermic effort hardly more useful than heterothermy. An evolutionary compromise is then responsible for the success of endothermic creatures: stable body temperatures are as high as possible given the danger of heat death and the energetic cost of metabolism, water loss, and insulation (Spotila and Gates 1975).

Disadvantages of smaller size in endotherms are obvious: convection losses and demands for insulation increase, as does the necessity of rapid metabolic response to falling temperature. Since the capacity to store fat is directly proportional to body mass, large mammals can maintain constant temperatures for months without eating. No hibernator heavier than about 5 kg has to reduce its winter body temperature by more than a few degrees, but small mammals would starve in a few days unless becoming torpid (French 1988).

Insulation value of feathers and furs can be stunning. Arctic mammals have skin temperatures comparable to those for a well-clothed man even at −32°C. Plumage of birds is about a third more insulative than is the pelage of similarly sized mammals. Color of the body cover makes no difference as far as the absorption of IR radiation is concerned, but light-colored species may reflect over 50% of visible wavelengths while the dark-colored species may reject just 15%. Actual effects on radiation balance are complex and the amount of solar energy reaching the skin may be higher in a lighter colored animal.

Coping with heat is no less a challenge. Small desert rodents as well as camels reduce their water loss by using countercurrent heat exchange in respiratory passages to cool the exhaled air to temperatures much lower than their body cores. Some desert rodents and birds resort to estivation (torpidation), a lethargic state similar to hibernation. Ability to live on air-dried food, production of concentrated excreta, hiding in the burrows, and tolerance of salt water are other common adaptations among small mammals (Schmidt-Nielsen 1979).

The link between endothermy and evolutionary success of birds and mammals is obvious. Possession of constant portable microenvironment had to be bought by many adaptations maintaining very high metabolic rates, but it conferred clear competitive advantages in benign environments and opened up the most inhospitable parts of the biosphere for colonization. Our species could not have succeeded without it: ectothermic sapience is unthinkable, at least with life as we know it.

4.2   REPRODUCTION AND GROWTH

Heterotrophs have a continuum of reproductive strategies ranging from maximizing birth and growth rates, often in a single prodigious reproductive bout (semelparity), to successive breeding (iteroparity) producing fewer but larger neonates of greater competitive ability. Borrowing the terminology of growth equations MacArthur and Wilson (1967) labeled the first strategy r selection (r being the intrinsic rate of increase), the other one K selection (K being the upper asymptote of population size).

Clearly, r selectionists are great opportunists, pouring a much larger part of their production into reproduction, which makes them into obnoxious pests and efficient colonizers. But this prodigious channeling of energy into offspring severely limits the survival of parents and the chances of repeated reproduction. Endoparasites are notable exceptions to this rule. In contrast, adaptation to limited resources makes the larger K selectionists long-term occupants in more stable settings. Most heterotrophs do not operate at these extremes: they tend toward the r or K end of the continuum.

Biochemical commonalities make possible a generalized calculation of conversion efficiencies achieved by growing heterotrophic organisms (Calow 1977). When assuming average energy density of zoomass polymers at about 23 kJ/g ATP must supply 860 J to synthesize this biomass, and with 65% efficiency of tissue respiration 1323 J are needed to generate the requisite ATP. Besides the 463 J lost in this transformation another 619 J will be lost during the energy transfer from ATP to the polymers so that the net efficiency of zoomass growth will be about 96%.

This is only a theoretical maximum. More realistic rates—considering digestive efficiency at 90%, molecular turnaround at least once in the species' lifetime, molecular transport losses, and mechanical inefficiencies—would be just over 70%. Actual efficiencies are easily measured in rapidly growing unicellular organisms: net values range between 50 and 65% for bacteria; protozoan and yeast efficiencies are about 40–50%. Estimates for invertebrates show gross growth conversion efficiencies of 30–65% for mollusca, 35–55% for crustaceans, and 25–60% for insects.

Gains (in g/day) for neonate mammals scale as $0.0326w^{0.75}$. Notable departures from the trend are the fast-growing pinniped carnivores and the slow-growing primates. Mammalian neonates average about 12% of protein and 2% of fat, but gray seals have 9 and humans up to 16% of fat, and hence an extraordinarily high energy density of 8.75 kJ/g at birth, compared to usual mammalian values of 2.9–3.6 kJ/g. Conversion efficiencies decline with age from 30–37% during the initial postnatal growth to about 15–20% when reaching 25% of adult weight and to well below 10% when nearly mature.

An evolutionary tradeoff works toward equalizing gross conversion efficiencies of different mammals: herbivores have low absorption efficiency (an understandable consequence of digesting such intractable polymers as cellulose and lignin) but high net conversion rates, whereas carnivores have high absorption rates but rather low net efficiencies, a result of a much more mobile way of life dictated by the search for and capture of prey.

4.3 LOCOMOTION

Much of the human fascination with the animal kingdom arises from our admiration of running ungulates, flying birds, or swimming whales. Many feats of animals in motion are extraordinary—none more than the annual long-distance migrations—but direct measurements of energy requirements of animal locomotion are either difficult or impossible and much of our knowledge must rest on theoretical considerations.

Running, flying, and swimming performance is limited by maximal power availabilities, that is, by metabolic scopes of heterotrophs. These scopes usually have been expressed as ratios between maximal and resting oxygen consumption. For running mammals the scopes are typically about 10 times the resting rate but for horses the range is 20 and it peaks at 31–32 in coyotes, wolves, and dogs. For birds the scopes do not appear to go beyond 15 (but they start from a higher basal level), a range that may be sustained for short periods of time by some fishes. Reptilian and amphibian scopes are no higher than 5–10, posing a severe limitation to activity of these heterotrophs (Huey et al. 1983).

Whereas the top aerobic power is around 150 mW/g in small birds and 50 in small rodents, it is no higher than 9 mW/g in toads and 3 in iguana. Rapid motion of many reptiles depends primarily on short, anaerobically energized bursts requiring long subsequent periods of recovery. Flying insects put out mostly 0.12–0.58 W/g, or up to 100 times their resting rates (moths can go up to 150 times).

Swimming needs little or no energy to support the neutrally buoyant fish and the effort is overwhelmingly channeled into overcoming the drag of the relatively dense medium. The product of drag (kgm/s^2) and speed (m/s) gives the metabolic power (kgm^2/s^3) needed for swimming; since drag is proportional to the square of speed, the power needs will go up with the cube of velocity. Specific energy cost of swimming appears to be very similar for all species, declining with size along a line sloping at approximately -0.3.

This physical imperative—larger animals having lower specific locomotion expenditures—is repeated with the runners and flyers. The number of steps a running animal must take per unit of distance is inversely proportional to its length (i.e., roughly to the one-third power of its mass), while the work accomplished for each step must be proportional to its weight. The work of running over a unit distance then is proportional to the 0.67 power of the mass, and that of per unit body mass per unit distance to the -0.33 power of the mass. Simply stated, smaller creatures must take many more steps, each one needing power in direct proportion to its mass.

Actual measurements confirm the expected economies of scale with near perfection as their regression slopes are between -0.28 and -0.35. This close match of simply derived expectations and actual needs is remarkable because the animals are far from isometric. If the cost of locomotion is related to one step, animals appear to be equally economical regardless of their size (Schmidt-Nielsen 1984). This means that the cost of running at higher speeds will rise much faster in the smallest mammals than in the more massive species—but also that there is little difference in power inputs among animals with similar body mass; consequently, such differently structured mammals as goats and cheetahs have similar locomotion energy needs!

Maximum oxygen consumption increases with body mass to the power of 0.85 while the exponent for the total running costs is only 0.67. Because the available power goes up faster than the cost of running, large mammals are able to develop speeds much higher (up to 10 times) than the smallest ones—and regardless of their apparent fitness for running (hence the high speeds of bears, hippos, or buffalo). Large mammals also save much energy (more than half at high speed) because of the elastic structures of their legs (Alexander 1984).

Most of the energy invested in locomotion is in the search for food. Searching strategies have been traditionally divided into "cruise" (or widely ranging) and "ambush" (sit and wait) modes. Many animals fit neatly into these categories—hawks and tunas in the first one, rattlesnakes and herons in the second—but an extensive survey of search strategies indicates that most foragers exhibit saltatory pattern of movement, alternating pauses and moves and falling somewhere between true cruisers and ambushers (O'Brien et al. 1990).

As they graze or hunt many animals seem to be trying to maximize their total energy gain per unit of foraging time. Optimal foraging theory initially perceived such behavior as one of the great natural laws and its proponents still believe that optimal foraging maximizes animal fitness and plays a critical role in natural selection (Stephens and Krebs 1987). Critics have pointed out that foraging behavior is shaped by more factors than just searching for and consuming feed; necessity to maintain constant antipredator vigilance is perhaps the most critical concern for most grazers.

More fundamentally, the critics argue that animals are not simply designed, that optimal strategies may or may not emerge from the randomness of mutation, and that, in any case, their existence is intestable (Pierce and Ollason 1987). Most important is the question "What does natural selection maximize?" The life of heterotrophs cannot simply be partitioned into independent activities—and optimal foraging is not one of the universal energetic imperatives of heterotrophic life. Consequently, it may be more rewarding to look at several important links between foraging and energy expenditure.

Garland (1983) reviewed critical relationships for mammalian species: daily movement distance (DMD), incremental cost of locomotion (ICL), daily energy expenditure (DEE), and ecological cost of transport (ECT). DMD scales as $1.038w^{0.25}$ (w in kg) but it can vary by almost two orders of magnitude. Carnivores move much more than herbivores (respective body weight multiples are 3.87 and 0.87), but the exponents are identical (0.22). ICL (in J/km, a constant independent of speed) scales as $10{,}678w^{0.7}$. DEE is a relatively uncertain variable averaging about $800w^{0.71}$. ECT, expressed as a percentage of DEE [ECT = 100 (DMD × ICL/DEE)] scales as $5.17w^{0.21}$ for carnivores and as $1.17w^{0.21}$ for other mammals. ECT of small noncarnivorous animals thus may be less than 1% of their DEE, whereas large carnivores may spend 10–15%, or even a larger share of their DEE, on locomotion.

Most insects excel at hovering and are competent, if rather slow forward fliers, but it is the flight of birds that has been so envied by humans. Men lack the radical avian adaptations: lengthening of forelimbs to produce wings delivering sufficient lift and thrust, muscles strong enough to sustain the flapping, respiratory system superior to mammals in its extraction of atmospheric oxygen. Most of our knowledge of bird flight is derived from aerodynamic theory supported by obviously difficult free-flight measurements of mostly smaller (30–300 g) birds (Tucker 1973; Greenewalt 1975; Kendeigh et al. 1977).

Larger birds must fly faster to generate sufficient lift, hence the wing loadings must go up with mass, approximately as the power of 0.33. Typical flight speeds are proportional to wing loading to the power of 0.55 and hence to body mass raised to the power of 0.18. Speeds minimizing the cost of transport are equal (in m/s) to $14.6w^{0.2}$, a relationship resulting in usual velocities of about 9 m/s for a starling and close to 20 m/s for large geese.

A simplified formula puts the power requirements (in W) directly proportional to body mass multiplied by a factor of 84.7. Since the power needed for flight goes up faster (exponent 1.0) than the available power (a multiple of the resting rate whose exponent is 0.72), there must be a maximum size for flying animals: no birds heavier than about 13 kg fly and larger flyers spend hours as gliders aided by wind currents rather than as active flappers. Similar limitations appear to restrict the size of hovering flyers to birds lighter than 100 g.

The energetics of long-distance migrations of tiny birds is satisfactorily explained by preflight accumulation of fat and by utilization of tailwinds (Baker 1981). Whereas a sparrow's ratio of lipid/lean dry weight is 0.34–0.44, this ratio was measured up to 3.42 for blackpoll warblers and as high as 3.5 for ruby-throated hummingbirds ready to embark on their long flights (Blem 1980). In-flight consumption of

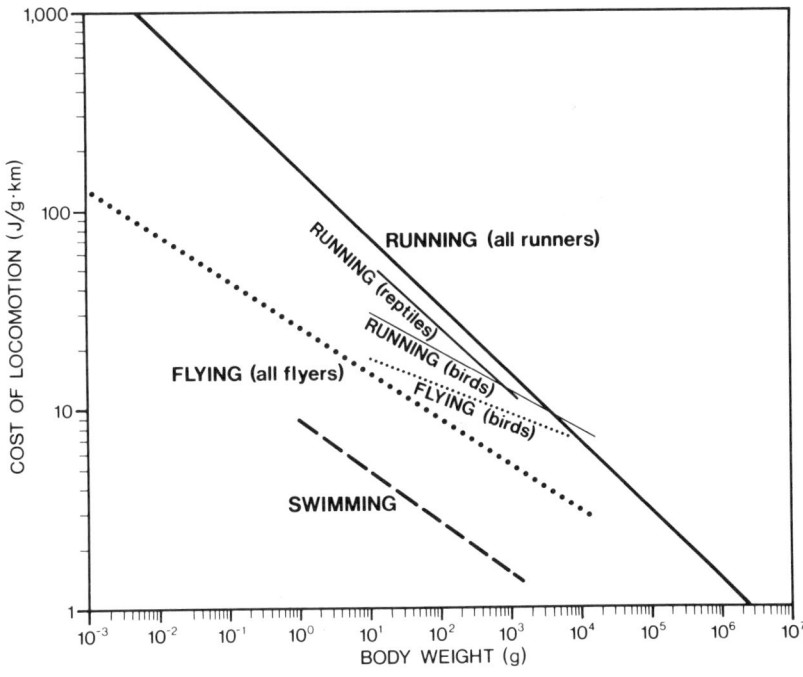

Figure 4.3 Significant differences separate energy costs of running, flying, and swimming. Running is always the most energy-intensive mode of locomotion but it can be done by heterotrophs, whose body mass spans nearly 10 orders of magnitude. In contrast, there are few fliers heavier than 10 kg. We have no reliable information on the energy cost of swimming among whales. Calculated from equations in Robbins (1983) and Schmidt-Nielsen (1984).

25% of body mass is common in long-distance migrants and 50% loss is possible.

Flying heterotrophs are by far the most important biotic pollinators, and their invaluable services must be energetically rewarding. The three key energetic variables of pollination include the nutritional content and metabolic availability of blossom-derived food; the work involved in extracting the food; and the energy spent in flying to blossoms (Faegri and Pijl 1979). Small ectothermic pollinators can live on dilute nectar and can harvest minute drops; larger endotherms must feed on more concentrated sugars present in larger volumes.

Comparisons of the three modes of locomotion done in terms of energy costs per gram per kilometer eliminate the speed as a variable and make it possible to generalize on energy requirements of unrelated species moving at speeds differing by more than an order of magnitude (Fig. 4.3). Running is clearly most costly. Its incremental cost is as high as 100 J/g·km for the smallest mammals. There is no substantial difference between bipedal and quadrupedal running, or between the bipedal running of birds and quadrupedal motion of mammals. Reptiles have nearly identical incremental energy needs for running; even cockroaches and ants fall close to the plotted line.

The flying line shows the total cost of the activity but the comparison is hardly affected as the flight metabolism is a large multiple of the resting rate: for animals between 1 and 1000 g flying is always cheaper than running. Insects need more than 20 J/g·km, the largest birds just a few joules. Swimming is the least energy intensive way to move but our knowledge of actual expenditures does not extend to the largest swimmers, weighing 10^8 g.

Before presenting a broad survey of animal biomasses and productivities I must recall the disparity between our knowledge of photosynthetic biochemistry and ecosystemic realities and note that the gap between our appreciation of animal metabolism, reproduction, growth, and locomotion on the one hand and zoomass densities and ecosystemic productivities on the other is even greater. Furthermore, rapid extinction rates in many locales will preclude the eventual closing of that gap.

4.4 BIOMASSES AND PRODUCTIVITIES

Mobility of animals is one of the principal difficulties encountered in assessing heterotrophic biomass and productivity. Monitoring methods are often tedious and usually expensive, and they have large error margins. The variety of heterotrophs and a considerable variability of their composition are additional complicating factors. There are more than a million different heterotrophs, even omitting bacteria, fungi and protozoa (Barnes 1989). Most of these species are arthropods (at least 800,000, dominated by insects) and there are only about 40,000 different vertebrates, 50% of them bony fishes, nearly 25% birds, 14% reptiles, and just about 10% mammals.

The third major difficulty arises from large variability of heterotrophic body compositions. For most vertebrate species there is at least a twofold difference of energy densities depending on their growth stage, seasonal intraindividual variations on the order of 20–30% are common for many birds and mammals, and energy contents of different developmental stages or castes of insects differ by up to 40%. Approximate means are 18.5 kJ/g for bacteria, 22 for fungi, 19 for molluscs, 23 for arthropods, 21 for fishes, and 23 for mammals and birds.

Soil heterotrophs represent the bulk of the zoomass in most terrestrial ecosystems where they provide many invaluable ecosystemic services, above all as decomposers. Bacteria are by far the most abundant microorganisms. In poor soils there may be less than 10 g of bacteria per square meter but an alfalfa field may have up to 900 g/m^2 (9 t/ha) of living cells. As the size of soil heterotrophs increases, their density declines: there are 10^{14} bacteria, 10^9 fungi, 10^7 nematoda, and just 10^2 earthworms per square meter of soil. Earthworms are the most conspicuous soil invertebrates. In Darwin's (1888) opinion, "It may be doubted whether there are many other animals which have played so important a part in the history of the world, as have these lowly organized creatures."

Total invertebrate soil biomass ranges from just a few to a few hundred grams per square meter, or about 20–1100 kJ/m^2 with values around 100 kJ/m^2 being perhaps most common. Total aboveground zoomass can only very rarely approach or surpass this value. The expected inverse relationship between size and density holds for all heterotrophs as their numbers per unit area are equal to the -2.25 power of their length (Fig. 4.4). Densities of the largest, strictly herbivorous aboveground heterotrophs are thus very low, but their combined zoomass can rival that of soil invertebrates in some exceptionally favorable circumstances.

Up to the early 1970s the Ruwenzori National Park in Uganda supported as much as 145 kJ/m^2 of large herbivores, whereas the Serengeti total is only about 60 kJ/m^2. These zoomass densities are not matched anywhere else. Ants and termites usually do not add up to more than 20 kJ/m^2; total avifaunas barely surpass 1 kJ/m^2 even in tropical forests; zoomass of insectivorous mammals is nearly always below 1 kJ/m^2 as their bodies are small. The feeding base is reduced in higher trophic levels and so is the zoomass of omnivorous and even more of carnivorous animals. The total for all predators in the Ngorongoro Crater, perhaps the world's best place for carnivores to hunt ungulates, is less than 0.7 kJ/m^2, roughly 1% of the herbivorous prey (Schaller 1972).

A plot of mammalian densities and body masses shows an overall negative slope of -0.75 (Damuth 1981), a finding corresponding to the -2.25 slope of the length–density line in Fig. 4.4 (mass is the cube of length, and the -0.75 power of this cube is equal to the -2.25 power of the length). More interestingly, since the BMR is proportional to the 0.75 power of the body mass, Damuth's rule

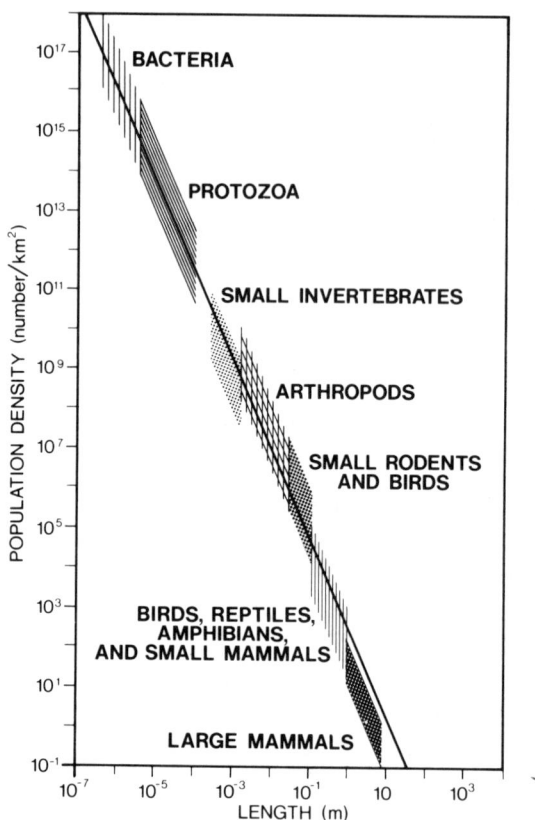

Figure 4.4 Population density of heterotrophs plotted against their body length on a double-log graph. The curve has a slope of −2.25; shadings indicate approximate positions of major groups. Based on Damuth (1981), Eisenberg (1981) and Mc-Mahon and Bonner (1983).

implies that energy harvested daily per unit area is a constant independent of the unit mass of the feeding animals.

No herbivorous species of mammals can thus outstrip another energetically only because of its larger size. But the standing zoomass of the smaller species will be considerably lower. Or as Van Valen (1973) put it, larger body size means that "there are fewer species and fewer species per genus, but not necessarily less participation in the energy flow." At the same time it must be stressed that density differences of up to three orders of magnitude can be found for identically massive animals with weights between 0.1 and 100 kg; ecosystemic and interspecific variabilities leave much room for departures from the general trend.

These realities are well illustrated by scaling analyses of home ranges whose extent is determined by a combination of metabolic needs and feed quality, density, and variability. Ranges of carnivorous passeriformes scale to 1.75, herbivorous galliformes to 1.39, omnivorous rodents to 0.97, and only herbivorous rodents fall close to the metabolic exponent with 0.81 (Mace et al. 1983). For many birds and mammals home ranges must overlap. Degree of this overlap scales to between $2.19w^{0.27}$ and $4.6w^{0.34}$ (small animals up to 2 kg are either solitary or in pairs and small groups, larger animals form herds) and the exponents are similar, but of opposite sign, to those scaling range for the turnover time of the standing zoomass.

This led Calder (1983) to generalize that in the long run each home area will support the same number of animals, either larger ones sharing the space and developing social structure or smaller solitary species with much faster rates of sequential replacement. This simplification ignores the ability of larger species to migrate or the common use of hibernation and estivation. Consequently, many animals need much larger or smaller areas than suggested simply by their size, but in general size is a critical determinant of life strategies.

Size has a much less reliable relationship with trophic levels. Elton (1927) was the first ecologist to recognize the declining numbers of heterotrophs in higher food-web levels while pointing out their often increasing size. Subsequently, Hutchinson's research opened the way for Lindeman's (1942) pioneering trophic–dynamic approach to ecology. Hairston et al. (1960) generalized this fundamental energetic concern by concluding that "herbivores are seldom food-limited, appear most often to be predator limited, and therefore are not likely to compete for common resources."

In this three-link food chain only the predators are energy limited and their pressure on grazers does not allow these primary consumers to regulate their energy intake. Fretwell (1987) extended this approach to a continuum of terrrestrial food chains with up to four links. Five links may be the limit in aquatic environments although the richest kelp beds and coral reefs may be exceptions going up to the seventh level. There is no simple progression of animal sizes along these chains.

If there is no simple pattern of sizes, there is an inevitable decline in energy transfers first quantified in Lindeman's (1942) paper in terms of progressive efficiencies [assimilation at level n/assimilation at level $(n - 1)$]. These efficiencies vary widely with different ecosystems and trophic levels. Shares of available NPP actually consumed by all herbivores can range from just 1–2% in some temperate forests and fields to 5–10% in most forest ecosystems, and they can reach maxima of 25–40% in temperate meadows and wetlands and extremes of 50–60% in rich tropical grasslands (Crawley 1983).

When the ratio is limited to aboveground heterotrophs it is rarely above 10% in any temperate ecosystem, and when it is restricted to vertebrates it would be mostly around 1% in these environments. With assimilation efficiencies ranging typically from 40 to 60% for ectotherms and 60 to 90% for endotherms possible values of Lindeman's efficiencies at the primary consumer level (excluding decomposers) would only infrequently fall outside the range of 1–15%.

Ecological efficiencies for all herbivores as well as for individual grazers can depart significantly from the frequently cited "typical" 10% transfers. The dominant range for ecological efficiencies of predators is higher, 5–20%, with maxima around 25%. There is no evidence of predictable taxonomic, ecosystemic, or spatial variation in these efficiencies for either grazers or predators. Efficiency split along the thermoregulatory divide is clear (Brafield and Llewellyn 1982). Ectotherms have growth efficiencies commonly around 20% and often over 30%, while for endotherms 2% is typical, 4% maximum.

Ecological growth efficiencies generally reflect this split (maxima of just 4% for endotherms, over 30% for carnivorous ectotherms) but the values are relatively low for some detritivore invertebrates (5–8%). Endothermy reduces massively the ecological growth efficiency, but for individual species this high price is adequately compensated by competitive advantages homeotherms of every size enjoy in every terrestrial and marine environment. But they cannot dominate the planetary heterotrophic biomass: its bulk is formed by mostly unseen or inconspicuous microbes and invertebrates. They could thrive without large vertebrates, without men, but we could not survive without their decomposition and nutrient cycling services.

Large, long-lived animals have high capacity to accumulate zoomass but they do so very slowly and the energy flux through their populations is relatively low. For elephants the annual consumption/zoomass ratios are less than 10, for the smallest rodents over 700. Small mammals and birds have relatively large energy throughputs and high capabilities to produce new zoomass: production/zoomass ratios (turnover rates) are 2–3 for rodents and 0.05 for elephants.

Arthropods are the most prolific producers with production/biomass ratios of up to 10 for termites and some spiders. The ratios are generally higher for ectotherms (mostly well above 1, although

Table 4.1 An Estimate of Planetary Heterotrophic Biomass[a]

Biota	Total Biomass (Mt)	Approximate Energy Content (EJ)
Terrestrial		
Decomposers	7000	140
Invertebrates	800	17
Wild vertebrates	10	
Domesticated vertebrates	200	5
Mankind	80	2
Total	8090	164
Oceanic		
Decomposers	1000	19
Invertebrates	500	10
Fish	100	2
Mammals	20	1
Total	1620	32
Grand total	9710	196

[a] For comparison, terrestrial phytomass stores just over 1 Gt, oceanic about 700 Mt.

ants rate only between 0.3–0.9), but show no obvious dependence on their body mass. In contrast, for endotherms there is the expected decline with increasing body mass: their production/biomass ratios range from about 5 for shrews to 0.5 for larger rodents to 0.05 for elephants, following the -0.75 slope (Bakker 1975).

Some arthropods also have aggregate consumption rates that actually rival those of the largest herbivores: many ant species consume 200–350 kJ/m^2 annually, as much as elephants. But all small heterotrophs have a limited capacity to accumulate new tissues in their tiny bodies and hence their densities remain low. McCullough (1973) noted that it is no accident that the livestock species combine the capacity to produce with that to accumulate zoomass. Such compromise mammals usually weigh 40–400 kg, birds between 1 and 10 kg.

Global assessments of heterotrophic biomass cam aim at just the right orders of magnitude. Conservative calculations result in a terrestrial total of roughly 8 Gt (Table 4.1). This is less than 1% of standing phytomass, with vertebrates accounting for only about 3% of heterotrophic biomass and decomposers for at least 80% of the total. Estimates for aquatic ecosystems are even more uncertain.

Data for freshwater fish (Gerking 1978) indicate biomass densities ranging between 50 and 500 kJ/m^2; densities of invertebrate herbivores and detritivores go up to 1 MJ/m^2.

Using a rough combined figure of 5 kJ/m^2 for all habitats (Mann 1984) results in some 300 Mt of fresh weight of standing fish biomass. For comparison, potential annual yield of major commercial species (herring, mackerel, anchoveta, cod, plaice, tuna, salmon) was put at at least 100 Mt/y, that of squids at up to 100 Mt, and the yield of all other potential catch at 100–200 Mt (Tont and Delistraty 1977). The multilayered nature of oceanic life, its great patchiness, and its migratory nature make the accounts of marine zoomass and their averaging per square meter difficult.

In any case, there is certainly a much smaller difference between decomposer and invertebrate biomasses in the ocean than on the land, and the inverted trophic pyramid means that, on the average, standing heterotrophic biomass is at least twice as large as the oceanic phytomass (Table 4.1). Total oceanic zoomass may equal about one-fifth of the terrestrial aggregate with vertebrates being relatively twice as abundant as on the land and decomposers accounting for less than two-thirds of all heterotrophic mass.

Global grand total of close to 10 Gt, or roughly 200 EJ, of heterotrophic biomass equals roughly 1% of all phytomass stores. Perhaps the most notable outcome of this exercise is the demonstration of a relatively high and steadily rising share of the anthropomass. At the beginning of this century biomasses of humans and wild vertebrates were about equal, but by its end there will be an order of magnitude difference. Moreover, zoomass of domesticated vertebrates, dominated by large ungulates, is as much as 20 times the wild vertebrate total. Few comparisons demonstrate so impressively the relentless ascent of the most adaptable as well as the most destructive heterotrophic species, whose metabolism, food requirements, and energy expenditures will be examined in the next chapter.

5

HUMAN ENERGETICS: NUTRITION, METABOLISM AND ACTIVITY

It is clear that, contrary to the general impression, the emperor of human nutrition has no clothes but at least he knows what clothes he should have, for his subjects have by now produced an almost complete list of the raiment, haberdashery and accessories necessary—but they cannot be worn with conviction while there is continuing dispute concerning the size and cut of every item in the list.

—Michael V. Tracey
Human Nutrition

Between 1840 and 1880, during the great pioneering decades of modern science, studies of human energy needs, conversions, and expenditures were in the mainstream of the new science of energy. The contributions of these studies were critical in formulating the canons of general energetics. The subsequent emergence of small, portable combustion engines and the diffusion of electric motors marked the beginning of the demise of man as a prime mover, a shift that has been largely responsible for removing human energetics from the core of energy studies and for its overwhelming relegation to specialized niches of physiological and nutritional interest.

Yet studies of human energetics have been advancing in many fascinating ways with new research illuminating everything from minutiae of cellular biochemistry to appraisals of biophysical limits of the body as a machine, from the roles of individual nutrients to constructions of nationwide food energy balances. But much remains uncertain and unknown. After looking at complexities of human nutrition, and before surveying the record of human energy expenditures and thermoregulatory capacities, this chapter highlights several major existing uncertainties concerning human energy requirements, a matter of fundamental importance for adopting rational nutritional and agricultural strategies.

5.1 NUTRITIONAL COMPLEXITIES

No energy conversion is more immediately essential for our existence than continuing oxidation of foodstuffs—yet our understanding of this process ramains surprisingly incomplete (Tracey 1977). About 50 essential nutrients, including not only preformed organic compounds but also numerous mineral elements, must be obtained in plant and animal foods, digested, absorbed, and distributed by the blood for the body to grow, live, and adapt. Energy-yielding compounds— carbohydrates, proteins, and lipids—are macronutrients, consumed on the order of 10^1–10^2 g/day, but adequate intakes of micronutrients—vitamins and minerals—are also essential. Deficiencies of calcium, phosphorus, magnesium, and zinc impede normal growth; vitamin inadequacies disrupt the functioning of metabolic and maintenance processes.

Most of the biosphere's huge carbohydrate stores are metabolically inaccessible to humans, who cannot digest wood's lignin, cellulose, and hemicellulose. Our dietary carbohydrates are complex polysaccharides and simple sugars, monosaccharides fructose and glucose, and disaccharide sucrose. All of these compounds contain 17 kJ/g and are almost always the principal energizers of human life. They are eaten as cereal products, tubers, legumes, and refined sugar. Complex carbohydrates are preferable to supply indigestible dietary fiber, a critical part of proper nutrition (Trowell et al. 1985).

Proteins, containing 23 kJ/g, supply amino acids necessary for normal growth. Synthesis of human proteins is impossible without concurrent digestion of 11 essential amino acids, which also must be present in correct ratios. Available complete in animal foods and mushrooms and incomplete (one or more amino acids not present in the desirable ratio) in plant foods (above all in legumes, cereals, and nuts), their qualitative importance far surpasses their possible energetic contributions.

Lipids (fats) are the most energy-dense nutrients (39 kJ/g) containing three essential fatty acids; like their amino counterparts, these must be digested preformed to become precursors of prostaglandins (to regulate gastric function and smooth-muscle activity and to release hormones) and parts of cell membranes. The other role of lipids is to carry fat-soluble vitamins (A, D, E, and K). Fats come concentrated in plant oils, butter, and lard as well as in meat, fish, and dairy products. Their average consumption in preindustrial societies was very limited; now they constitute as much as 40% of total energy intake in rich nations.

Digestion in healthy people on balanced diets has a very high efficiency for carbohydrates (99%) and a slightly lower efficiency for fats and proteins (95 and 92%); more than 20% of dietary protein (about 5.2 kJ/g) is lost through urine. Shares of energy actually available for body maintenance and growth are thus virtually equal to the gross energy content of carbohydrates (17 kJ/g), marginally lower for lipids (38 rather than 39 kJ/g), and appreciably lower for proteins (17.23 kJ/g). Most food composition tables list actually available energy. Finally, alcohol (ethanol) is a peculiar energy source of relatively high density (29.3 kJ/g); it can be utilized only by the liver at hourly rates not surpassing 0.1 g/kg body weight (i.e., about 190 kJ/h, or 53 W, for a 65 kg man).

5.2 REQUIREMENTS AND UNCERTAINTIES

Fundamental relations are easily stated. Human energy needs are determined by expenditures. Basal metabolism—that is, energy required for the maintenance of vital functions and homeothermy—is nearly always the largest component of these energy expenditures, and it is a function of body size, composition, and age. Activity needs—energy spent at work and during leisure and varying with the nature, intensity, and duration of the tasks—account for most of the remaining expenditures. Requirements of growth—the sum of the energy value of the newly formed tissues and products and the cost of their synthesis—are of minor importance except in infants and pregnant and lactating women.

Accurate quantifications of these links are difficult. Calculations of expected requirements must rely on information derived from large-scale measurements of basal metabolic rates or actual food intakes, although typical activity expenditures come from much smaller sets of respirometric observations. Given their considerable variabilities, these measures are only approximate indicators of individual or population needs whose actual levels are also affected by the quality of diets and by health.

International consensus on food energy needs has undergone a number of adjustments since the periodic standard-setting meetings of nutritional experts begun shortly after World War II. The latest consensus meeting expressed total energy requirements as multiples of the BMR, almost always the largest component of energy expenditure that can be fairly accurately quantified under standardized conditions (Joint FAO/WHO/UNU Expert Consultation 1985). The most accurate determinations of the BMR would require a precise knowledge of body composition in order to ascertain the total mass as well as the shares of all metabolically active tissues.

Both of these values are highly age-dependent, and sexual differences are of importance owing to the principal role of fat, which accounts for most human dimorphism (Bailey 1982). The difference is very small at birth (newborns average about 14% fat), but it increases with age: by age 60 Western men average 23%, women 36% fat. An average Western adult weighing 65 kg, with lean body mass of 53 kg, 12 kg of fat, 10.5 kg of bones, and 25 kg of muscles, would store about 500 MJ. For global estimates a value about 20% smaller (lighter stature of non-Western populations and their lower fat content), that is, roughly 400 MJ, would be more appropriate. Global anthropomass would then equal about 2 EJ.

The insidious storage flux of fat with aging—at about 6–16 MJ/y, or 0.2–0.5 W—has its counterpart in the loss of lean body mass. At birth only around 20% of a baby's mass is muscles, muscle tissue averages 52% of weight in young men and 40% in young women, and autopsies show people over 70 years with about 40% less muscle than they had had as young adults. Rapid buildup of the metabolizing lean body mass in childhood and adolescence, its adult stabilization, and its later gradual decline are reflected in variations of metabolic rates. But even in adulthood the four metabolically most active organs— kidney, heart, liver, and brain—account for nearly two-thirds of the BMR while a newborn's brain, just one-tenth of its body weight, will account for almost exactly half of resting metabolism.

By the mid-1980s about 11,000 BMR measurements in healthy individuals of both sexes and all ages were available (Schofield et al. 1985). The latest consensus consultation used these measurements to find that simple linear equations give the best correlations in all categories. Inclusion of body height in the equations has no effect on improving the fits, which are rather poor for adults: for men aged 30–60 (daily BMR $= 0.0485w + 3.67$ MJ), that is, for most of the economically active population, they explain only 36% of the group variation.

Basal metabolic rates per kilogram of body weight are about 2.3 W at birth, peak at around 2.7 W three to six months later, and then start sharply declining (Fig. 5.1). By the time body growth stops in the late teens the rates are only slightly above the adult level, and after staying fairly stable for four decades they renew their decline after 60. By the age of 70 the fire of life, a mere 1 W/kg, burns at less than half the rate at birth.

Energetics of human growth is complicated by considerations of body composition: proteins are always more expensive to synthesize. There is no shortage of values on energy cost of growth, ranging from as little as 14.6 kJ/g for some infants recovering from malnutrition to over 34 kJ/g for adults during overfeeding experiments. Average of 21 kJ/g, representing basically the cost of growth in young children, is a generally accepted value (Spady et al. 1976). Energy cost of growth is highest during the first month of life, claiming about a third of food intake. Then it declines to 5–6% by the end of the first year and about 2% by the end of the first decade; the growth spurt in the early teens pushes it up to 3–4% but soon comes the swift descent of the late teens to small fractions of 1% needed for the maintenance and renewal of adult tissues.

Pregnancy among well-fed women of rich countries has a higher relative energy cost than childhood growth. There is not only the addition of new tissues—fetus, placenta, and associated maternal growth—but also a rise of the BMR and the cost of increased cardiovascular and respiratory effort, so that the total average weight gain of 12.5 kg (baby's birth weight of 3.3 kg and mother's 4–5 kg of

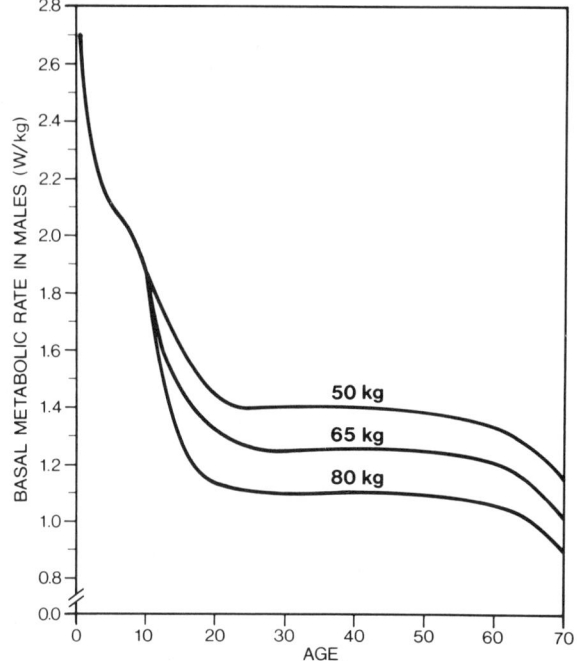

Figure 5.1 Lifetime of basal metabolic rates for males: great uniformity in infancy and a fast decline in childhood are followed by considerable adult differentiation. Calculated from equations in Joint FAO/WHO/UNU Expert Consultation (1985).

fat reserves account for most of it) costs about 335 MJ (Hytten 1980), or almost 27 kJ/g. But most of the world's pregnancies are actually far less energy intensive—and many of them appear to carry no extra cost at all.

Average energy content of human milk is 2.9 kJ/ml; its median production rises from about 720 ml/day during the first month to 850 ml/day in the third month and declines afterward. Conversion of food energy to milk has a high efficiency of about 80%. With a daily mean of 800 ml lactation would cost 2.9 MJ a day. As a typical Western woman would start breastfeeding with some 150 MJ of fat reserves that will be converted to milk during the following six months, the actual additional energy cost is only about 2.1 MJ/day. Still, this is nearly twice the extra energy need in pregnancy and 370 MJ needed for half a year of breastfeeding represents a higher energy cost than carrying a baby to delivery after 280 days.

In poor countries, where smaller women give birth to smaller babies, the costs should be lower. Astonishingly, several sets of detailed energy balance studies show that many women in poor countries have, not as exceptional individuals but as groups, extremely low energy needs during pregnancy and lactation. Prentice (1984) found that, compared to standard metabolic expectations, pregnant rural Gambian women appeared to have energy shortfalls of at least 600 kJ and as much as 2.15 MJ/day, even if they just sleep and rest, and up to 2.50–4.14 MJ when performing their heavy work duties.

The unavoidable conclusion is that these women can maintain genuine energy balance on what seem to be incredibly low levels of food intake—not merely 5–10%, but easily 20–40 and even close to 50% lower than the expected requirement. Adair and Pollitt (1982) found a similar situation in Taiwanese mothers bearing healthy children. And Norgan et al. (1974) found that among the Kauls of New Guinea there was no difference in energy intakes of nonpregnant and nonlactating and pregnant and lactating women.

Whole populations have been shown to live with surprisingly low food energy intakes. Among the Senegalese Ferlo a large seasonal food deficit (nearly 1.25 MJ/day compared to standard recommendations) is not accompanied by any significant increase in malnutrition or by any clinical signs of food deficiency (Benefice et al. 1984). And in New Guinea the average energy intakes of the whole coastal Kaul tribe were just 27% (males) and 13% (females) higher than their expected basal metabolic rates. Not only some exceptional individuals but whole populations can use their food energy much more efficiently than the standard expectations would have it. Reduced activity, weight loss, or lower milk production would be undesirable, but clear evidence of harmless adaptations means that there is a range of long-term averages of energy intakes compatible with regulation of expenditures without exceeding the homeostatic limits.

Recommendations of energy intakes thus are pursuits of a moving target whose position is determined not only by differences in individual or population-wide energy conversion efficiencies but also by culturally conditioned work habits and attitudes, seasonal fluctuations in staple diets, and a host of genetic and environmental factors controlling the adaptive process. In the short term there is no pattern relating intakes and expenditures of individuals; in the long run the two values coincide but large individual differences in homeostatic levels make any a priori applications of standard calculations even more error-prone than in the case of populations.

Looking back, Widdowson (1983) noted that the fundamental question she had posed in the late 1940s—"Why can one person live on half the calories of another, and yet remain a perfectly efficient physical machine?"—has never been satisfactorily answered. Because of these basic uncertainties, estimates of national or even global prevalence of malnutrition based on comparisons between known means of food energy supplies and standard expected requirements are highly questionable. And even in protein-energy malnutrition, the most acute form of nutritional deficiency afflicting many children in poor nations, energy intakes in some areas may be adequate, or even excessive, in relation to body weight or age (Bhattacharya 1986).

All we know is that *Homo sapiens* is a flexible converter of food energy responding with altered metabolic efficiencies to different diets, environmental conditions, specific tasks, and health states. The question about food requirements is not simply "how much?" but must include "for what?" and "in what context?" These questions remove the search for food requirements from the realm of quantifiable considerations to the much larger and largely unquantifiable setting of cultural preferences and social expectations.

Human energetics is so contextual and so value-laden precisely because it concerns humans. Borrini and Margen (1985) summed up its challenge well: before defining specific food requirements it is imperative to appreciate the perceived needs and wants of the people and their customs, the structure and dynamics of their society, and the ecology of their environment; in synthesizing these determinants we may discover that there is not only no single answer—but no simple answer as well.

5.3   ENERGY EXPENDITURES

Energy expenditures can be measured directly and with great accuracy: classical experiments confirmed the equivalence of food intakes and energy losses and little can be done to improve their design and accuracy. Indirect calorimetry, measuring the volume and gas content of expired air, is much more practical. Oxygen utilization and carbon dioxide production can be easily calculated and converted into energy equivalents (1 liter O_2 equals 21.1 kJ/g of starch, 19.6 kJ/g of lipids, and 19.3 kJ/g of protein). Thousands of respirometric investigations are now available, and Durnin and Passmore's (1967) work remains an unsurpassed comprehensive summary of these efforts.

The simplest way to express the energy requirements of various activities is to contrast them with the BMR, that is, with energy flows of about 70–90 W for adult males (78 W for the reference man of

65 kg) and 55–75 W for females (64 W for the 55-kg woman). Sitting and standing require the deployment of numerous muscles to maintain desirable postures and even the former may require as much as 60% more energy than a person's BMR. Typical markups for sitting are 15–20%, while standing usually requires 1.3–1.5 times BMR. Studies of walking on a level surface indicate lower efficiencies at speeds both below and above the optimum range of 5–6 km/h.

Gross energy cost of walking varies not only with speed but also with sex and age: as a group, adult slim women have the lowest requirements and heavier women the highest requirements, but the differences are not very large, varying only about ±10% from the mean. Uneven surfaces, muddy roads, or deep snow will raise the costs of level walking by up to 25–35%. Walking with light loads (up to about 15 kg) on the level is not too demanding as such loads are carried with about the same effort as similar additional body weight. Heavier loads can be very taxing, especially when carried uphill. Walking uphill is a function of both the gradient and the speed, and detailed studies show nearly linear increases across a broad range of speeds and inclines.

Running usually requires 700–1400 W (about 10–20 times BMR for adults). Compared to other mammals, energetic cost for running humans is relatively high, but humans appear to be unique in virtually uncoupling the cost of transport from speed (Carrier 1984). Whereas quadrupeds have optimum speeds for different gaits (walk, trot, and canter for horses), human costs of running per unit distance are essentially independent of the speed between about 2 and 6 m/s. Two factors explain this extraordinary capability: bipedalism (in quadrupeds ventilation is limited to one breath per locomotor cycle as the thorax bones and muscles must absorb the impact on the front limbs) and efficient heat dissipation.

Thus Tarahumara Indians of northern Mexico could run down deer and Paiutes and Navajos pronghorn antelopes, Kalahari Basarwa could chase to exhaustion duikers, gemsbok, and during the dry season even zebras, as did some Australian Aborigines with kangaroos: in hot weather these excellent animal runners simply cannot match man's ability of long-distance variable-speed running and heat dissipation. Carrier (1984) believes that this was a notable evolutionary advantage that served man's ancestors well in appropriating a new niche—as diurnal hot-temperature predators.

Man's excellence in running also can be illustrated by steadily improving record speeds for every distance. Since 1910 the annual rate of improvement has averaged less than 1 m/min, a gain imperceptible in sprints now run at speeds over 10 m/s but a reduction of about half an hour in running the marathon (42,200 m), which is now run by many of the world's best athletes at a faster pace than the record 10 km run as recently as 1945. Ryder et al. (1976) believe that the historic rate of improvement could continue for several decades.

An enormous variety of occupational activities is best treated in groups dividing work into exertion categories first proposed by Christensen (1953). For men light activities involve exertions demanding 1.8–4.3 times BMR, moderate 4.3–6.6, heavy 6.6–8.8, and very heavy 8.8–11.2; for women the respective multiples times are 1.6–3.8, 3.8–5.9, 5.9–8.1, and 8.1–10.3. Most of the jobs in the now huge service sector fall into the light work category. These include virtually all office, education, retail, food, and health services, as well as repairing cars or driving trucks.

Although the brain's high metabolic rate is responsible for up to 20% of adult BMR, thinking hard makes a barely discernible difference: long periods of intensive mental effort may leave a person tired, even exhausted, but respirometry studies showed long ago that the additional maximum energy cost of complicated brainstorming is no more than about 4 W; science is, energetically at least, light work. Light expenditures also characterize a multitude of tasks in modern manufacturing and the construction industry. But as soon as carrying, lifting, and loading of materials is involved the averages rise into the moderate or heavy expenditure range. Farming and extraction industries, often even when modernized, require plenty of moderate exertion and frequently heavy work.

Classification of most recreational activities is critically dependent on their intensity. Light workouts usually demand moderate expenditures and normally moderate exertions may be readily pushed into the heavy category. Mass-participation sports usually requiring heavy energy investment are basketball and soccer, which include a good deal of running. Among the less widespread high-energy pastimes are many track and field disciplines, mountain climbing, cross-country running, rowing, and squash.

Extremes of energy expenditures among active adults thus range from as little as 6 MJ (about 70 W) for a slim, older woman taking care of a small household to over 30 MJ (230 W) for a lumberjack working a long shift. On a given day, or for a few days, expenditures and intakes need not balance, but in the long run they have to tally. Energy requirements of a person or a population are calculable and predictable with a fair degree of accuracy—but only once all the necessary demographic and anthropometric data are available and once there is sufficient information to make sensible estimates about typical levels and durations of occupational and leisure activities.

This is an extraordinarily challenging task even for a small population and simplifications are unavoidable. The FAO regularly prepares such approximations and during the 1980s their daily per capita values ranged from 9.2 MJ in India to 11.1 MJ in the United States; averages for poor nations range mostly between 9.6 and 10.0 MJ and for the rich they are about 10.8 MJ. Climatic differences should be considered in such calculations but their effect on energy needs is not easily generalized.

What is much easier to do is to outline the limits of humans as machines. Not surprisingly, individual performances will vary rather widely but the maximum power outputs and the best achievable efficiencies can be determined fairly accurately.

5.4 LIMITS OF PERFORMANCE

An impressive amount of experimental work (McGilvery 1975; Shephard 1978; Bouchard et al. 1981; Nadel 1985) allows solid generalizations about the peaks and limits of human energetic performance. During outbursts lasting seconds (lifting heavy objects, sprinting) as well as during sustained taxing effort lasting tens of minutes or several hours, limits of human performance are set by the rates of hydrolysis of high-energy compounds. ATP is stored in muscles at minuscule levels, averaging 5 mmol/kg of wet tissue. With 20 kg of active muscles and 42 kJ/mol of ATP this is an equivalent of just 4.2 kJ, the total sufficient to energize contractions for just 1/2–3/4 second of maximal effort.

The most rapid way of replenishment is through the breakdown of phosphocreatine (PC)—but PC's muscle stores are also rather limited (20–30 mmol/kg of wet tissue) and the recharge will last no more than a few seconds. Outburst performances rely heavily on these limited ATP and PC stores. Maximum metabolic power rating achievable by the ATP–PC route is large, 3.5–8.5 kW for a typical Western man (65–70 kg with 20 kg of active muscles) and as much as 12.5 kW for a trained man, but the overall capacity averages just 20–40 kJ and reaches no more than 65 kJ for the best adapted bodies.

Somewhat longer exertions can be energized by muscular glycogen. As with ATP–PC recharge, this process is anaerobic and there is the expected power–capacity tradeoff. Maximum metabolic power of anaerobic glycolysis is 1.8–3.3 kW for average men, but top performances can reach as much as three of four times the phosphagen-derived total. Exertions lasting 30 seconds to 3 minutes are largely energized in this way, with maximum contribution coming after about 1 minute of strenuous work. All prolonged efforts are powered primarily through aerobic recharge. As the body's oxygen store of about 1 L can support moderate exertion for no more than 1/2 minute, all subsequent higher energy demands require linear increases in pulmonary ventilation: human power is largely a function of maximum oxygen intakes.

Eventually higher exertions fail to elicit higher rates of ventilation as oxygen pumping reaches a limit. This maximum aerobic power ranges between 20 and 55 ml/kg·min, or between 350 W and 1350 W in adults. A metabolic range of 600–900 W would include most mildly active people, but the rate can go over 90 ml/kg·min for elite endurance athletes. This is equivalent to more than 2 kW and a flux 25 times higher than the BMR, an impressive metabolic scope in comparison with most mammals. Aerobic rates are slightly lower in women than in men of the same age and after adolescence the annual decline is around 0.5 ml/kg·min.

Peak aerobic capacities range from 1.5 to 3.5 MJ for healthy adults but can easily surpass 10 MJ for good athletes, with maximum metabolic outputs at an astonishing 45 MJ. Both glycogen and fat are substrates of oxidative metabolism, with fatty acids' share rising up to 70% during prolonged activities. Yet almost every longer physical exertion has its share of anaerobic breakdown of glycogen. Resulting pyruvate is converted to lactate (its accumulation in active muscles leads to the well-known weakness and pains), and a convenient indicator of anaerobic threshold is the breaking point when lactate blood levels start increasing exponentially. In sedentary people this point comes invariably when the workload is 50–70% of the maximum aerobic power, but in endurance athletes this threshold may rise to 85% of peak aerobic power.

Limits of human performance can change. Exceptional genetic endowment is a prerequisite but even average individuals can increase their aerobic power by 20% with training; maximum trainability may be around 40% of the initial level. How efficient are these performances; how good a machine is the human body? The answers can be only conditional. Complex energy transformations in the human body result in highly idiosyncratic outcomes. Metabolism of each mole of glycogen-derived hexose (containing 2.81 MJ) conserves 39 mol of ATP (42 kJ/mol), a nearly 60% conversion efficiency; about 40% of this energy powers muscle contractions for a total gross efficiency of 23%. Rates of ATP conservation can be as low as 40 and as high as 70%, giving a range of 16–28% for the final gross efficiencies.

This range has been repeatedly confirmed by careful ergometric studies. Anaerobic conversion is much less efficient (about 10–13%) and so the long-duration efficiencies will be highest for elite aerobic performers deploying their muscles at higher power rates in activities most conducive to peak kinetic power outputs. Almost invariably this is cycling (Whitt and Wilson 1982). Bursts of 1 kW are possible in pedaling for just a few seconds, rates of 300–400 W can be sustained for up to 10 minutes, and exertions lasting more than 1 hour are limited to less than 200 W in untrained men.

Regardless of the length of a race, cycling is the fastest mode of human locomotion (Fig. 5.2) and its high sustained power outputs and high efficiencies have energized all human-powered record-setting machines not only on land but also on water and in the air. In 1984 an MIT Monarch B pedal aircraft completed a triangular 1500 m course at nearly 10 m/s (Drela and Langford 1985); Flying Fish II, a hydrofoil-equipped watercraft ridden like a bicycle outpaces, at a maximum speed of about 6.5 m/s, a single rower in a shell (Brooks et al. 1986).

The question of how much labor a man could be expected to do in a day became prominent during the late seventeenth and eighteenth century when larger numbers of laborers were being employed in expanding industrial establishments (Ferguson 1971). Notable estimates began in 1699 with Guillaume Amontons's observations of glass polishers, whose exertion during a 10 hour workday he equated with raising continuously a weight of 25 lb at 3 ft/s. Augustin de Coulomb's 1798 figure of a fair day's work at 205,000 kgm derived from climbing the 2923 m high Tenerife Peak in the Canary Islands in slightly under 8 hours. The former calculation is an equivalent of 3.66 MJ of work at a rate of 102 W, the latter of 2 MJ at about 72 W. We can do little to improve on these values.

A healthy adult should easily tolerate many hours of work at 40–50% of his or her maximum aerobic capacity. Putting this conservatively, at 40 ml O_2/kg·min, this translates into metabolic power of 380–490 W for men of 60–70 kg, or, with typical kinetic efficiencies around 20%, to anywhere between 75 and 100 W of useful work. Ultimately, the limits of human performance are a matter of adequate food supplies. The body's carbohydrate stores can release just 8.4 MJ, enough for a day's minimal survival needs. Only 2–2.5 kg of protein (33–42 MJ), or no more than 20–25% of total stores, can be

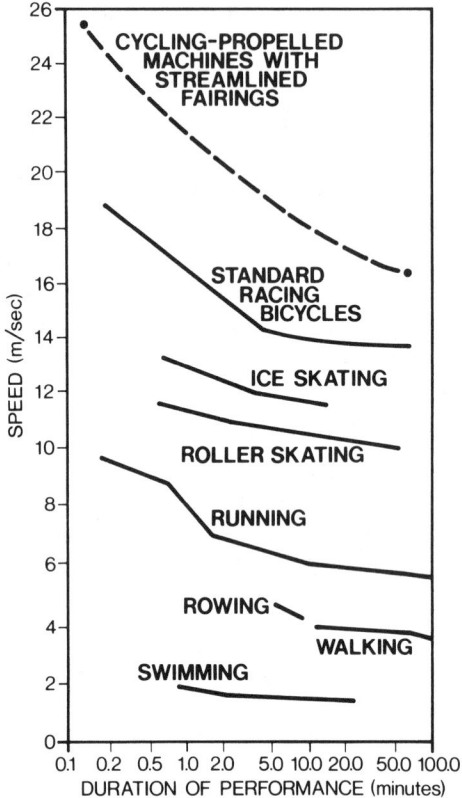

Figure 5.2 World record speeds for various modes of man-powered locomotion. Cycling is fastest; rowing and swimming are surprisingly slow. (From Whitt and Wilson 1982).

metabolized without endangering life. Clearly, starving people must derive most of their energy from fat: 2–3 kg of lipids is a structural part of cells but the rest—6–18 kg (226–678 MJ) in adults—can be converted to last at least 40–50 days.

This chapter ends with a few pages on human thermoregulation. This could have been included in the general survey of heterotrophic metabolism, but human thermoregulatory abilities are so impressive —and they have contributed so decisively to our complexification as a species—that they merit a separate discussion.

5.5 THERMOREGULATION

As in any homeotherm, the normal range of human temperatures is very narrow. The lowest mean is at 35.1°C in Andean Indians, and

37°C is the standard physiological datum; values up to 40–41°C are tolerated for short periods of fever or hard work in a hot environment, and brief hypothermic spells as low as 29–27°C are also survivable (Hardy et al. 1971). In contrast, the Earth's inhabited regions have extremes of around 50°C in subtropical deserts and −60°C during Arctic winters. A hairless heterotroph has only four choices in order to survive in temperatures below the zone of thermal comfort (for resting unclothed people between 24 and 29°C): to insulate by becoming sufficiently obese, to boost metabolism, to cut thermal conductance through vasoconstriction— or to create a portable micro-environment.

The first adaptation is not a very practical evolutionary option for an originally tropical, long-limbed, upright walker–runner even if the food to sustain it could be found. Marked differences of BMRs —they are highest in northern Europe and among native North Americans—confirm the usefulness of the second thermoregulative approach. Also, in response to surface chilling Africans raise their metabolism slower than Europeans, who are, not surprisingly, bested by the Inuit. That vasoconstriction works has been convincingly documented with sleeping unclothed Australian Aborigines (Scholander et al. 1958). Whereas white Australians had recurrent erratic elevations of their metabolic rate (up to more than twice their mean), the Aborigines slept undisturbed, their metabolism not only steady but in some cases even below the BMR.

This would be grossly inadequate to handle seasonal deprivations even in temperate latitudes where clothing and heating must be used to protect against the cold. Interestingly, the Inuit, whose harsh environment forced them to develop extraordinary multilayered clothing protection replicating the furry adaptations of Arctic animals, appear to have the poorest naked body insulation, that is, the highest thermal conductance, of any population studied. The microclimate of the Inuit's body was constantly an almost tropical one and hence there was no need for metabolic adaptations.

Still, boreal populations have evolved minor but important adaptations: sweat glands, which are more abundant on the face and greatly reduced on the body trunk and legs, and higher finger temperatures, which are advantageous for outdoor activities. But the bulk of the Arctic and Subarctic thermoregulation was always a matter of intelligence, of effective mimicry of the hunted mammals. Animal precursors can be seen in one common heat adaptation—the wearing of dark-colored clothes in desert environments. As in the case of dark bird plumage such robes absorb a very large part of incoming radiation, which is subsequently lost by convective heat transfer without impinging on the body.

Adaptation to hot climates is not only fast but also very effective for all healthy people, and it is retained even by populations that have lived for generations in temperate or cold climates. Hanna and Brown (1983) include the ability to tolerate heat as one of the distinctly human characteristics alongside bipedalism, hairlessness, a large brain, and a symbolic linguistic ability. Both the rapidity and the effectiveness of the process can be perfectly illustrated by Wyndham's (1969) revealing experiments with South African white males with no recent exposure to heat during work. They had to perform moderate work in a hot (33°C) and humid environment, and the acclimatization period consisted merely of 10 consecutive days of such a regime. Their responses were remarkable. Rapidly rising core temperatures reaching a danger zone, heart rates approaching tolerable maxima, and low sweating rates were swiftly transformed to levels nearly identical to those of highly acclimatized native Bantu.

Other studies have demonstrated that such an adaptive training required just 100 minutes a day regardless of the thermal environment for the rest of time. Initial response to heat is the dilation of peripheral skin blood vessels, compensated by vasoconstriction elsewhere, and shifting of large volumes of blood into hands and feet. The starting point is highly variable but at 28–32°C of skin temperature perspiration begins, first on the trunk, then on the extremities.

Eccrine sweat glands, whose density ranges from $52/cm^2$ on the thigh to $240/cm^2$ on the dorsum of the hand, are highly effective heat dissipators. Without active sweating an average person loses about $12 \ W/m^2$ of body surface, equally split between respiration and skin diffusion. Above 28°C evaporative losses climb exponentially so that even in a resting man in still air they will remove about $130 \ W/m^2$ at 48°C, or a total of up to 230 W for an adult with a resting metabolism of 80 W. With work the perspiration rates go up to surpass greatly those of sweating mammals: whereas a horse can lose every hour $100 \ g/m^2$ and a camel up to $250 \ g/m^2$, a man can perspire more than $500 \ g/m^2$ (Folk 1976).

Perspiration of $500 \ g/m^2$ translates to heat loss of 550–625 W for most adults, sufficient to regulate temperatures even in extremely hardworking individuals. Designers of heating and ventilating plants use 586 W (or about $325 \ W/m^2$) as the maximum heat output of a manual laborer, and under normal free-conversion conditions, with air moving just at 0.45 m/s, virtually all of this load can be lost through evaporation of sweat. Similarly, a pedaler on an ergometer in a free-convection environment can sustain output of nearly 600 W without any noticeable increase in body temperature no matter how long he works (Whitt and Wilson 1982).

The best acclimatized individuals can produce up to 1100 g of sweat per square meter per hour, an amount adequate to remove 5 MJ of metabolic heat. At a rate of 1390 W this exceeds all but the most strenuous athletic exertions. The highest reported short-term peak sweating rates are 4000 g/h. At such levels rehydration becomes an acute necessity as humans can neither tolerate substantial dehydration nor store large volumes of water. Voluntary dehydration (drinking less than is perspired) is common during heavy exertions, with the deficit gradually replaced within a day. This partial un-coupling of thirst from actual water requirements is also of inestim-able adaptive value.

There is little doubt about the early hominid origins of this effec-tive adaptation to heat, most likely in hot, dry Plio-Pleistocene environments of East Africa. Without this ability *Homo sapiens* could not diffuse as a hardworking pantropical species engaged in extensive, and often highly demanding, collection of plants and pursuit of animals. And without retaining this ability in northern climates there would have been no post-1500 global European expansion, which has been responsible for much of the economic, technical, and social transformation fo mankind.

6

MAN AS A SIMPLE HETEROTROPH: GATHERING, HUNTING, AND FISHING

Through many lustres of the sun rolling through the sky they passed their lives after the wide-wandering fashion of wild beasts. No sturdy guider of the curved plough was there, none knew how to work the fields with iron ...

—Lucretius
De rerum natura

No matter what the latest anthropological consensus on the phylogenetic record, evolutionary transitions, and species dating may be, one reality remains constant: for more than 99% of its existence the genus *Homo* survived as a fairly simple heterotroph, with no permanent abode, a gatherer, hunter, and fisher, an omnivorous user of basic tools. This existence was not too dissimilar from that of less advanced hominids whose record now goes back as far as 4 million years, to the East African *Australopithecus afarensis*.

Obviously, chances of any meaningful appraisal of the energetics of *afarensis* are nonexistent. And, unfortunately, such chances do not improve even after bypassing *Australopithecus boisei*, *Homo habilis*, and *Homo erectus*, skipping nearly 4 million years of evolution, and focusing on the much more abundant archeological record of the past 50,000 years. Remains of implements used in hunting and fishing mark the technical improvements: first bows and arrows about 25,000 years old; line fishing about half as old; nets made from twisted fiber along with hair or thongs from about 8000 B.C. An abundance of bones from numerous caves and pits around the world adds up to a long list of animals, ranging from mollusks to elephants, eaten around the fires.

Isotope ratios (C^{13} and N^{15}) can be used to separate plant remains—charred or as organic residues scraped from the inside of potsherds—into legumes, nonleguminous C_3 species, and C_4 and CAM varieties; their determinations in human bone collagen can uncover relative amounts of terrestrial and aquatic foods in prehistoric diets (De Niro 1987). This archeological evidence is most interesting—but it is insufficient for any meaningful energy balance calculations. Compared to durable bones and shells, plant remains have been preserved only infrequently, and we have no idea what fraction of killed animals was brought to the excavated sites, if the sites were in constant or intermittent use, and how many people they served. Moreover, extreme spatial and temporal variability of the fragmentary evidence leaves little room for generalizations.

But the long span during which humans were foragers makes it imperative to do better than simply pointing out the inherent weaknesses of our data base or dismissing the energetics of foraging as a matter of specialized archeological or anthropological interest. Except in a handful of cases studied in modern energetic anthropology, estimates and deductions are unavoidable. The best records are available for several groups of the Basarwa (Bushmen, or the San) gatherers-hunters during the late 1950s and the early 1960s (Lee 1979; Tanaka 1980; Silberbauer 1981).

There is enough evidence for reviewing the lives of traditional gatherers-hunters (Lee and De Vore 1968; Wadsworth 1984; Price and Brown 1985) as well for systematic analyses probing the general existential modes and dynamics of these now virtually extinct societies (Service 1979; Hayden 1981; Kirk 1981; Winterhalder and Smith 1981; Testart 1982; Kelly 1983; Clark and Brandt 1984). Naturally, information on groups that survived in extreme environments to be studied by modern science offers only limited insights into the energetics of foragers active in more equable climates and more fertile areas.

6.1 ENERGETIC IMPERATIVES AND SINGULARITIES

By far the most important conclusion from the recent crop of forager studies is the absence of a "typical" hunter–gatherer pattern. Studied groups have varied greatly in virtually every measured parameter, from food sharing to fertility (Hill and Hurtado 1989). Consequently, forager energetics is also a matter of countless peculiarities and exceptions. Large differences in forager's habitats and diets translated into population density means differing commonly by up to a factor of 100. These variations displayed no obvious relationships with energy fluxes, such as decreasing poleward or increasing proportionately to the primary productivity of surrounding ecosystems.

Population densities ranged over at least two orders of magnitude as the various groups, numbering from just 10 to many hundreds of people, exploited territories of 10^2–10^3 km² (Fig. 6.1). Kirk (1981) stresses this disparity for the precontact Australian Aborigines: the continent's average population density was about 5.5 persons per 100 km², for one of the largest tribes it was merely 1.1 persons per 100 km², and along the coast of today's Sydney it was 2 people per 1 km².

Density means offer no information about the intensity of land use. Kelly (1983) suggested a coverage index (total exploited area divided by total residential mobility distance) to indicate the intensity of land utilization. Predictably, it would be highest for gathering societies and up to an order of magnitude lower for hunters. Identification of general energetic imperatives behind these density disparities must consider above all a variety of biomass and productivity attributes.

Most phytomass in tropical and temperate rain forests is in indigestible cellulose and lignin; desirable fruits, seeds, and nuts are a

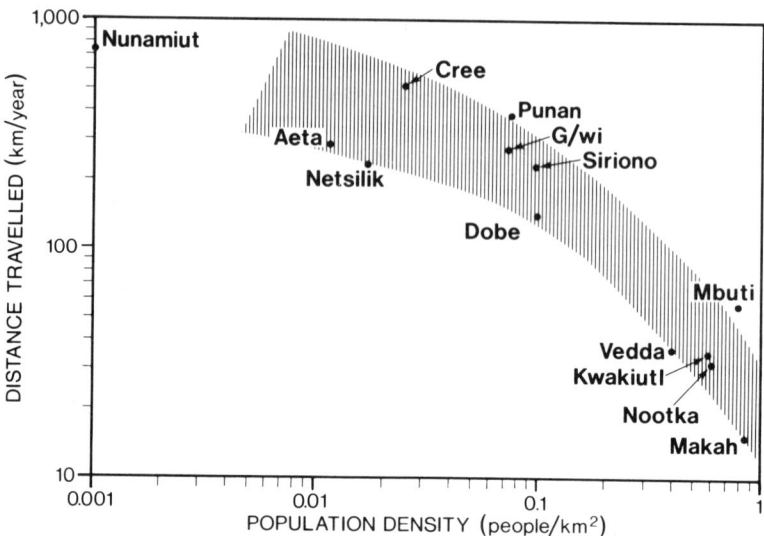

Figure 6.1 Relationship between annual mobility and population density plotted for several foraging societies. Data from Kelly (1983) and Murdock (1967).

very small portion of total phytomass and are commonly inaccessible and well protected by hard coats. Tropical species diversity makes gathering energetically more taxing owing to often considerable distances between the plants of desirable species. In temperate and boreal forests seasonal productivity at and near the ground level and snow cover make for very low winter returns. In contrast, grasslands and woodlands often produce an abundance of easily collectible fruits, seeds, and roots; and concentrated patches of tubers and rhizomes are a relatively more important part of semidesert phytomass.

Consequently, forest gatherers had to relocate their camps as many as 40–50 times a year, whereas foragers in grasslands and semideserts moved fewer than a dozen times—but their moves were usually longer. Ethnographic evidence confirms the expected inverse link between the total annual mobility and population density. Not surprisingly, average densities of foragers were always lower than those of similarly sized herbivorous mammals, which can digest a much wider range of available phytomass. According to Calder's (1983) allometric equation there would be typically five 50 kg individuals per square kilometer but densities of large Hominoidea prorate to about two animals per square kilometer (Bernstein and Smith 1979). The documented densities of tropical rain forest hunters are below one person per square kilometer; those in boreal forests are mostly an order of magnitude lower. This extensive roaming, sharpening the adaptive responses, was certainly an important factor in human evolution.

Herbivores were always the most important hunted animals. Going after carnivores was not only dangerous but energetically unprofitable given their low densities and thus poor energy returns in their pursuit. But most tropical forest zoomass is folivorous, that is, arboreal, small, and inaccessible—in contrast to large grassland and tundra grazers of grasslands. But since these ungulates are highly mobile, the groups dependent on their hunting had to cover larger areas increasing poleward.

Exploiting such resources becomes impractical without frequent camp shifts. As the distances of daily hunts grow, residential change is a must. The same is also true in snowbound boreal forests where gathering is reduced to negligible levels during winters. Not surprisingly, exploitation ranges of North America's Indians and Inuit extended over many thousands of kilometers even for small groups. Dependence on fishing was another important key factor: in the Pacific Northwest, with its massive runs of salmon, fishing allowed the establishment of permanent settlements and a sharp reduction of mobility needed for food acquisition.

Energetic imperatives are also easy to discern behind the minimum gatherer-hunter group sizes, which were determined by the requirements of successful hunting. As an individual hunter's daily success rate in killing large mammals was rarely higher than 15–30%, at least 3–6 hunters were needed to assure daily meat supply for their families. With an average of 6–7 people per family this translated to minimum sizes of 18–40 people. A minimum number of adults was also required to track and butcher a large mammal and to transport its pieces to a camp. Benefits of group hunting are clear from Harako's (1981) studies of Mbuti bands in the Ituri tropical rain forest. Archers shooting monkeys had a daily success rate of 10% and averaged just 110–170 g of meat per person per day, spear hunters averaged 220 g, and group net hunting yielded 370 g for each band member.

In common with chimpanzees, the genetically closest primate species, all foragers were omnivorous. Everything digestible was eaten, from dwarf willow leaves and the contents of caribou stomachs mixed with seal oil in the Arctic to termites captured during nuptial flights and roasted and ground ungulate hides in Africa's semideserts.

Groups highly dependent on gathering ate various parts and products of scores of different plants—but a small percentage accounted for the bulk of food intake. Preference for energy-dense seeds—grasses (15 MJ/kg), legumes (17.7 MJ/kg), and nuts (up to 27 MJ/kg)—was universal.

The high value of meat in foraging societies is well documented quantitatively by the willingness to spend much energy in its acquisition. Some energy expenditures were extraordinarily large. A reliable report has the Kalahari hunters running for up to 30 km without pause and catching up with the wounded animal with an all-out spurt (van der Post and Taylor 1984). The optimizing approach—selecting the prey giving the greatest food reward for the lowest expenditure of energy—was common, but much of hunting was opportunistic.

Meat of herbivorous animals, an excellent source of protein, was largely a marginal supplier of fat: at only about 6 MJ/kg and with just 10% lipids even its abundant consumption left people feeling hungry and craving fat. Could it be, asks Hayden (1981), "that the meat was actually valued little, whereas fat contained in the prey was the real prize"? This is an energetically unexceptionable position—but energetic explanations are not always applicable. For !Kung Basarwa easy-to-gather, energy-dense *mongongo* nuts provided the best energy return and most of their food—but /Aise, another Basarwa group with access to the nuts, did not eat them because to them they did not taste good (Hitchcock and Ebert 1984).

We will never be able to reconstruct long-term food balances for representative foraging societies at the height of their global diffusion. More fundamentally, I agree with Hayden (1981) that even the most elaborate environmental approaches will not make it possible to obtain adequate data for the calculations of carrying-capacity levels for foraging populations. All we can do is use the few detailed accounts available for twentieth-century foragers and some descriptions of earlier subsistence to reconstruct approximate energy returns of individual activities.

These net energy returns—calculated by adding up energy costs of particular foraging activities (using the detailed descriptions in the anthropological literature and standard assumptions of energy expenditures) and dividing them into the edible energy content of the obtained food—are as high as 30–40 for collecting energy-dense roots, 10–20 for all gathering, in the same range for individual hunts of large ungulates, barely above the break-even point for bow and arrow hunts of small mammals, and, astonishingly, over 2000 for coastal whaling.

6.2 ROADS TO COMPLEXITY

Perhaps the most important outcome of recent inquiries into foragers' way of life has been the growing realization, considerably helped by archeological advances, that many gatherer-hunter societies reached levels of complexity—including sedentism, high population densities, large-scale food preservation and long-term storage, social stratification, elaborate rituals, and incipient crop cultivation—usually associated only with farming societies. This cultural complexity had its energetic foundations in exploitation of extraordinarily productive environments, the development of substantial storage economies, and incipient agricultures.

The image of "small, ephemeral encampments occupied by a few people eating, sleeping, scraping hides and only occasionally reproducing" (Price and Brown 1985) did not fit some foragers as far back as the Upper Paleolithic, when the mammoth hunters in the Moravian loess region lived permanently in well-built semisubterranean dwellings, made a variety of excellent tools, and could fire clay and sculpt. Zvelebil (1986) argues that delayed acceptance of farming in Europe's northern and eastern forest zone was the result of high foraging productivities.

Farming appeared to have no inherent superiority, thus only the severe depletion or disappearance (often climatically induced) of a key intensively exploited resource or the competition with neighboring farmers pushed the foragers into regular agriculture. In almost all cases of complex foraging the highest productivities were associated with exploitation of marine resources. Yesner (1980) generalized the attributes of these adaptations to include high biomass and high resource diversity, utilization of migrating species, sedentism, technical complexity and cooperative resource exploitation, high per capita productivity, and high population density, as well as territoriality, resource competition, and warfare.

Although the typical sizes of foraging groups ranged between 20 and 50 people, the Pacific Northwest's semipermanent settlements commonly housed (literally: in well-built wooden structures) several hundred people. Seasonal abundance of salmon and its preservation by smoking were the energetic basis of this extraordinary population density and resultant social complexity: compared to cod's 3.2 MJ/kg, chinook salmon has about 9.1 MJ/kg, largely owing to its high (15%) fat content. Effects of the high food energy value of migratory marine species are even better demonstrated in the case of northwestern Alaskan Inuit. In spite of extremely harsh environment and low and unpredictable density of large land mammals, these groups were able to secure a food energy surplus in less than four months of near-shore hunting of baleen whales accessible in the leads and open waters during their migration (Sheehan 1985). Even a minimum estimate of the baleen whale harvest could—together with subsidiary exploitation of other marine species—result in food surpluses.

The baleen whale's huge mass (even the most commonly landed immature two-year-old animals averaged nearly 12,000 kg), incomparable food energy value (about 36 MJ/kg for blubber), and easy storage in underground permafrost cellars offered an unmatchable net energy return and supported large (over 500 people) permanent settlements and social complexity. This adaptation appeared to be self-amplifying: more people could field more whaling crews, resulting in more sightings and in higher chances of hunting success.

The reliance on seasonal food flows in sedentary foraging societies required extensive and often elaborate storage (Hayden 1981). Testart (1982) argues that such large-scale, long-term food storage brought a total change in foragers' mentality, new attitudes toward time, work, and nature, new perceptions of being more in control. Undoubtedly, intensive food storage was critical in supporting population expansion and enabling the groups to stabilize at a higher level of density. The necessity for planning and budgeting of time was perhaps its most important evolutionary contribution: tool making and maintenance as well as the preparation of storages had to be concentrated in slack periods preceding the times of maximum seasonal food availability, which had to be spent in intensive foraging to maximize the time-limited harvesting opportunities.

Once this pattern was mastered there was no turning back without sharp reduction of prevailing population densities. Large-scale storage is not only incompatible with mobility but it is also conducive to development of new activities which might further enrich the society. Accumulation of goods can become unlimited, and it is natural to incorporate gradually incipient agricultural practices; thus the stage is slowly transformed to a fundamentally different way of subsistence.

The process was evolutionary and multifocal, its onset spread over several millennia in different parts of the world. Man ceased to be a simple, opportunistic, omnivorous heterotroph and became—through crop selection and cultivation, irrigation and nutrient recycling—an increasingly refined manipulator of solar energy flows, an overwhelmingly herbivorous producer of large quantities of a few staple crops, a rapidly learning organizer of accumulating social and technical complexity.

7

MAN AS A SOLAR FARMER: TRADITIONAL AGRICULTURES

Rain in the village must be plentiful.
I dream of fragrance with the rice-plants full.
Since Heaven's impartial in its overflow
Of grace, strong reeds and tares will likewise grow.
Men find such growths unwelcome from the harm
They always do those who work a farm.
Hence none of the good villagers can shirk
In seasonable tasks of weeding work,
Piling tares by the river in defense
Of cleaner crops. Grain is life's sustenance.
. . .
Whatever grows will rise in mad confusion
And toil must guide the crop to its conclusion.

—Du Fu
Directing farmers

Du Fu's verse defines agriculture very well. On the most abstract level it can easily be seen as a prescient description of negentropic effort: agricluture as periodically strenuous energy investment combating the natural tendency toward entropic disorder and producing orderly harvests. An ecologist might point out that stressing separation of crops from other phytomass and extolling grains as life's sustenance perfectly characterize farming as a manipulated ecosystem ultimately heavily reliant on cereal monocultures and hence inimical to the maintenance of species diversity and systemic stability. An anthropologist might focus on the inevitability of practically universal participation in seasonally demanding labor alternating with periods of extended rest.

So we see that definitions of agriculture abound. But explanations —why the practice started, why it diffused so widely, and why it was adopted at what is in evolutionary terms a fairly rapid rate—are hard to come by.

Of the many published causal theories (reviewed in Reed 1977 and Pryor 1983) it is easiest to reject the idea of agriculture as a technical innovation or the result of an invention (Carter 1977), as well as the notion of farming's revolutionary nature (Childe 1951); the well-documented evolutionary nature of the process makes these ideas untenable.

It is equally easy to refute the necessity of such prerequisites as skilled tool making, sedentism, or storage habits: simple tools (wooden digging sticks, tiny stone-chip blades) suffice; there are many records of sedentary foragers who made no move toward farming while the Tehuacan Valley had no permanent settlements but thousands of years of crop cultivation (Bray 1977); and there are numerous examples of food storage in foraging societies and its absence among gardening societies. Many foraging societies lived for a very long time side by side with the agriculturalists. Indeed, farming had no universal appeal; there were no compelling reasons for its early adoption.

Most of the evidence points to the combination of population growth and environmental stress as the principal factor in adopting and intensifying agriculture. Cohen's (1977) work is perhaps the best example of this ecological approach. Gathering and hunting will eventually bring diminishing returns, leading to gradual extension of incipient cultivation techniques and slow adoption of a greater variety of small-scale gardening practices. This is not a response to acute food shortages but rather an adaptation to changing demands and resource availability.

This basically energetic explanation should be enlarged by pointing out the social advantages of agriculture. Crop cultivation efficiently fosters association, clearly a desirable goal for our sociable species; farming promotes individual ownership and accumulation of material possessions; it makes it easier to have larger families; and it facilitates warfare. Besides, the net returns of agricultural endeavors were in no way superior to those of foraging and hence the adoption of farming cannot be seen basically as a quest for minimization of energy inputs. Quality considerations were certainly important: concentration on plants whose parts were formerly gathered in the wild could be readily explained by their particular nutritional qualities (high protein or oil content) rather than by any generalized striving to minimize search and transport.

Whereas the transition from foraging to farming was initiated and sustained by a complex of energetic, nutritional, and social impulses, further evolution of agriculture can be seen as a matter of clear energy imperatives. Boserup (1965, 1976) elaborated this link with great clarity when looking at the evolution of peasant societies. As a particular food production system reaches its limits the affected population can migrate or stay and stabilize or decline—or it may adopt a more productive subsistence arrangement. The last option may not be initially any more appealing or probable than the other solutions. When it comes, the shift requires higher energy inputs so that even with higher food production the net return ratio may decline—but the higher edible energy flux will support a larger population.

Not surprisingly, there will be a natural tendency to postpone the switch as long as a particular society could subsist within a less intensive arrangement. Intensification advances in several fairly universal stages from long forest fallow (with just one or two crops followed by a regeneration of 15–25 years), to bush fallow (with 4–6 crop years eventually equaling the rest spell), to short fallow (a crop or two followed by a year off), regular annual cropping (with fallow reduced to fall and winter months), and, finally, to multicropping (often irrigated) with two or three grain, seed, or forage crops, or up to five or six vegetable crops, planted in rapid succession.

Each of these successive steps recovers more of the site's potential photosynthesis as food and supports more people per hectare of arable land as it demands higher energy inputs, first for forest clearing, planting, and digging, and eventually for repeated plowing, harrowing, seeding, weeding, construction of terraced fields, wells, and dams, and irrigation. These activities also require further energy investment for making or buying more sophisticated tools and implements. Moreover, as a large part of energy inputs is in the form of long-term investments (in cleared or terraced fields, irrigation systems, roads) and as intensive cropping demands planning, storage, and trading, intensification of agriculture has been a key ingredient of civilizational complexification, promoting innovation, specialization, interdependence, and exchange of goods and techniques.

Agricultural intensification also led inevitably to growing reliance on sources of energy other than human muscle. Plowing, even on a relatively small scale, was either enormously taxing or outright impossible without draft animals; manual threshing and milling of grains was so labor-intensive that inanimate power was necessary to process concentrated harvests; long-distance distribution of grain to cities also had to rely on animal power and wind, and ironmaking consumed charcoal for smelting.

Except for draft animals, these energy subsidies into traditional farming prorated to fairly small amounts per hectare. During the nineteenth century traditional farming in Europe and North America received rapidly growing fossil-fueled subsidies in the form of steel implements, cleverly designed machines, and steam threshers. However, as animate power (excepting uncommon steam tractors) remained its sole kinetic energizer and as fertilization was still overwhelmingly dependent on recycling organic waste and planting leguminous crops, Western agriculture remained traditional until the early decades of twentieth century—and most of the poor world's farming continues to be a hybrid of traditional and modern practices. Again, studies of traditional agriculture are far from being of historical interest only.

7.1 EXTENSIVE FARMING

Nomadic pastoralism and shifting farming are two very dissimilar modes of traditional agriculture sharing the intermittent and extensive use of land. Sustainable population densities of many nomadic pastoralist societies were not any higher than those of the less widely roaming foragers, while in suitable environments even the moderately productive shifting cropping with long regeneration cycles could support populations an order of magnitude larger than most of the settled foraging communities. Shifting agriculture is a part of the evolutionary sequence running from foraging to incipient farming to permanent cropping, but pastoralism followed domestication of animals as an adaptation to arid regions or as a response to desiccation.

For millennia both of these ways of life dominated huge areas of all continents (except Australia). In Africa, they blended into mixtures of seminomadic agropastoralism, sometimes retaining a significant bit of foraging. Current pastoralist retreat follows rising population densities, which led to unsustainably large animal herds resulting in ruinous overgrazing. And a few groups among tens of millions of pastoralist and shifting cultivator families in Asia, Africa, and Latin America still conform to pure types of traditional subsistence: there is much trading, outside indirect energy inputs, and transitional modes of existence.

There is no shortage of general reviews of pastoralism and its transformations (Monod 1975; Helland 1980; Salzman 1981; Khazanov 1984), but studies of traditional pastoral energetics are rare. Only the human inputs are easy to approximate. As these pastoralists engaged in no improvement of their grazing lands, their labor was confined to herding the animals, guarding them against predators, milking regularly and butchering infrequently, and sometimes also building temporary fenced enclosures. These tasks usually required only light to moderate exertion for two to four hours a day and made child labor common. Large numbers of livestock—up to 100 camels, 200 cattle, and 400 sheep and goats in East Africa (Helland 1980)—could be managed by a single herder. This energetic appeal is one of the key reasons for pastoralists' reluctance to be converted to farmers.

Monod (1975) correctly states that a numerical analysis of the relationship of livestock and grazing has little significance unless one specifies a host of local factors determining the grazing potential. Water availability is critical in all arid settings. In cool season with good pastures camels can go up to 90 days without water, sheep 30—but cows only 3 and calves just a single day. Sustainable size of the herd thus can be partially manipulated by changing its species composition. Recent densities in East Africa were 2–8 livestock units per capita, 0.8–2.2 people/km^2, and 0.03–0.14 heads/ha (Helland 1980; Evangelou 1984; Coughenour et al. 1985).

Shifting cultivation alternates variable but always short cropping periods with variable but commonly fairly long periods of fallow. The practice was once ubiquitous on every continent except Australia and in the twentieth century it has remained of major importance for tens of millions of families in Africa (Allan 1965), Latin America (Watters 1971), and Southeast Asia (Spencer 1966). There are many fundamental similarities in all regions (Grigg 1974; Okigbo 1984).

The cultivation cycle starts with clearing of natural vegetation. The most labor-intensive phase of field preparation is felling of large trees, trimming and pollarding of smaller trees, and slashing of younger growth. After a period of drying the cut phytomass is burned. Fire clears away the slash litter, prepares the surface for planting, reduces the regrowth rate of forest species and attack rates of pests, and, although most nitrogen is lost, mineral nutrients in the soil are recycled.

A variety of edible, fiber, and medicinal species, dominated by grains (rice, corn, millet), roots (sweet potatoes, cassava, yams), and legumes (various beans, peanuts), is grown in often unruly-looking gardenlike arrangements, typically with high degrees of interplanting and intercropping and staggered harvesting. Although two to five staples may provide most of the food energy, the number of cultivated crops is rarely less than a dozen and often there are 30–50 species crowded in a small area; in humid tropics, this bears a conspicuous resemblance to the variety that prevailed in the original forest.

Profuse gardens are commonly fenced to keep domestic or wild animals away and a good deal of time must be spent guarding the crops against mammal and bird predation and keeping the herbaceous and ligneous competitors in check; weeding may have to be repeated as many as five or six times. Except for once- or twice-a-year harvests of grain crops, there is a continuous digging of roots and picking of seeds, leaves, and stems. Total labor inputs vary between as little as 600 and as much 3200 h/ha, and the net energy returns cluster between 11 and 15 for small grains and between 20 and 40 for most root crops, bananas, and also for good corn yields; they reach the highest levels (maxima close to 70-fold gain) for some roots and legumes.

One person requires as much as 10 and as little as 2 ha of land in fallow and under the crops, with the actually cultivated area ranging from just 1/10 to 1 ha. Compared to pastoralist population densities of mostly 1–2/km^2, shifting cultivators' densities—with extremes of 15–55/km^2 and most common values of 30–40/km^2—are typically an order of magnitude higher. Their energy returns are almost invariably superior, as is the security of their food supply.

In comparing the practices of shifting agriculture on the three continents many expected energy relationships recur: location of clearings close to the settlements to minimize the walking distance with harvested crops (more remote locations also have an energy explanation: to save labor on building fences against roaming domestic animals); preference for an easier task of clearing the secondary growth; men felling the trees but women carrying disproportionate shares of less strenuous but still taxing, repetitive chores, such as weeding and harvesting.

On the other hand, no form of agriculture can be governed by simple maximization of energy returns. Nutritional imperatives are easily discernible. Beans always have a higher energy return than corn—but legumes are less palatable than cereals or tubers and that is why the less energy-efficient corn or cassava will be the lead staple. Similarly, Rappaport's (1968) data from New Guinea show that energy expended to grow feed for highlanders' pigs is at best equal to their total food value and often can surpass the energy returned in pork. These pigs, which may be net energy losers, are not grown for energy but for protein and fat, two nutrients very scarce in the dominant energy-rich roots. Widespread human desire for at least an occasional fatty and meaty feast is a sufficient justification for a seemingly inefficient practice—as well as another excellent example of a quest for specific qualities superseding general quantitative considerations.

No form of extensive cultivation could produce enough food to lay the foundations for a high culture: incipient urbanization and emergence of states could be supported only by intensified modes of cropping. Widespread use of draft animals is a key characteristic of intensive farming (although not the precondition: none of the New World high cultures used them); diverse implements and increasingly more complex machines followed, as did the spread of fertilization, irrigation, and multicropping.

7.2 ANIMALS, IMPLEMENTS, AND MACHINES

Plowing opened the soil for planting of small cereal seeds on scales vastly surpassing those of hoe-dependent farming. All of the Old World's high cultures were creations of grain surplus, and regular plowing was their energetic hallmark. First, as early as 5000–3000 B.C., came the simple light, symmetrical wooden stone- or metal-tipped ard (scratch plough) opening a shallow furrow for the seed without inverting the soil; millennia later there followed a much heavier asymmetrical turn-plow with a metal moldboard inverting a ridge of soil and leaving behind deep (10–20 cm) furrow (Leser 1931).

Primitive ards scratching the surface of light soils were often pulled by people but in heavier soils their traction was inordinately taxing, offering no energetic advantage to hoeing. Real plowing required traction and endurance superior to those deliverable even by a gang of strong men: the first known remnants of plows coincide with the time of cattle domestication both in Europe and in Asia. Animals later came to be used not only in various field tasks but also in raising water, crop processing, and transportation.

Since draft animals, the dominant tractive force for about five millennia of human history, are still so important throughout the poor world, their energetics deserves—but rarely gets—closer attention. Power of working animals is roughly proportional to their weight; their actual field performance, however, is determined also by the sex and age of an animal, its health and experience, soil and terrain conditions, and efficiency of the harness, so that calculations of actual contributions as well as their energy cost in traditional farming can be approximations only.

The three principal working species—oxen, horses, and water buffalo—have a profusion of physiques and performance capacities (Rouse 1970; Cockrill 1974). There are twofold to threefold differences between light (Chinese Sanhe horses at 350 kg) and heavy (European Percherons or Clydesdales at 1 t) varieties. With sustainable pulls equivalent to about 14% of body weight for horses and 10% for other species, typical drafts range between 20 and 80 kg. Figure 7.1 superimposes the performance fields of common working species on power isolines, showing most horses below 1 hp and donkeys often no better than a man.

Mechanical considerations actually favor smaller animals: with the identical type of harness their line of pull is lower, and the more parallel this line is with the direction of traction the greater the efficiency of work. The lower pull line will also lower the uplift on drawn implements, resulting, most notably, in much less strain on a plowman. Actual draft required for field work will vary above all with the task and soil type. Deep plowing needs drafts of 120–170 kg; shallow plowing, heavy harrowing, and mowing require 80–120 kg; cereal harvesting with a mechanical reaper and binder demands about 200 kg.

Well-fed horses have greater endurance than oxen and can also provide much higher power during brief exertions. Pulls up to 35%

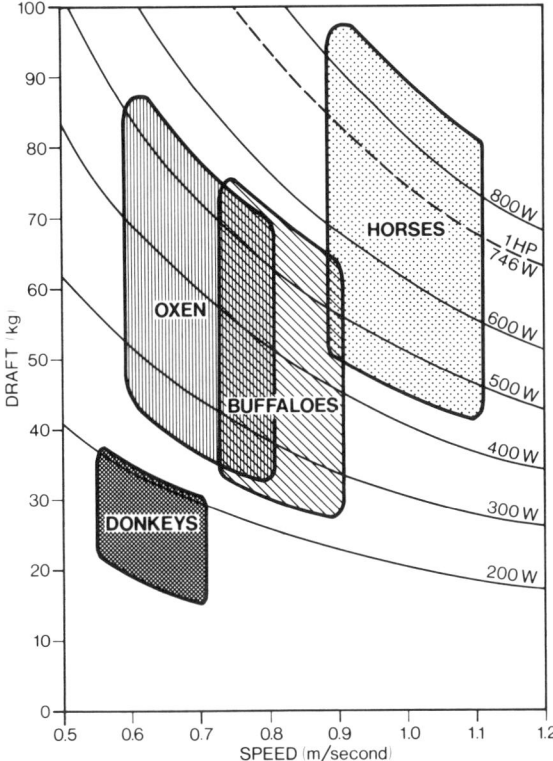

Figure 7.1 Draft, speed, and useful available power for the four common species of working animals. Superiority of horses is obvious, although most could never work at a sustained rate of 1 hp or nearly 750 W.

of body weight are possible, equivalent to working rates of 2.2 kW or 3 hp! And horses, unlike cattle or humans, are also unique in not requiring any additional energy for standing. Horses' unusually powerful suspensory and check ligaments mean that they can rest and even sleep without any energy markup. This is an important energetic advantage because it allows for resting in harness and for lower metabolic expenditures while grazing.

Consequently, the traditional cropping of relatively large cereal fields of temperate climates was always done best with horses, and proper harnessing was a decisive factor in utilizing most efficiently their large draft capabilities (des Noëttes 1931; Haudricourt and Delamarre 1955; Needham 1965). The throat-and-girth harness, present in all ancient horse-using cultures, was inefficient: the point of traction was placed too high and the throat strap used to prevent the

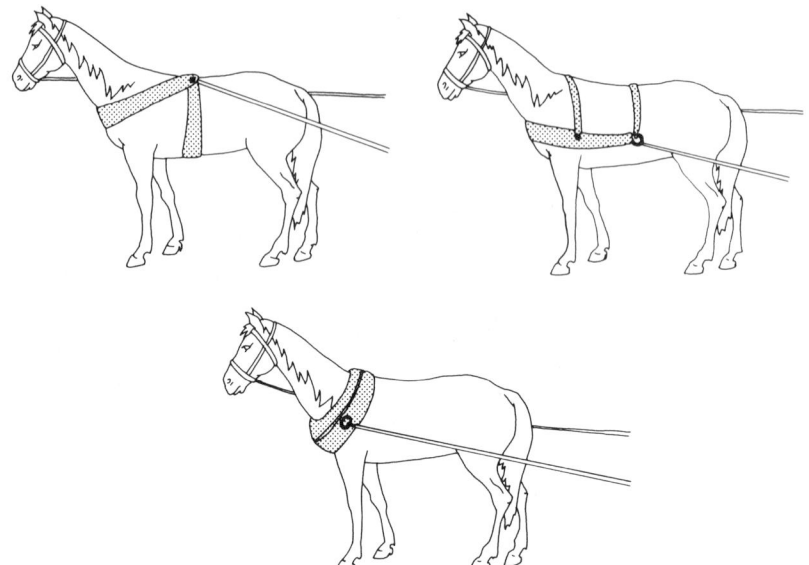

Figure 7.2 Evolution of the horse harness from inefficient throat-and-girth arrangement (top left) to breastband (top right) and fitted and padded collar (bottom). (Based on des Noëttes 1931).

backward slippage of the girth pressed on the trachea, choking the animal during heavy exertion as it lowered and advanced the head (Fig. 7.2).

The breastband harness, a Chinese invention, had the point of traction too far back from the most powerful shoulder and breast muscles and the harness was suitable only for lighter work (Fig. 7.2). The collar harness also originated in China. By the fifth century A.D. its initial form was depicted on the Donghuang frescoes, it reached Europe by the ninth century, and it was universally adopted by the end of the twelfth century (Fig. 7.2). Usually made in one piece, with an oval wooden frame, softly lined to fit the horse's shoulders, and with attachments just above the animal's shoulder blades to connect the draft traces, the collar harness provides the most comfort and the most efficient pull for heavy exertion. A pair of horses with collars could pull loads at least 4 times, and as much as 10 times, as heavy as horses with throat-and-girth harnesses.

Working bovines on all continents traditionally have been harnessed by neck or head yokes. The double-neck yoke has been the most widespread type throughout Africa, the Middle East, and the Indian subcontinent. A beam, often shaped to fit the necks, is held in place by wooden sticks, chains, or ropes. Its throat fastenings tend to choke the animals, its point of traction is high, animals must be of the same size, and a pair must be used even where one would suffice. The single-neck yoke was dominant in China and East Asia as well as in Central Europe.

No draft animal offers a universal fit. Horses—powerful, fast, smart, and easy to handle—were often too light, expensive to harness and to feed properly and in the tropics (besides their susceptibility to trypanosomiasis) they tired readily. Oxen, difficult to train and slow in work, compensated by stolidity, simple harnessing, and easy feeding.

A working animal that best fits its environment is undoubtedly the water buffalo (Cockrill 1974). These heavy animals are surprisingly nimble, moving easily on steep, narrow earth bunds dividing rice fields as well as in the slippery, deep mud of the fields; their large hooves and flexible pastern and fetlock joints are of great help. They browse readily on hedgerow and oddland grasses and they can even graze on aquatic plants while completely submerged. During slack periods grasses and rice straw are enough to keep them fit; they mature fast and are superior converters of feed, needing about 43 MJ/kg of gain compared to cattle's 78 MJ/kg. They are docile (children are often in charge), are easily trained to work (a week may be enough), and can perform reliably for at least 10–15 years, although 25 years is not uncommon.

The number and distribution of working hours for draft animals are determined by the pulse of seasonal cropping, with totals of 60–140 days most common. The amount of work accomplished in a day varied widely—as did the day itself. In plowing, invariably the most difficult task of traditional farming, the daily performance ranged from as little as 0.15 ha for a single buffalo in wet fields and 0.2–0.3 ha for a pair of African or Asian oxen in dry soils, up to 0.5–0.8 ha for a good pair of horses during a long day of stubble plowing or grassland breaking. In late nineteenth-century America, gang plows pulled by a dozen horses could finish 1 ha in just 2.5 hours.

In intensive traditional farming animals usually worked no more than 1100–1400 hours. With average power ratings about 500 W for oxen and buffalo and 700 W for horses their annual useful work would be equivalent to 2–3.5 GJ for every healthy animal, or 10–20 MJ for every day of work. The energetic value of working animals in comparison with manual labor is clear: compared to maximum sustained human exertions at 50–100 W (mean 75 W), draft animals commonly used in fieldwork can deliver 400–800 W (average around 600 W), typically an eightfold, and usually not less than a sixfold, difference.

Moreover, during numerous critical periods marking the course of every cropping year when speed is essential for timely planting, harvesting, or storage, a well-fed animal can work long hours for a few days with rests and can accomplish close to 30 MJ of useful work a day, more than 13 times the work of a good laborer. The obvious question to ask is about the working animals' energy cost: how much of a burden are they for a traditional farmer? According to the Subcommittee on Horse Nutrition (1978), maintenance requirements of a mature 500 kg horse will be about 70 MJ/day of digestible energy. Depending on the shares of highly digestible concentrates (corn, oats) and less digestible roughages (hays, straws), this may represent a very broad range of required gross feed energies but values of 80–100 MJ would be most common, and 90 MJ/day may be a slightly liberal average.

For actual working requirements we can rely on Brody's (1945) impeccable metabolic measurements on a 500 kg Percheron working at a rate of about 500 W expending about 10 MJ/h and a 700 kg horse delivering 750 W metabolizing 14 MJ/h. With 8 hours of work and 16 hours of rest (at 3.75 MJ/h), this translates to 140–170 MJ/day, the range closely confirmed by traditional recommendations used by horse-owning farmers. Calculating the annual feeding burden of draft animals in a traditional agroecosystem is more complicated. Usually only about two-thirds of the animals work and in many cultures there is a mixture of animals, most of them of lesser working capacities and needs than good horses.

Rather than preparing hypothetical accounts of possible requirements I will offer two detailed calculations for the two extremes of traditional farming: the United States in 1910, dependent on horse-and-mule farming energized by lavishly feeding grain and good hays, and China in the early 1950s (before the introduction of the first small tractors), the world's largest traditional agriculture powered by more than 50 million oxen, horses, water buffalo, and donkeys.

The year 1910 is a perfect year, the eve of rapid mechanization and declining horse power: there were 24.2 million horses and mules on American farms—and only 1000 tractors and no trucks. I assumed two-thirds of all animals working and eating a standard ration of 110 MJ/day; for the other six months and for other animals I assumed maintenance feeding (requirements of young growing animals were expressed in adult feeding units). The result: an annual requirement of about 50 Mt of feed (at 16 MJ/kg) equivalent (with average yields of 1.5 t/ha) to some 35 Mha. This means that roughly 20–25% of farmland was needed to feed the working animals.

In contrast, it is obvious that the Chinese could not afford to devote at least a fifth of their land to feeding their draft animals. In the early 1950s, before they started making small-sized and walking tractors, they had about 55 million working animals, mostly oxen and water buffalo, and fewer than 10 million horses, mules, and donkeys. Typical workday needs were about 125 MJ, and with 140 days worked on average by two-thirds of the animals and with concentrate feeds accounting for just 25% for working and 10% for other animals (the rest being roughage), the country needed only about 20 Mt of grain feeds.

That would have equaled a bit more than 10% of all unprocessed cereal harvests, but as the Chinese fed their animals a large portion of their grain-milling residues (e.g., 50–75% of wheat bran) and oil-pressing wastes (including highly proteinaceous soybean and peanut cakes), whose total output at that time amounted to over 20 Mt/y, no more than about 8 Mt of unmilled grain (including legumes) was needed for all draft animals, claiming about 7% of farmland.

A different and revealing way of comparing the two extremes is to note that an American horse needed at least about 1.2 ha of farmland, whereas a Chinese draft animal had to do with as little as 0.13 ha, nearly a tenfold difference resulting from a combination of smaller sizes, less work, and poorer feeding in China. Fundamentally, this was a matter of farmland availability as in 1910 Americans had about 1.5 ha per person, Chinese in the early 1950s merely 0.16 ha.

Whereas the 1910 American share of 20–25% of farmland needed to feed draft animals can be safely taken as the maximum land claim in relatively rich traditional agricultures with an abundance of land, it would be a mistake to see the Chinese share as a minimum. Draft animals of the Indian subcontinent have been even less demanding: in some regions there was virtually no competition between cattle and people for land, carbohydrates, or protein. Elsewhere bullocks and buffalo would claim on the order of 500–600 m^2 of cultivated land per animal. With traditional dryland grain harvests not surpassing 1300 kg/ha, these land demands translate to preempting production of anywhere between 65 kg (India) and 1560 kg (United States) of grain, or roughly 1–25 GJ. In edible terms this equals (with at least 10% milling rate) 0.9–22.5 GJ of food energy. The maximum was enough to support about six people on the typical preindustrial,

Figure 7.3 Twine-binding harvester was the first completely mechanical field machine; its diffusion opened up the plains of North America, Argentina, and Australia for grain farming. This model was manufactured during the 1880s by D. M. Osborne & Co. in Auburn, New York, as a very successful modification of Appleby's invention. (From Ardrey 1894.)

grain-dominated diets. But the large, well-fed horses requiring so much feed were working at a rate at least tenfold higher (easily at 750–850 W) than an average man.

Even when considering the necessity to feed the nonworking horses (about one-third of the total stock), the difference would scale to more than sixfold. Clearly, even a horse claiming the land capable to produce food for six people makes energetic sense on the basis of a narrow efficiency gain alone. Moreover, these heavy horses were able to provide traction for tasks that were both energetically and logistically beyond the practical means of human labor, tasks ranging from deep plowing of clay soil to pulling wheat combines, breaking up large expanses of natural grasslands, and performing critical tasks in a timely and efficient manner.

Early key innovations in plowing and harvesting included inter-changeable cast iron moldboards (1819) soon superseded by steel walking plows (1833); two- and three-wheel riding plows followed after 1864 and gang plows, with up to 10 blades, drawn by a dozen horses, were used on the plains by the late 1880s. The first mechanical grain harvester was patented in 1858, and after several decades of unsuccessful attempts, the twine knotter was introduced in 1878 perfecting the first fully mechanical machine (Fig. 7.3), which was adopted with great rapidity making possible large expansion of wheat-growing areas before the end of the century (Ardrey 1894; Rogin 1931).

The peak of horse-drawn harvesting mechanization was reached when a combined header and thresher was marketed during the 1880s. The largest combines needed up to 40 horses and could harvest a hectare of wheat in less than 40 minutes. But harnessing and guiding 16 to 40 horses to pull these large machines was a logistic nightmare. The time was ripe for a much more concentrated source of tractive power. Enormous time savings brought by horse-drawn gang-plowing, end-gate seeding, harrowing, and combining in the most efficient large-scale operations can be illustrated by a comparison with small-scale oxen-powered cropping of the early ninetenth century: the most productive combination of horses, implements, and machines required merely 5% of the human labor invested into the earlier operation.

Energy costs are not so easily appraised. Net human energy investment—energy spent above the basic survival rate equal to about 1.5 times BMR—is fairly closely approximated by assuming an average exertion in traditional farming tasks equal to 4 BMR: 1 hour of farming then costs 700 kJ. Since draft animals are kept solely for the traction, their total annual feeding costs must be charged against the hours actually worked; a horse then costs about 30 MJ, an ox 25 MJ/h. With these realistic assumptions the comparison shows little increase in overall energy cost.

This near-identity is clearly fortuitous but the great similarity is not surprising. The useful work to be done was the same, harvesting of about 1300 kg of wheat, and the animate power, with its best sustained efficiencies between 15 and 20% for both men and beasts, was the only direct energizer. But where 1 worker-hour was aided by draft work costing about 9 MJ in the early decades of the nineteenth century by its end 1 hour of human labor could control some 220 MJ of horse work. Farmers ceased to be important energizers of the process and became controllers of large energy flows.

7.3 CROPPING INTENSIFICATION

Agricultural intensification has three key ingredients: water, nutrients, and crop diversity. Relationships between crop yields and water supply are complex (Doorenbos et al. 1979). For common grain crops the lowest water utilization efficiencies (in kilograms of harvested yield per cubic meter) range between 0.6 and 0.8 and the best returns are 1.0 for wheat, 1.1 for rice, and 1.6 for corn. Gravity-fed irrigation is energetically most desirable but in river valleys and on cultivated plains it has always been necessary to lift large volumes of surface or underground water.

Even if only half of evapotranspiration requirements were to be supplied by lift irrigation, if would be necessary to raise at least 3000 m^3 of water per hectare for a typical grain crop: irrigation efficiencies were hardly above 50%, demanding a doubling of the theoretical need. Lifting this volume just 1 m requires about 30 MJ or, with about 20% efficiency, an input of some 150 MJ of energy. A man steadily working at 60 W would need almost 700 hours to accomplish the task.

This is an extraordinary burden in the context of farmwork: 1 ha of a wheat field can be hoed by a single person in 12–20 days of steady work and cut with a cradled scythe in 8 hours—but supplying half of its evapotranspiration need would take nearly 3 months of 8 hour days. Not surprisingly, traditional agricultures tried to do with as little irrigation as possible, or they employed a variety of ingenious mechanical devices (Forbes 1956; Needham 1965). The most primitive were tightly woven or lined scoops, baskets, or buckets slung on ropes and handled by two men facing each other, dipping the device, and swinging it over a ridge never higher than 1 m.

The oldest and simplest water-lifting mechanism achieving widespread diffusion was the counterpoise lift (swape or well-sweep), perhaps best known as the Arabic *shaduf*. Its effective lift was usually 1–3 m. Whereas the shaduf required downward pulling on a rope to lift the counterweight, the Archimedean screw needed tiresome cranking to rotate a wooden double helix inside a cylinder with which only low lifts were possible. Hand- or foot-operated paddle wheels were commonly used in India, Korea, Vietnam, and Japan to irrigate small paddies. In China the same function was done largely by various water ladders, commonly known as dragon backbone machines (*long gu che*), basically square-pallet wooden chain pumps with a horizontal pole trodden by two or more men, who supported themselves by leaning on a pole.

The rope and bucket lift, especially common in India (*monte* or *charsa*), uses one or two pair of oxen walking down an incline while lifting a leather bag fastened to a long rope. An ancient device, best known by its Arabic name of *sāqīya*, carried clay pots on two loops of rope upside down below a wooden drum to fill at the lower end and to discharge into a flume at the top. Devices with clay pots, bamboo

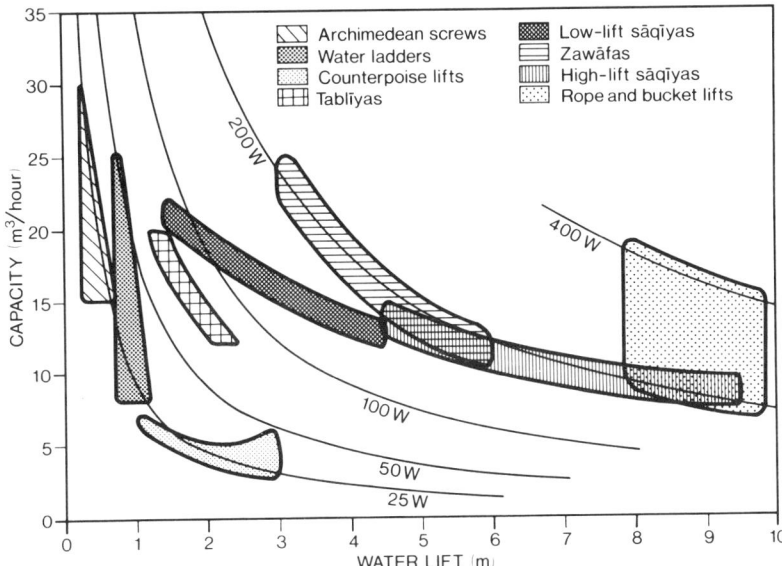

Figure 7.4 Performance fields of traditional irrigation devices ranging from relatively high-capacity, low-lift Archimedean screws to rope and bucket lifts drawing water from depths greater than 8 m. The 50 W isoline delimits the extent of continuous human labor. Based on data in Molenaar (1956), Forbes (1965), and Needham (1965).

tubes, or metal buckets fastened to the rim of a single wheel were driven either through right-angle gears by animals or, if equipped with paddles, by water current. Again, an Arabic name, *noria*, is most famous, although some of the largest batteries of stream-driven norias, with diameters of 15 m, are known from China (*hung che*). The Egyptian *tablīya* has a double-sided all-metal wheel scooping up water at the outer edge and discharging at the center.

Energetic imperatives of traditional irrigation are best revealed by plotting the usual ranges of lifts and hourly capacities (Fig. 7.4). A single man working in 2–4 hour spells at rates close to 100 W will easily power all low-lift Archimedean screws, low-capacity water ladders, and counterpoise lifts; lifts over 3 m require a pair of animals, as do all other high-lift methods; high-volume deep-well bucket lifts may require three to four oxen putting in up to 1.6 kW.

Energy costs of irrigation, calculated in the same way as those of tillage, range from just 100–250 kJ/m^3 of water for human-powered low lifts to up to 4.5–6.5 MJ/m^3 for animal-powered medium and high lifts. Cost–benefit generalizations are impossible owing to differences in crop sensitivities to water supply. A single specific calculation will demonstrate the considerable energy returns. Spring wheat would suffer a 23% yield reduction with a 20% water deficit spread over the whole growing period. Supplying this need of about 80–100 mm would bring an additional yield of 300 kg/ha of grain. When irrigated with human-powered devices the cost, assuming 50% irrigation efficiency, would be 200–500 MJ/ha of additional food intakes—only about 5–10% of the yield gain. With animals energy return in terms of edible cereals would be almost as favorable.

Together with adequate water supply, nutrients are the critical inputs opening the photosynthetic work gates and none is more important than nitrogen. Traditional farming resorted to three basic replenishment strategies: returning a part of the phytomass to soil by plowing in crop residues; recycling of animal and human wastes and other organic materials; and planting of leguminous green manures to be incorporated into the soil in order to provide nitrogen for the subsequent grain crop. These practices also carry other important agroecosystemic benefits, above all improving soil moisture-holding capacity and tilth.

Crop residues, above all cereal straws and stalks, are a large reservoir of recyclable nitrogen. In traditional cultivars nitrogen was about equally divided among the grain and the residue but because residues were needed for household use and animal feed and bedding, only a small fraction was returned directly to the soil recycled in manures.

Manures can be recycled to support high yields, but only with much repetitive heavy labor. Low nitrogen contents (mostly just around 0.5%) necessitate voluminous applications (up to 40 t/ha·y), and a variety of field losses (commonly surpassing two-thirds of the initially available nitrogen) greatly reduce nutrient availability. As with irrigation returns, there is no possibility of simple generalizations, but in cereal cultivation investments of animate labor in manuring returned easily as much as 20 times more energy in grain, clearly an excellent benefit–cost ratio.

Green manuring has relied mainly on the nitrogen-fixing legumes, above all on vetches (*Astragalus*, *Vicia*) and clovers (*Trifolium* and *Melilotus*). These plants can fix 100–300 kg N/ha·y and where the climate allowed their winter growth (3–4 months) added at least 30–60 kg N, enough to produce a good summer cereal crop. In the long run the provision of adequate nitrogen is of such importance that intensive farming cannot do without the legumes, and they are planted as edible species.

This desirable practice represents perhaps the most admirable energetic optimization in traditional farming. It formed the core of all intensive agricultural systems relying on complex crop rotations: China's soybeans, beans, peas, and peanuts alternating with millets, wheat, and rice; India's lentils, peas, and chick peas with local grains, wheat, and rice; Europe's peas and beans with wheats, barley, oats, and rye; and West Africa's peanuts and cowpeas with millets.

Food legumes were self-fertilizing crops leaving large quantities of nitrogen (10–40 kg/ha) for the subsequent grain crops. They yielded at least 50% more protein per hectare than dryland cereals, so although their harvests were lower, their protein content is two to three times higher than that of grains. Moreover, they have plenty of lysine, complementing the cereals deficient in this amino acid and enabling healthy growth on meatless mixed-plant diets. Some of them (soybeans, peanuts) also provided edible oils needed to increase the low energy density of largely vegetarian meals, and oilseed cakes, as well as vines and stalks, make excellent feeds or fertilizers.

This appraisal of traditional agricultures concludes with a few pages on their persistent energetic limitations and on the performance of three important and influential farming systems: Egyptian and Chinese multicropping based predominantly on human labor, and western European farming based on progressively heavier draft animals pulling more complex implements and machines.

7.4 TRADITIONAL AGRICULTURES

For most of recorded history increases in agricultural production came from expansion of cultivated lands, the trend evident in the Old World until the eighteenth century. Cultivated areas grew but yields showed hardly any upward trend as the agronomic practices, animate energy requirements, and mechanical devices changed only very slowly—and as the threat of massive famines was an ever-present reality. This is the key paradox, the principal advantage versus the fundamental weakness of traditional agricultures, or subsistence peasant societies.

Unlike the simpler survival cultures of foragers, traditional agricultures have been able to support much higher population densities as the area of cultivated land was extended, its productivity gradually improved, and some products or services were bartered for those of nearby villages. Subsistence food was ensured by equalized sharing of cultivation rights and harvests among village households. In normal years there was a small but important storable surplus. And, as is best shown by the history of eastern China or northern India, these arrangements were maintainable for indefinite periods of time.

Yet these peasant societies, with their high fertility rates and often low per capita food production, achieved material welfare barely above the minimal needs and were always vulnerable to periodic peacetime famines. This pattern of existential misery and recurrent hunger and massive death persevered in parts of Europe right into the nineteenth century (e.g., the Irish famine of 1846–1851). As the traditional subsistence system was undergoing a rapid transformation into a new, commercial relationship, it continued on a grand scale throughout the poor world until the 1950s and it is still prevalent in less advanced regions and localities of Asia, Africa, and Latin America.

The most important difference between commercial agriculture and the vulnerable peasant farming is in their divergent energy conversion strategies. It is perhaps best elucidated, as Seavoy (1986) argues, by posing a seldom asked question: why do peasant societies increase their populations to the maximum carrying capacity during normal crop years and expose themselves periodically to seasonal hunger or famine during consecutive harvest failures? Moreover, why has this happened even to the societies with low population densities, high soil fertilities, and fairly elaborate farming techniques?

In spite of enormous cultural differences traditional peasant societies shared a strong preference for subsistence compromise where minimum levels of material welfare and food safety were acquired with the least expenditure of physical labor. This predilection can be illustrated on the one hand by the persistence of shifting agriculture and on the other by reluctance to expand permanently farmed lands and to adopt more intensive cultivation. As already stressed, shifting cultivation, which requires relatively low and largely nonspecialized energy inputs, has been a preferred way of food production in all thinly populated forest regions. There it included even those populations that had longstanding contact with settled farmers, be it in the Southeast Asia or Latin America or even in Europe, where the last recorded instances of the practice in Scandinavia and northern Russia date to early decades of the twentieth century. Clearly, permanent farming held little appeal until higher population densities demanded more intensive uses of smaller areas of farmland to maintain the accustomed nutritional levels. Increased energy expenditures were also needed to clear new lands for permanent fields or to create new fields on the slope lands by terracing or by building irrigation canals. Again, these steps were taken reluctantly.

The other important strategy to reduce labor inputs was to spread them as widely as possible, that is, in practice to transfer them from adults to children (Caldwell 1976). Even with children, women—the low-status persons in peasant societies—have to do a dispropor- tionately large share of heavy work, but having a large family is the easiest way (available also to infertile couples, who commonly resorted to adoption) for parents to minimize their future labor exertions. The energy cost of having an additional child is negligible (even the pregnancy may be "free") compared to its labor contributions, which start with children as early as four to five years of age and assure much less heavy work for parents in their old age.

Seavoy (1986) sums it up well:

> Having many children (an average of four to six) and transferring labor to them at the earliest possible age is highly rational behavior in peasant societies, where the good life is equated with minimal labor expenditures, not with the possession of abundant material goods. There is no need to invoke mysterious terms like traditional peasant behavior ... ignorance of human reproduction, masculine prowess, survival strategies, tragedy of childlessness, or joys of family life to explain why peasants desire large families.

This process of very slow intensification of food production can be best illustrated by focusing on three traditional agricultures of out- standing importance: Egyptian, Chinese, and European farming. Minimum populations supportable by predynastic Egyptian farming were just 1–1.3 people per hectare cultivated land. Butzer's (1976) reconstruction of ancient Egypt's demographic history has the Nile valley's population density rising from 1.3/ha of arable land at 2500 B.C. to 1.8 at 1250 B.C. and 2.4 at 150 B.C. when the country was the Roman empire's largest food surplus area. Then came centuries of decline and stagnation. Only the spread of irrigation boosted yields. By the mid-1920s the still basically traditional farming, helped by inorganic fertilizers, was feeding 6 people per hectare.

China's traditional farming was considerably more innovative than Egyptian agriculture. Chinese contributions to the art of irrigation have been outstanding and other improvements included the horse collar harness and an integrated dryland cultivation tool complex consisting of curved iron moldboard plow, multitube reed drill, and various horse-drawn weeding implements. As an enduring paragon of traditional agriculture, China's farming has been studied from many angles and in much detail (Perkins 1969; Ho 1975; Bray 1984). Buck's

(1930, 1937) studies offer a unique portrayal of China's traditional practices. Typical size of fields was only about 0.4 ha, nearly 50% of all land was irrigated, more than 90% of it was planted in grains, and there was a profusion (more than 500) of cropping systems.

Counting just the human labor, energy return ratios for staple grains and sweet potatoes were around 25, for corn about 40, for grain legumes around 15, and for plant oils about 10. Inclusion of animal labor does not substantially change these relative proportions. Virtually all of the edible crop harvests were consumed directly as the grains provided 90% of all food energy.

Such diets could support large population densities: the average was about 4.5 people per hectare of sown land, or at least 5.5 persons per hectare of arable area. This is nearly as much as in Egypt during the 1920s. Rice-growing southern China could do even better, surpassing 7 persons per hectare by the late 1920s. And in the most intensively farmed areas total yields were at least 36 and up to 48 GJ/ha, enough to feed 12–18 people. But even during the 1920s peasants recalled an average of three crop failures within their lifetime serious enough to cause famines.

European experience was somewhat less harsh owing to a more equable climate in the west and to generally lower population densities. But otherwise the continent's farming experienced similar prolonged productivity stagnation, periodic deep declines, and centuries of gradual intensification punctuated by minor and major famines (Slicher van Bath 1963; Duby 1968; White 1970; Fussell 1972). Oxen were the principal draft animals, plows were wooden, sowing was by hand, harvesting with sickles, and threshing with flails; yields were low and highly variable. All of this changed only very slowly during the millennium after the demise of the western Roman Empire. Notable innovations included scythe instead of sickle for harvesting, shoulder collars for horses, and horses as principal draft animals (since the thirteenth century in the West, although oxen remained important elsewhere).

Subsistence capacities of European farming doubled from 2–2.5 people per hectare of arable land in the Middle Ages to 4–5 ha by the beginning of the nineteenth century. In the most intensely cultivated parts of the continent they doubled again before its end when the farming was already benefiting from many external energy subsidies, changing into a new energetic hybrid where harnessing of solar radiation is inextricably tied to substantial inputs of fossil fuels.

8

PREINDUSTRIAL COMPLEXIFICATION: PRIME MOVERS AND FUELS IN TRADITIONAL SOCIETIES

Nature is so subtil and so penetrating in her ways, That she cannot be used except by great Craft; for she does not openly reveal that which may be completed within her, this completion must be accomplished by man.

—Paracelsus
Das Buch Paragramum

Complexification of human societies between the time of the first agricultural settlements and the explosion of industrial civilization rested on the harnessing of three classes of primary energy: muscular exertions of humans and animals, movements of water and wind, and the burning of biomass fuels. Some of these conversions eventually reached outputs high enough to power incipient mass production processes of evolving industrial society. Roads to higher performance led through multiplication of small forces and technical innovations. The two approaches often were combined: monumental architecture of antiquity required massed labor as well as the ubiquitous application of labor-saving devices.

In the Western world the dominance of muscle, water, wind, and wood ended only in the second half of the nineteenth century. Traditional energies are the foundation of the current affluence of industrialized countries. More importantly, in large parts of the poor world these energies remain essential existential necessities. Their study and understanding are thus more than matters of historical interest (Singer et al. 1954–1984 (8 volumes); Needham 1954– (6 volumes); Forbes 1964–1972; Klemm 1964; Daumas 1969; Lindsay 1974; White 1978; Landels 1980).

Energetic bases and performance of preindustrial societies are surveyed in this chapter initially through a detailed look at animate power in transport, construction, mining and manufacturing, followed by an examination of the advances in harnessing of water and wind, and finally through an appraisal of the availability and use of biomass fuels.

8.1   ANIMATE POWER

Societies deriving their kinetic energy solely from animate power had little physical security and affluence to offer. Intensification of animate energy inputs was achievable only through mass concentrations of labor and through deployment of clever devices, but ancient civilizations never used mass manufacture based on slave or free labor. Atomization of production (Christ 1984) remained the norm, and massed labor left its most impressive legacy in buildings of unsurpassed esthetic appeal.

Direct applications of animate power have obvious logistic limitations. Without mechanical aids easing the effects of gravity and friction, human capacities in lifting and conveying are limited to weights of several tens of kilograms. Nepali Sherpa porters, recognized as the ablest load carriers, usually shoulder 30–35 kg. In the traditional Chinese sedan chair two men conveyed a customer, again a load of 30–35 kg per porter.

More efficient utilization of manpower depended on simple mechanical devices—levers, inclined planes, pulleys, windlasses and capstans, treadwheels, gearwheels, and wheels. Horizontal windlasses (and winches) and vertical capstans enabled continuous power transmission by ropes or chains through simple rotary motion, a task made easier by cranks, a Chinese invention unknown in European antiquity (Fig. 8.1).

Treadwheels allowed much better performance than turning of windlasses or capstans and efficient transmission of their power via gearwheels enabled a wide variety of final uses (Fig. 8.2). Similarity of action with cycling, the most efficient form of human locomotion, is obvious: powerful leg, abdominal, and back muscles do nearly all the work. There were also external vertical wheels, horizontal and inclined treadwheels. Regardless of the particular mechanical arrangements, maximum inputs to power these devices were small: with a single man no more than 150–200 W during brief spells of strenuous exertion, at best 100 W during hours of sustained effort. The largest, eight-man, treadwheels received brief inputs of about 1500 W and steady flux of up to 700–800 W.

For millennia horses were very uncommon working animals. Unlike bovines (running on grasses and crop residues), they needed either good pastures (too few in the dry Mediterranean region or in densely populated eastern China) or concentrate feed (oats, legumes), and they were more difficult as well as more expensive to harness. Bovines—above all oxen and water buffalo—thus dominated not only field work but also transportation of goods and provision of stationary power. There were also working yaks, mules, donkeys, camels, goats, and dogs (in wheels turning kitchen spits and pulling carts or wheelbarrows).

A typical ox or horse, often old and weak, turning a whim (a circular crank pulling a beam attached to a central axle), could not put out more than 400 W, and even when two or four stronger animals were so harnessed their combined power would not be higher than 1–2 kW. On commonly used treadwheels it was impossible to fit more than two animals and hence to secure more than 1 kW. Transport capabilities were dependent not only on the animal but also on the ability to reduce friction, that is, on the quality of roads and vehicles.

Wheels varied from heavy, primitive, segmented solid disks or

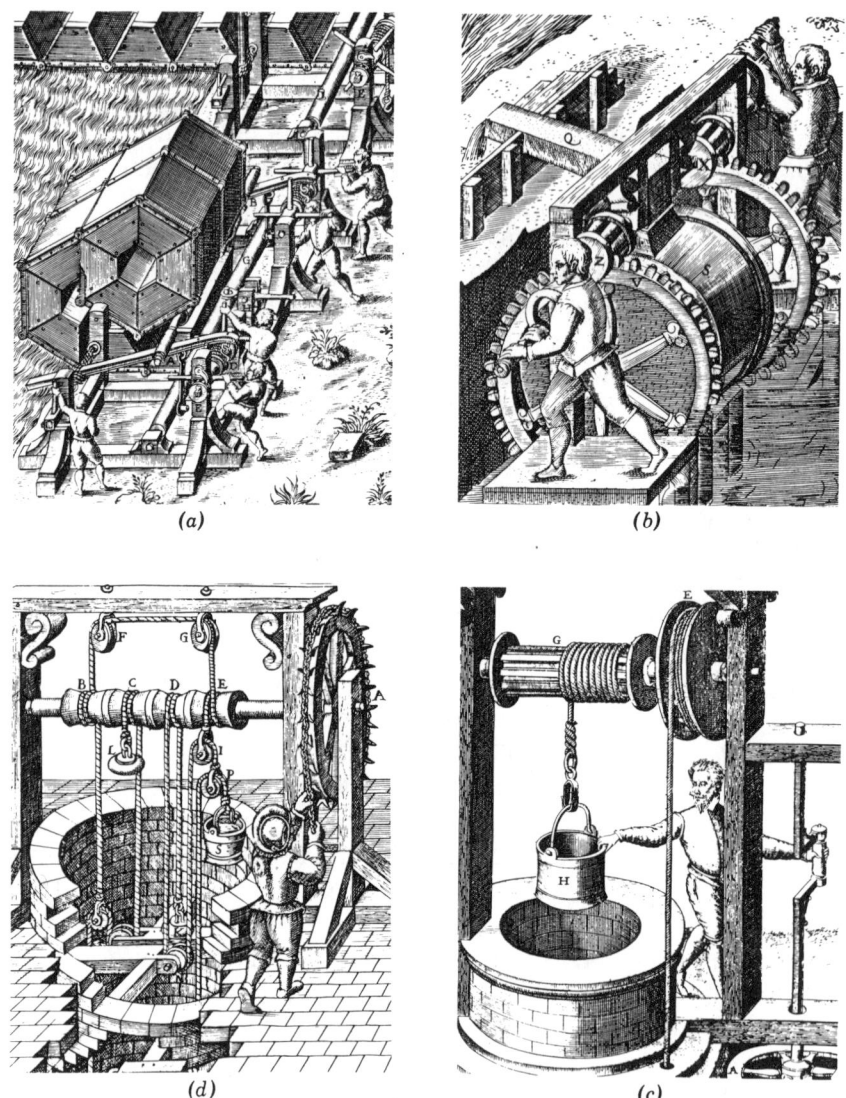

Figure 8.1 Hand-operated rotaries in Ramelli's (1588) book of diverse machines include (*a*) spiked winches, (*b*) horizontal cranks, (*c*) vertical doubly supported cranks, and (*d*) inertia chains. Note also the compound pulleys in the last image.

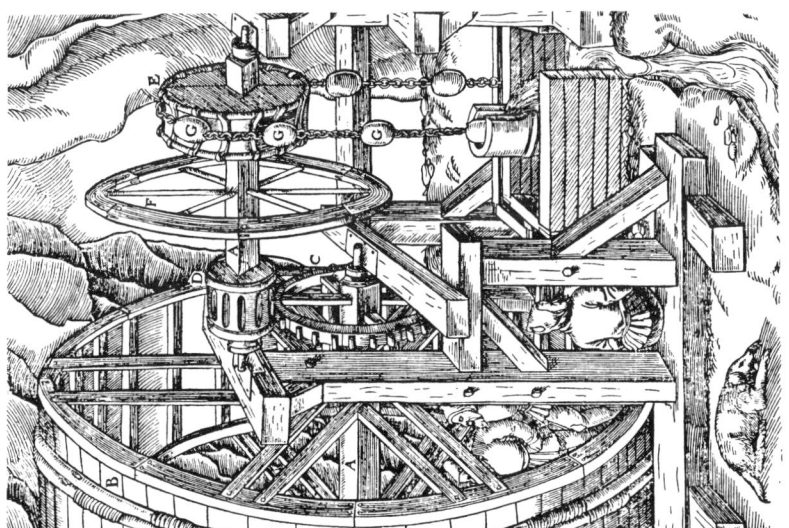

Figure 8.2 Internal vertical treadwheels pictured in Agricola's (1556) famous book were indispensable for lifting water from deep mines (left, powering a chain-and-ball pump) and were also used in milling (a goat-powered version on the right).

drums fixed to a rotating axle, to light, multispoked arrangements rotating on a fixed axle. The front axle itself was either pivoted (ancient Persia, Celtic Europe) or fixed (Roman era). Roads ranged from muddy ruts and sandy trails to hard-top *viae*. Whereas a smooth, hard, and dry road requires force of only about 30 kg to wheel a 1 t load, a loose gravel surface may easily need 150 kg and the force needed on sandy or muddy roads can exceed 200 kg. Small, poorly fed animals were often unimpressive performers. A pair of early twentieth-century draft horses could pull a 3–4 t rubber-tire cart on hard-top roads, but Forbes (1965) estimated tractive effort at no more than 680 kgm/s for a typical preindustrial horse cart.

In contrast, a man aided by a wheel could move at best a load of around 160 kg with a Chinese barrow bearing the load above the large wheel's axle (sometimes lightened by installing a small sail) and a load of 60–100 kg with a European barrow—but with no more than half of the horse cart's speed. The effort thus came only to 30–70 kgm/s. Low speeds of animate transport limited the daily range to no more than 70 km for passenger horse carts, 30 km for horse-drawn good wagons, and 10–15 km for men with wheelbarrows.

Where the loads had to be carried the lower human performance was greatly compensated by flexibility of loading, unloading, and carrying on narrow paths. Similarly, for carrying loads mules and donkeys with panniers were preferred to horses. Typical loads were about 30% of the animal's weight (i.e., generally 50–120 kg) on the level, 25% in the hills, and speeds did not exceed 5 km/h. Animate land transport was never suited to a large-scale movement of goods: it cost more to move grain just 120 km on Roman roads than to ship it across the Mediterranean. Land transport was the most ubiquitous energetically determined hindrance to complexification of preindustrial societies.

Horse-drawn transport rose to an unprecedented, though brief, dominance in the rapidly growing cities of nineteenth-century Europe and North America with multiplication of private coaches, cabs, omnibuses (first in London in 1829), and delivery wagons (Dent 1974). Late 1890s London had some 300,000 horses (Fig. 8.3), and some New Yorkers were thinking about creating a large suburban belt of pastures to accommodate the swelling horse numbers. Within a decade streetcars and cars ended this equine dominance.

Man-powered waterborne movement had much higher power rates than land transport, and oared vessels offer fine examples of clever design and mass labor integration (and, frequently, of terrible suffering). Assuming 80% rowing efficiency, 100 W sustained and 200 W

Figure 8.3 Victorian London Bridge at noonday—a horse-drawn traffic jam in a growing city. (From *The Illustrated London News* 16 November 1872.)

peak burst power, *penteconteres* (vessels with 50 oarsmen) taking Greek troops to Troy were propelled with 3.5–7 kW of power. Triple-tiered *triremes* with 170 rowers could fly with up to 24 kW, enough power to go over 20 km/h into a devastating ramming attack. Large seventeenth-century galleys with 56 oars were crewed by five men per oar—and up to 200 warriors oared a large Maori dugout canoes. Limits of aggregate useful human power in sustained rowing were thus 12–20 kW.

These totals were surpassed in numerous cases where large numbers of men were skillfully integrated into smoothly working power systems to do some extraordinary construction tasks. A well-known Egyptian painting of transporting a colossus from a cave at el-Bersheh (1800 B.C.) depicts 127 men pulling a sledge whose path is being lubricated by a worker pouring water from a vessel (Fig. 8.4). Short bursts of over 30 kW could easily move the 50 t load if the lubrication reduced the sledge resistance by about 50%. Protzen (1986) determined that pulling the heaviest (140 t) Inca stone block at Ollantaytambo up the ramp required the concerted participation of some 2400 men. Peak power of this group would have been around 600 kW, but we know nothing of the incredible logistics of this effort.

Similarly, construction of the three great pyramids at Giza, the foremost ancient example of integrated use of labor, remains a matter of conjecture. There are no extant contemporary images or descriptions of how it was done. Two millennia later Herodotus reported on a grand causeway of polished stone over which sleds pulled the blocks from the edge of the Nile to the Giza plateau. Barber's (1900) calculations showed that 900 men harnessed in double rank on four draft ropes could drag a 60 t stone up such a greased causeway. The best one can do is to determine the minimum energy cost of building a pyramid.

Herodotus remains the earliest credible source on the total labor involved—64,000 men for 80 days a year for 20 years—and Mendelssohn (1974), using basic physical considerations, came up with 70,000 seasonal laborers and up to 10,000 permanent masons. Assuming 80,000 men averaging 900 h/y at a net cost of 700 kJ/h for two decades results in the total human energy investment of 1 PJ and an average net power input of about 15 MW during the working hours.

In contrast, most of the European cathedrals required total energy investments two orders of magnitude smaller. Their frequently interrupted construction commonly stretching over decades, even centuries, needed 10^2–10^3 workers for a full-time equivalent of about

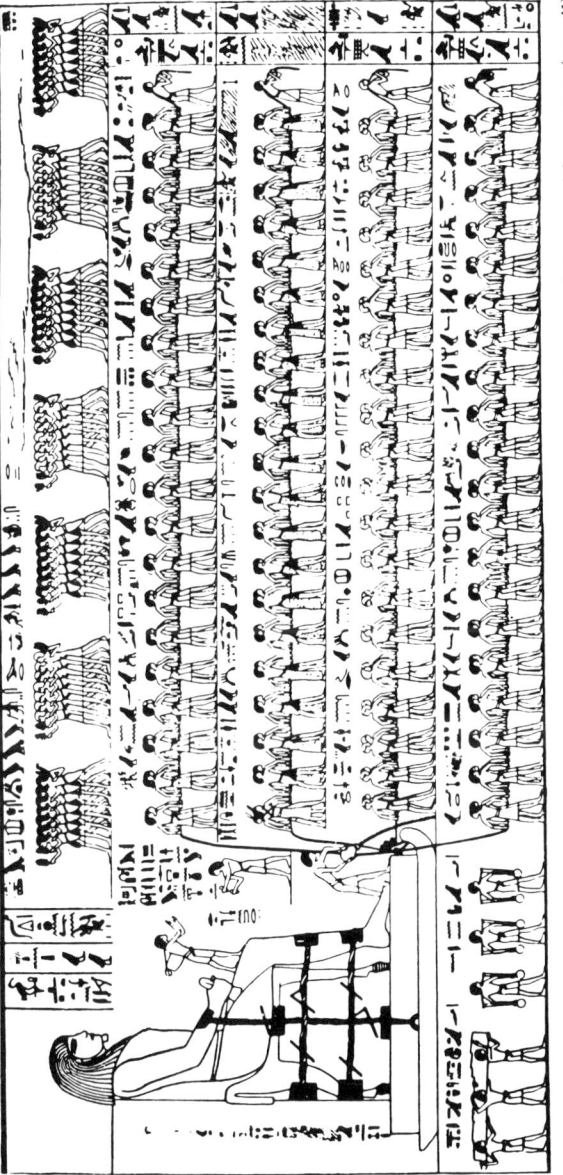

Figure 8.4 Egyptian technique of transporting huge loads shows the path of towing lubricated by water (or oil) by a man taking a ride with the colossus. (From Rühlmann 1962.)

10 years and the total energy investment mostly between 10 and 50 TJ and peak power flows of a few hundred kilowatts. Medieval societies did not surpass antiquity in skillful use of mass labor but they were far ahead in their widespread reliance on the kinetic energies of water and wind.

8.2   WATER AND WIND

While animals helped lighten human labor for millennia, the history of water and, with the exception of ships, wind as important prime movers is much shorter. The first literary reference to a primitive watermill is from Greece of the first century B.C.; one of the first descriptions of simple vertical shaft windmills in the windy region of today's eastern Iran dates from the middle of the tenth century. After their diffusion throughout the Old World both kinds of machines had gradually assumed decisive roles in the life of the preindustrial Europe and eventually helped to energize the beginnings of large-scale industrialization (Wilson 1956; Moritz 1958; Reynolds 1970; Fox et al. 1976; Torrey 1976; White 1978).

The horizontal wheel (Fig. 8.5), whose place of origin remains untraceable, has persisted in many regions of Europe and everywhere east of Syria, but vertical arrangements, turning the millstones by right-angle gearing, offered better performance and became dominant in a variety of later applications. The basic typology of vertical waterwheels is pretty straightforward (Fig. 8.6). In low falls (heads of 1–3 m) the impulse of flowing water was exploited by undershot wheels, with diameters roughly three times as large as the head for paddle wheels and two to four times for a Poncelet wheel with curved blades. Traditional undershots had efficiencies of just around 20%, nineteenth-century wheels up to 35–45%. Location on swift currents was desirable as the theoretical power of undershots is proportional to the cube of the water speed. Breast wheels combined the impulse of the flowing water with gravity for heads of 1.5–5 m; their efficiencies were no better than those for well-designed undershots.

Traditional overshots exploited heads greater than 3 m (diameters equal to about 75% of the head), usually by feeding water through troughs or flumes into the buckets; the impact is relatively unimportant as the weight of descending water generates the bulk of the energy. Top efficiencies reached 85%, although 60% was more common. The advantage of using relatively sluggish flow to operate the overshots was commonly weakened by the necessity of a carefully

Figure 8.5 Horizontal waterwheel was simple to make but rather inefficient; it was best suited for small-scale grain milling. (From Ramelli 1588.)

regulated water supply often requiring construction of ponds and races. Still, overshots became favorites for many applications, ranging from wood sawing to blacksmithing, besides efficient grain milling. There were also floating wheels and tidal mills (Minchinton and Meigs 1980).

Unprecedented continuity, reliability, and magnitude of power delivery by waterwheels afforded new productive possibilities, especially in mining and metallurgy. In many ways the beginnings of European and American industrialization lay clearly in these specialized waterwheels, which were introduced between the tenth and the thirteenth century and subsequently were enlarged and improved. Power of the simplest small machines was hardly impressive but even they made a great difference. Whereas in an hour two slaves (200 W)

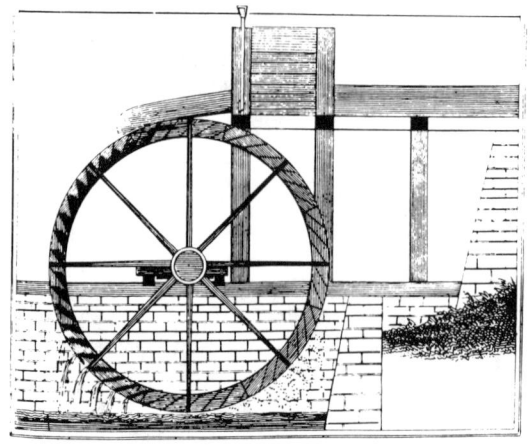

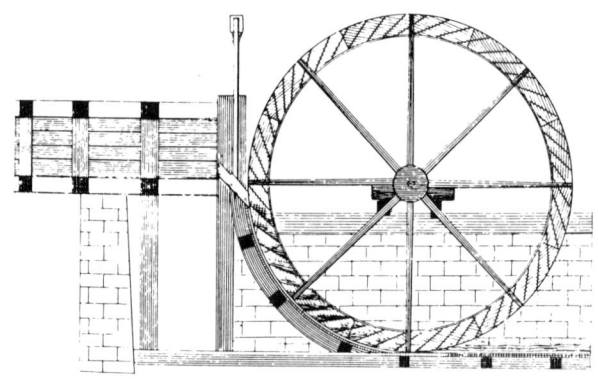

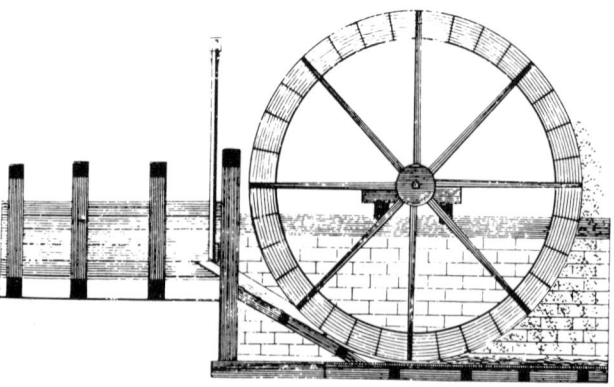

Figure 8.6 Late-eighteenth-century renditions of overshot, breast, and undershot wheels from *The Young Mill-Wright and Miller's Guide* (Evans 1795).

with hand querns ground 7 kg of grain and a donkey-driven mill (300–400 W) produced around 12 kg of flour, typical output for a set of waterwheel-driven (2.2–2.5 kW) millstones was 80–100 kg. If flour could have provided half the average food intake, a single small mill could have produced enough of it in a 10 h shift for about 3500 people, a good-sized medieval town.

The growth of unit capacities was slow: even during the eighteenth century waterwheels averaged only 3.7 kW, rarely exceeding 7.5 kW, and poor gearing led to low efficiencies. Larger power outputs were achieved by multiple installations of smaller units in suitable location and more recently also by building some sizable wheels. An exceptional Roman mill-line at Barbegal near Arles had 16 wheels with a total of 24 kW. A water-pumping installation on the Seine at Marly, built between 1681 and 1685 to supply the Versailles fountains, had 14 wheels (12 m in diameter) raising 3200 m³ a day 162 m in three stages, potential over 90 kW, and actual output of at least 60 kW (Klemm 1964). Aggregrate capacities reaching into megawatts came only with industrialization requiring concentrated power supply. By the early 1830s Shaw's water works on the Clyde near Glasgow consisted of 30 mills rated about 1.5 MW with a possible extension to 2.2 MW.

The largest British and American waterwheels had diameters around 20 m and capacities in excess of 50 kW. *Lady Isabella*, the largest wheel ever built, was a remarkable example belonging to the Great Laxey Mining Company on the Isle of Man (Reynolds 1970). This pitchback overshot machine had diameter of 21.9 m; its rotation was transmitted to the pump rod, reaching 451 m to the bottom of the lead-zinc mine shaft, by the main-axle crank and 180 m of timber connecting rods. The theoretical peak power was about 427 kW and normal operation delivered about 200 kW. Built in 1854, the wheel worked until 1926 and was restored after 1965.

The first radical improvement of water-driven prime movers came only in the first half of the nineteenth century with the development of the water turbine. First, in 1832, came Benoit Fourneyron's reaction turbine with radial outward flow built to run forge hammers at Fraisans. Under the head of 1.3 m and with the rotor diameter of 2.4 m it delivered 38 kW. In 1837 two of Fourneyron's machines for the Saint Blaisien spinning mill worked under heads of 108 and 114 m, and the power of 44.7 kW was large enough to run 30,000 spindles and 800 looms (Smith 1980).

Fourneyron's turbines, with high performance limited only to particular flow and pressure conditions, were soon replaced by better

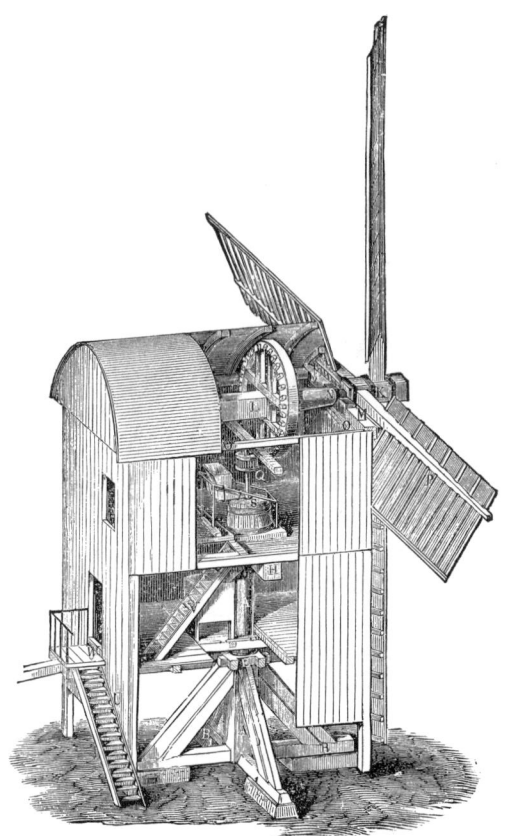

Figure 8.7 Cut-out view of a sturdy post mill (Wolff 1900).

machines, above all the inward-flow turbine of James B. Francis in 1849, axial flow turbines of the Kaplan type, and jet-driven Pelton wheels of the 1880s. Gradually, turbines replaced waterwheels as the prime movers in manufacturing but soon they stopped turning complex gears and started to rotate electricity generators. Two millennia of water power as a direct prime mover were nearly over.

The windmill, the other inanimate prime mover contributing decisively to the gradual intensification of Western economic life, was first recorded in Europe during the last decades of the twelfth century. Windmills were vertically mounted and had the ability to turn the driving shaft into the wind. All of the early Western European machines—except for the low-power Iberian octagonal sail mills with triangular cloth transferred from the Middle East—were post mills (Fig. 8.7). Their gears and millstones were housed in a wooden structure pivoting on a sturdy central post. Their relatively low height

Figure 8.8 Typical late-19th century American windmill widely used to pump water for households, cattle and steam locomotives (Wolff 1900).

meant a limited performance: extractable power goes up with the cube of the windspeed—and the typical aboveground windspeed profiles show exponential rise.

Postmills were gradually displaced by tower mills or smock mills. Both of these structures had a fixed body and only the top cap was turned into the wind. With the introduction of fantail to power a winding gear in 1745 the sails could automatically be kept to the wind—but it took another century for this innovation to spread to Holland and Denmark, the countries with the largest numbers of mills. Development of American windmills took a new course after the middle of the nineteenth century with westward expansion. Instead of a few large and wide sails, great numbers of narrow blades or slats were put on solid or sectional wheels, most commonly equipped with governors and independent rudders, placed on top of lattice towers (Fig. 8.8).

During the eighteenth and nineteenth centuries windmills reached the peak of their global importance. By 1650 United Provinces of

Netherlands had at least 8000 machines, there were perhaps as many as 10,000 mills working in the early nineteenth-century England, a German count in 1895 showed 18,232 in use, and several million American windmills (in 1889 there were 77 manufacturers) were sold in the United States between 1860 and 1900.

There is no information on energy output of early windmills. Forbes (1958) estimated that a typical large eighteenth-century Dutch mill with 30 m span could, when equipped with improved sails and with 8–9 m/s winds needed for efficient operations, develop 10 hp (about 7.5 kW) at the windshaft. When equipped with the best modern sails and gearings its output could be raised to as much as 20–30 hp (15–22.5 kW). Measurements at a preserved 1648 marsh mill showed actual output of just 11.6 kW for a transmission loss of 61%. This confirms Rankine and d'Aubuisson's comparison of eighteenth-century prime movers in which they assigned 2–8 hp (1.5–6 kW) to post windmills and 6–14 hp (4.5–10.4 kW) to tower mills (these values are obviously for useful power). Wolff's (1900) measurements of American windmills indicate useful power rates from 30 W for a 2.5 m wheel to about 160 W for common 3.6 m machines and up to 1000 W for large 7.6 m devices.

Consequently, an acceptable generalization would be to assign 0.1–1 kW of useful power to nineteenth-century American wheel mills, 1–2 kW to small medieval and later post mills, 2–5 kW to large post mills, 4–8 kW to common smock and tower mills, and 8–12 kW to the largest eighteenth- and nineteenth-century devices. Typical medieval windmills thus had capacities roughly equal to common watermills but by the early nineteenth century many hydraulic installations were four to five times more powerful than even the largest windmills.

Finally, the ancient way of exploiting wind power—the sailship— has a rich, fascinating history (Torr 1964; Chatterson 1977), but only innovative medieval designs made it superior to oars in terms of maneuverability and reliability. Square-sail ships were efficient only with the wind astern and only the late medieval combination of square rigging and triangular sails made it possible to sail close to the wind. Together with better hulls, stern-post rudder, and magnetic compass, it gave the Europeans a highly efficient ship—and the first means of global travel, trade, and military projection.

Comparisons of record tonnages between antiquity and late medieval and premodern eras do not show any huge gains: ships with capacities in excess of 1000 t had been built by the Romans, but such sizes were always exceptional. A common Roman cargo vessel carried less than 100 t, voyages of great European discoveries started with ships nearly as small (*Santa Maria* at 165 t), a Great Armada vessel averaged 515 t, and typical late eighteenth-century India fleet ships were about 1200 t. Speed extremes are spanned by Roman cargo ships, which usually went no faster than 2–2.5 m/s—and the sleek clippers of the mid-nineteenth century that surpassed 9 m/s (*Lightning* in 1853) and averaged 8 m/s a day.

But tonnages and speeds are inadequate to calculate energies needed to move individual ships on long voyages or to come up with aggregate annual contributions of wind power harnessed by merchant or military fleets: hull designs, sail areas and cuts, cargo weights, and utilization rates are far too heterogeneous to come up with meaningful values. Still, Unger (1984) calculated the contribution of Dutch sailships during the seventeenth century at about 200 TJ (6.2 MW) a year, roughly equal to the total energy output from all Dutch windmills (DeZeeuw 1978).

Muscles, water, wind—the choice of prime movers in preindustrial societies was limited. Yet these alternatives were more varied than fuel supplies, which were dominated nearly everywhere by wood.

8.3 PHYTOMASS FUELS

Wood—stems, branches, bark, roots, or, preferably, charcoal made of this phytomass—and crop residues—straws, stalks, vines, and roots—has fueled the subsistence as well as the complexification of all preindustrial societies. Fossil fuels have displaced all but a small fraction of these phytomass fuels in all industrialized countries, but throughout the poor world they still cover major shares of national primary energy use and they commonly dominate rural cooking and heating needs.

Widespread interest in the energetics of industrializing economies has brought much new information about the rates and efficiencies of fuelwood and crop residue utilization and charcoal production (Earl 1973; Openshaw 1978; National Academy of Sciences 1980b; Hall et al. 1982; Smil 1983). As the basic subsistence needs—in most societies cooking two or three meals a day, in many also heating water or a room in winter months, preparing animal feed, or drying food—have changed little over time, this contemporary evidence can be extended to the preindustrial era.

But first the essential values on wood and crop residues as fuels must be considered (Tillman 1978; Smil 1983). Although thousands of woody species are available for combustion, their chemical composition is relatively uniform: about 43% is cellulose, 28–35% hemicelluloses, the rest lignin; carbon makes up 49–56%, oxygen 40–42%, and ash varies from less than 0.5 to 2%. In contrast, density differences can be large, from less than 0.4 g/cm^3 for poplars to nearly 0.7 for oaks up to 1.0 for some eucalyptuses.

Energy content goes up with the proportion of lignin (26.5 MJ/kg compared to 17.5 MJ/kg for cellulose and hemicellulose) and extractives (up to 35 MJ/kg). Differences among the common woods are not large; they are 17.5–20 MJ/kg for most hardwoods and 19–21 MJ/kg for softwoods. In comparison with standard coal, energy density of typical wood (19 MJ/kg) is about a third lower and it surpasses most crop residues by only about 5–10%. Cereal straws have heat values of 17–18 MJ/kg.

All of these energy densities refer to absolutely dry phytomass. In reality, wood as well as crop residues always have fairly high moisture content. Fresh mature wood averages 30% water for hardwoods and 46% for softwoods, and even air-dry wood (cut, stacked, sheltered, and let dry for at least two months) still contains about 15% moisture. Dry, ripe straws have 7–15% water. Net heating values of phytomass vary widely, from around 20 MJ/kg for very dry to about 15 MJ/kg for air-dry wood to as little as 5–6 MJ/kg for fresh wood and grasses.

In contrast, charcoal is a high-quality fuel with energy density (29.7 MJ/kg) equal to that of very good coal. Traditional production of this smokeless and virtually sulfur-free fuel has been very simple but also very wasteful. In primitive earth or pit kilns the partial combustion of the heaped wood provides the heat necessary to initiate carbonization. Charcoal yields are 15–25% of the dry wood charge by weight and a 1:5 ratio is the best approximation. Assuming net heat value of 14.5 MJ/kg of air-dry wood, energy loss in traditional charcoal-making would be 60%—but it is only half that in modern steel furnaces or retorts.

Considering the generally low levels of final energy demand, the consumption of phytomass fuels in traditional societies has been relatively high; often dismally low conversion efficiencies are the reason. Transition from uncontrolled and inefficient open-air fires to enclosed, regulated, and efficient burners has been very slow. Moving the open fire inside made little difference: fireplaces allowed fires to burn unattended overnight but otherwise performed quite poorly. Modern measurements show their heating efficiencies reaching at best about 10% and not uncommonly causing net energy loss.

Efficiencies of traditional stoves varied widely with design, fuel, and cooking practices but they rarely surpassed 20%. Benjamin Franklin's stove, introduced in 1740, marked the first step toward more efficient designs.

Judging by the best available studies of those remaining traditional societies that rely almost solely on biomass fuels for their heat requirements, the annual consumption of woody matter and crop residues averaged less than 10 GJ per capita for the poorest communities in the warmest regions and it could have been 30–50 GJ per capita in relatively rich mid-latitude regions with many larger cities and wood-based manufacturing.

For nineteenth-century western and northern Europe and North America the needs were much higher. In places between 40 and 50°N households relying solely on wood combustion in modern stoves need annually 50–200 GJ (15–50 GJ per capita). Lower combustion efficiencies and additional needs for cooking and water heating easily doubled these rates for wooden-age societies: the best per capita fuelwood consumption estimate for the United States in 1850 is about 97 GJ (Schurr and Netschert 1960).

Estimates of the worldwide total of biomass fuel consumption in the preindustrial age can aim only at the right order of magnitude. Assuming an average of 20 GJ per capita for the 1850 population of around 1 billion would give a total of some 20 EJ, and values up to 30 EJ seem plausible. Gross energy value of preindustrial biomass consumption thus would have been an order of magnitude smaller than the mid-1980s energy content of fossil fuels. Considering the superior modern conversion efficiencies, the difference of effective energies would be at least twentyfold.

But there were ingenious traditional ways of using wood and straw efficiently. The best examples are the three interesting space heating systems providing an uncommon degree of comfort: Roman *hypocaust*, Korean *ondol*, and Chinese *kang*. The first two used combustion gases to heat raised room floors before leaving through a chimney; the third involves just a large (typically about 4–5 m^2) heated platform in the main room to serve as a resting place during the day and bed at night. Advantages of these arrangements are obvious in comparison with brazier heaters, which have been common in both Eastern and Western societies (the British House of Commons was heated by large charcoal fire pots until 1791). They offered only a spot source of warmth and produced dangerous levels of carbon monoxide.

Metallurgical demand for charcoal rose with the replacement of bronze by iron starting about 1000 B.C. Iron melts at 1535°C, unaided charcoal fire can easily reach 900°C, and even with simple forced air supply its temperature can be increased to close to 2000°C. Besides its high energy density charcoal has the advantage of high purity, but its friability limited the mass of smelting charge. This presented no difficulties as long as the furnaces remained small. The simplest furnaces were just partially enclosed hearths built on hilltops to maximize natural draft; they produced small (up to 50–70 kg) masses (blooms) of iron. This low-carbon wrought iron had high tensile strength, it was malleable and easily forged, but it could not be cast.

Blast furnaces, originating in the Rhine–Meuse region in the fourteenth century, produce cast, or pig, iron, an alloy with 1.5–5 percent of carbon which cannot be directly forged or rolled and is weak in tension but sturdy in compression. Until the middle of the eighteenth century all ironmaking furnaces had very low heat efficiencies: they needed 3.6–8.8 times more fuel than the mass of charged ore (Johannsen 1953). Assuming ores averaging 60% of iron, 75% of which ended in the molten metal, smelting of wrought iron needed 8–20 kg of charcoal per kilogram (240–600 MJ/kg).

By 1900 the typical rate was down to just around 1.2 kg (36 MJ/kg) and the best rates were as low as 0.77 kg (Campbell 1907; Greenwood 1907). During the 1970s Brazil, the world's largest consumer of metallurgical charcoal, needed about 29 MJ/kg (Thibau 1978). Very high energy intensities of pre-1800 smelting caused extensive deforestation. A single English eighteenth-century furnace using wood from coppice hardwoods (15 year rotation, 7.5 t/ha·y) would have required about 1600 ha/y, or no less than 24,000 ha of trees for sustainable supply. Total annual wood needs in ironmaking added up to about 1100 km^2 of coppice growth in early eighteenth-century England, and to about 2600 km^2 of forest in early nineteenth-century North America.

Metal output of all pre-fossil fuel societies was primarily limited by the availability of fuel. When a single enterprise can strip up to 40 km^2 of trees a year it is easy to imagine the cumulative effect of scores of furnaces over a period of few centuries: preindustrial European deforestation was primarily a matter of horseshoes, mail shirts, nails, and axes.

Blast furnace locations reflected the limited radius of practical animal-drawn transport to bring in charcoal, the necessity of continuous rapid water flow to power bellows, and the requirements of minimum area for sustainable wood cutting. Inevitably, the sites were mountainous, forested locations, and the furnaces were at least 8–

10 km apart. Naturally, proximity to the ore was also essential, but ore made up only a fraction of charcoal's charge and it would have been easier to transport.

Finally, the maximum height of blast furnaces, about 7.5 m by the mid-eighteenth century, was limited not only by charcoal's friability (a taller charge would have crushed it to dust) but also by the maximum air blast available from water-powered bellows. Both obstacles were overcome when coal was turned to coke capable of supporting a heavy charge and converted to mechanical energy in steam engines always ready to run large bellows. Unprecedented power densities unlocked by combustion of coal enabled rapid intensification and widespread diffusion of mining and metallurgy, which laid the foundation of a new energy era.

9

FOSSIL FUELS:
HEAT AND PRIME MOVERS

Nature, in providing us with combustibles on all sides, has given us the power to produce, at all times and in all places, heat and the impelling power which is the result of it.

—Sadi Carnot
Reflexions sur la puissance ...

In all preindustrial societies utilization of energy flows—as direct radiation, wind, water, feed, food, and animate energies—either was virtually immediate or commenced after just a few months needed for crop maturation or a few years required for domestic animals and children to reach their working ages. Only in large trees cut for fuelwood or charcoal was the conversion delayed by decades, even centuries. Yet preindustrial civilizations depended on practically *instantaneous solar energy flows*. In contrast, industrial civilization has been tapping *stores of fossil fuels* transformed from biomass or created abiogenically mostly between 10^6 and 10^8 years ago. Extraction and combustion of these fuels could not change the irreplaceable reliance on solar flows in photosynthesis but made industrial societies dependent on resources that are not renewable on civilizational time scale.

We are living in an energetic interlude: the stores powering our way of life are finite and even the best conversion efficiencies and conservation measures cannot extend their life beyond several hundreds of years. But it is highly improbable that we will actually go on to exhaust all of their recoverable reserves. Long before reaching that point we will either go back to flows harnessed in ways superior to preindustrial practices, or we will become dependent on another class of stores. But the past advances of global industrial civilization and its prospects for many generations to come are defined by its consumption of fossil fuels.

The history of fossil fuel use can be traced in Singer et al. (1954–1984) and in Forbes (1964–1972) and the place of fossil fuels among the world's resources is appraised in McLaren and Skinner (1987). Discussion of their resources, extraction, transportation, combustion, and conversion to and transmission of electricity in this chapter is followed by a broad analysis of fossil-fueled civilization as a high-energy system in the next chapter.

9.1 COALS, OILS, AND GASES

Plurals are essential to convey heterogeneities of fossil fuels. Although there is a good case for abiogenic origin of some hydrocarbons (Gold 1987), most fossil fuels are clearly organic mineraloids with minor quantities of inorganic contaminants. They fit into one of two broad categories, coals or kerogens, and their physical state at normal temperatures divides them naturally into solids, liquids, and gases (Francis and Peters 1980; Tiratsoo 1980; Meyers 1981; Tsai 1982; Ward 1984; Royal Dutch–Shell 1984; Ikoku 1984).

Coals are sedimentary rocks formed largely by lithification of peats. Extreme values of proximate analyses span a wide continuum: carbon content ranges from 15% in the poorest lignites to 98% in the best meta-anthracites; volatile matter may be totally absent or as high as 85%; moisture and ash content can each vary between 1 and 45%; sulfur may be virtually absent or surpass 7% (1.5–3% range is most common); and there is about a fourfold difference between the heat value of the worst and the best coals (8.3 vs. 36.2 MJ/kg). Energy density of poor lignites is thus greatly inferior to that of air-dry wood (about 15 MJ/kg).

Major coal categories recognized worldwide are anthracites, bituminous and subbituminous coals (hard or black coals in Europe), and lignites (Europe's brown coals). International energy production and consumption comparisons have been using for decades a common denominator of the standard coal equivalent, a fuel containing 29.3 MJ/kg (7 Mcal/kg). No run-of-the-mine bituminous coal is this good; most steam coals have 18–25 MJ/kg.

Coal conversions can produce secondary fuels in all three states. Cokes have been critical for the advancement of iron smelting. The outlook for liquid synthetic hydrocarbons, produced for the first time on a large scale by Germany during World War II, improved after 1973, but by 1990 South African SASOL plants using the classical Fischer–Tropsch process were the only major liquefaction facilities. Coal-derived gaseous fuels range from low-energy (16–19 MJ/m^3) town gas from distilling of coal in closed retorts, frequently used for lighting during the nineteenth and early twentieth century, to a high-energy (30–38 MJ/m^3) synthetic fuel made by a variety of gasification processes.

Crude oils are physically distinct fluids of organic origin, but their ultimate analysis shows remarkable consistency owing to the dominance of a few homologous hydrocarbon series. Carbon accounts for 84–87% of total weight and hydrogen for 11–14%; the small remainder is sulfur (commonly 1–2%), nitrogen, oxygen, and traces of heavy metals. The energy content of most crude oils is 42–44 MJ/kg. Shares of constituent hydrocarbons and specific gravities are the basis for their classification. Different shares of paraffins, cycloparaffins, and aromatics result in a wide range of specific densities commonly measured in °API [°API = (141.4/specific gravity) − 131.5] and pour points.

Most of the major world export streams are rather heavy oils (°API 28–33 for Saudi crudes and 30–35 for other Persian Gulf oils; only Algerian, Libyan, and Nigerian oils are light, with °API 37–44) with low pour points (mostly between −21° and −36°C) but with undesirably high sulfur concentrations (1.4–2.8 for the Persian Gulf crudes). Natural gases are largely mixtures of the three simplest alkanes, methane, ethane, and propane. Methane dominates with 73–95%; propane is just 0.1–1.3%. Butane, pentane, and sometimes a few higher homologs are also present and CO_2, H_2S, N, He, and water vapor are found in many gases. Extreme heat contents for raw natural gas are approximately 30–45 MJ/m^3, with pure CH_4 at 35.5 MJ/m^3.

Besides liquids and gases there are also nonconventional hydrocarbons ranging from very heavy crude oils, oil shales, and tar sands to frozen gas hydrates in the Arctic. Heavy crudes require special methods of recovery and shales and sands yield only small amounts (less than 10%) of oil per unit of parental rock. Their commercial exploitation thus far has been negligible. Recoverable global totals are about 6.8 Gt of oil from sands (with 75% of this in Athabaska tar sands) and 5.5 Gt from more widely distributed shales (World Energy Conference 1986).

All fossil fuels fit into a broad category of mineral resources traditionally labeled nonrenewable. Their exploitation has been seen as an inexorable march toward physical exhaustion. These are incorrect assumptions. Fuels are forming steadily but their extraction is obviously vastly surpassing the rates of replenishment, so that nonsustainable resources would be a more fitting term. More importantly, these resources should not be seen within a rigid stock-exploitation frame because their magnitudes evolve with prices, input costs, and advances in exploratory, production, distribution, and conversion techniques.

Traditional classification of resources spans the extremes of an enormous resource base, the total (unknown or only very roughly estimable) quantity of the mineral in the Earth's crust, and relatively small reserves, that tiny part of the base whose spatial distribution and recovery costs at current prices and with existing techniques are known in detail. As Tilton and Skinner (1987) argue, this may be useful but it is not enough. What happens as we move toward and past the prescribed conditions? A superior measure of resource availability is the cost of producing additional or marginal units of a resource. This dynamic approach takes into account improvements in our techniques and ability to pay the price of recovery. Exhaustion is then not a matter of actual physical depletion but rather a burden of persistent and eventually insupportable real cost increases resulting in declining availability of a resource.

This is a gradual process allowing for adjustments and counter-measures including innovative techniques, various conservation efforts, and often surprisingly sweeping resource substitutions. Consequently, there are no sudden ends, just variable and gradual shifts onto new supply planes. This understanding is critical in appraising the rise and prospects of fossil-fueled civilization. Although its energetic base is the recovery of unsustainable stocks, the practical nonrenewability of fossil fuels does not imply any fixed dates for physical exhaustion of coals or hydrocarbons. Nor does this mean an early onset of unbearably rising real costs of recovering these resources.

Reserve estimates—global and national—are now readily available and resource appraisals have at least a sufficient degree of basic consensus, but these facts are not enough to offer any reliable speculations about the future recovery of fossil fuels. Coals have been explored longer than any other fossil fuels and are generally easier to find than hydrocarbons. The first detailed global summary of their reserves prepared in 1913 for the Thirteenth International Congress of Geologists (McInnes et al. 1913) ended up with 6.402×10^{12} t of resources and 671×10^9 t of recoverable reserves. Subsequent World Power Conference summaries (in 1936, 1948, 1960, 1962, and 1968) and triennial World Energy Conference tabulations since 1974 have not really change these totals in any meaningful ways.

As errors of $\pm 25\%$ cannot be considered excessive on a global scale, the 1913 value would be bracketed by approximately 500 and 840×10^9 t. All but one of the post-1913 reserve totals fell within this range; the latest sum is 838×10^9 t (World Energy Conference 1986). Differences in reporting criteria and in accuracy of national estimates make the resource totals highly unreliable. Fettweis (1979) offers an excellent discussion of these weaknesses. Knowing the precise total would make little difference as most of the resources will always remain undisturbed. The much more meaningful category of measured, recoverable reserves (in seams exceeding 30 cm and no deeper than 1200 m) is an order of magnitude smaller but even so it is a formidable total, over 700 Gt, a mass equivalent to about 250 years of the mid-1980s annual extraction of 2.8 Gt.

Both the reserve/production (R/P) ratio in excess of two centuries and the relatively low proportion of recoverable reserves to total resources indicate a relaxed outlook for availability of coal as a major source of primary energy on the global scale. Coal's uneven spatial distribution and the environmental consequences of its conversion will continue to be much more important in determining the extent of production than will concerns about its physical availability. At least 80% of all resources and about 70% of recoverable reserves are in just three countries, the Soviet Union, the United States, and China. The Soviet Union, thanks to its huge but low-quality Siberian deposits, has nearly 50% of the estimated resources and 25% of global reserves; the United States has about 17% of resources but some 30% of reserves.

Energy density of recoverable coal reserves varies greatly with the thickness of seams and heat content of the fuel. A 1 m thick seam of a very poor lignite (such as Germany's Braunkohle with 8.5 MJ/kg and specific gravity of 1.2) will store no more than 10 GJ/m^2, the same thickness of standard coal (29.3 MJ/kg) will contain up to 40 GJ/m^2, and the best anthracite will have about 50 GJ/m^2. Seams may be single or multiple, ranging from 30 cm to over 10 m, with modal values of 0.5–2 m. Typical energy densities of large coal basins are tens of gigajoules per square meter. The Pittsburgh bituminous bed as well as the huge (nearly 100,000 km^2) North Plains deposits of lignites and subbituminous coals in the Dakotas, Montana, and Wyoming have average energy density around 50 GJ/m^2.

The best mine sites are one or even two orders of magnitude higher: Arizona's Black Mesa, about 200 GJ/m^2; Montana's Ashland, 415 GJ/m^2; Fortuna, West Germany, the largest brown coal mine in the Rheinland, 430 GJ/M^2; parts of Victoria's Latrobe Valley (brown coal seams up to 100 m), over 800 GJ/m^2; and Queensland's Blair Athol (with 30 m of bituminous coal), about 900 GJ/m^2.

Estimates of ultimately potentially recoverable oil resources are much more meaningful than the oil-in-place totals: standard recovery methods get out only about a third of the oil present in the parent rock. Since 1945 there have been many estimates of global oil-in-place and ultimate recovery totals, feeding controversies regarding the probability of extreme values (Grossling 1977; Odell and Rosing 1983; Riva 1983). A summary analysis is more revealing than case-by-case comments. White (1987) arranged 47 estimates from 29 different sources published between 1965 and 1985 in an increasing order of a probability graph (Fig. 9.1). The median estimate of 0.95–0.05 range is 334 Gm^3 (2.1×10^{12} bbl). By 1985 one-quarter of this total was already extracted and one-third of it was in proved reserves.

The *R/P* ratio for all the remaining reserves of the median estimate and for the mid-1980s annual rate of extraction (about 20×10^9 bbl) would be about 80 years. Low (0.05) and high (0.95) estimates would

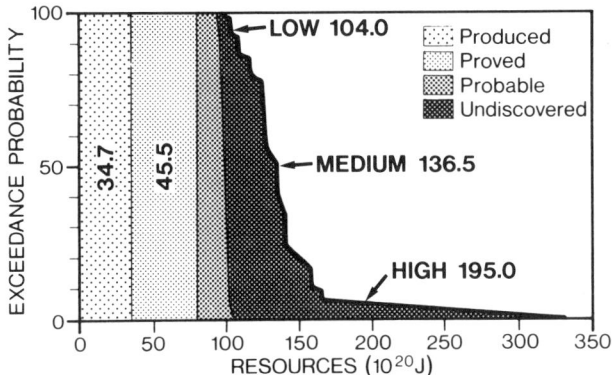

Figure 9.1 World oil resources at a glance: produced, proved, and probable resource totals as of January 1, 1985, are augmented by expert estimates of undiscovered deposits. There is almost exactly twofold difference between extreme assessments with exceedance probabilities of 0.95 and 0.05. (From White 1987.)

bracket the ratio between roughly 50 and 120 years. Unless the consensus is much too pessimistic, the oil era will have to be considerably shorter than the coal era. But its demise will certainly lie much further ahead than repeatedly forecast during the 1970s. A major downturn of the R/P ratio was reversed by a combination of reduced consumption and new discoveries and during 1987 also by some radical reevaluation of the known Middle Eastern and Venezuelan reserves.

This reassessment was largely responsible for a surprising 27% jump in the global total of recoverable oil during 1987 and for pushing the R/P ratio to 43 years, higher than it stood at any time since 1945, the year when the oil industry started to publish annual worldwide reserve surveys (Fig. 9.2). The post-1973 developments warn against any confident predictions of extraction rates but there is no doubt that the dominance of oil in the global energy supply will continue for at least two more generations and that the exit from the oil era will be gradual.

Estimates for ultimate potentially recoverable natural gas resources, presented in Fig. 9.3, show a greater spread between the extremes and a much longer R/P ratio for the remaining reserves: the median is 215 and extremes are 130 and 380 years at the mid-1980s annual rate of extraction. But both hydrocarbon resources share distributional unevenness far surpassing the patchiness of coal deposits. Although there have been some 30,000 hydrocarbon fields discovered worldwide, more than 70% of recoverable oil and gas is

Figure 9.2 Global crude oil reserve/production ratio for the years 1945–1987 calculated from annual year-end worldwide surveys in *The Oil & Gas Journal*. In spite of rapidly growing demand between the early 1950s and 1980 the ratio never dipped below 25 and during the late 1980s it reached new record highs. Clearly, the crude oil era will continue far beyond the year 2000.

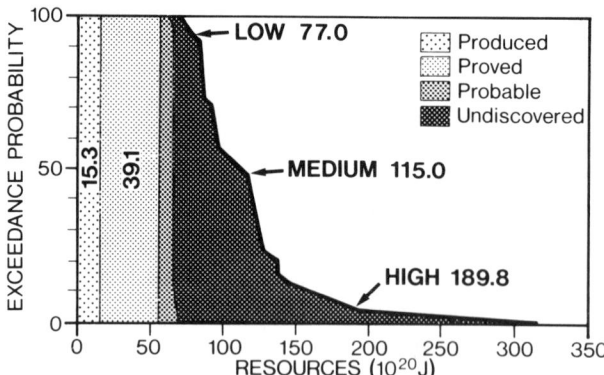

Figure 9.3 White's (1987) graphic summary of global natural gas resources using the same approach as in the crude oil assessment (Fig. 9.1). In this case the difference between extreme estimates of undiscovered resources is nearly 250%.

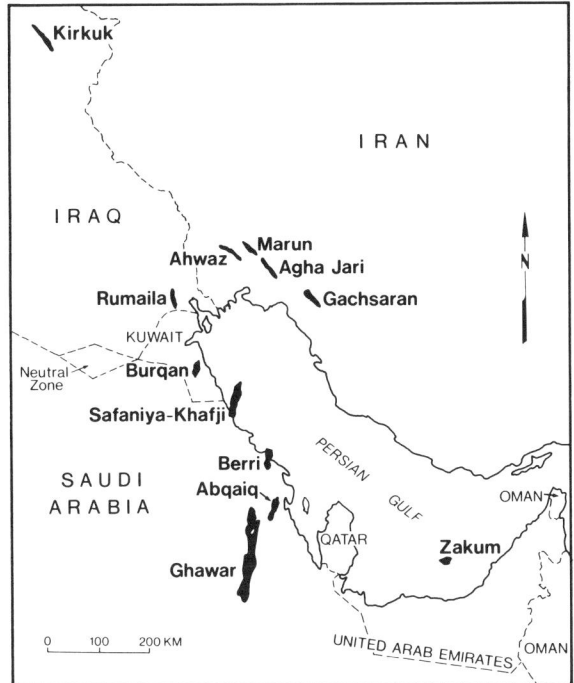

Figure 9.4 Perhaps the world's most important inequality of fossil fuel endowment: 12 out of the 15 largest oilfields concentrated in five countries around the Persian Gulf. Based on the CIA (1979) map of Persian Gulf oilfields.

in just 500 giant formations (each containing at least 80 Mm^3 of oil or 85 Gm^3 of gas or any combined energy equivalent) located overwhelmingly in five (of 260) producing basins: Persian Gulf–Zagros, West Siberia, Gulf of Mexico, Volga–Ural, and Maracaibo (Nehring 1978; Perrodon 1985). Of the world's 15 largest fields, containing more than half of all recoverable hydrocarbons, 12 are in the Persian Gulf–Zagros basin (Fig. 9.4).

Frequency of giant field discoveries has been steadily declining and the bulk of the new reserves will have to come from smaller finds, often in hostile environments and in offshore areas that may contain as much as 40% of the world's undiscovered oil resources. The Persian Gulf-Zagros basin will almost certainly remain an astounding singularity and its two supergiant fields, Saudi al-Ghawar with reserves of at least 550 EJ and Kuwaiti al-Burqan with the minimum of 470 EJ, will retain their extraordinary position as the world's largest deposits of conventional hydrocarbons. The Urengoi field in western

Siberia with some 270 EJ of proved reserves is unlikely to be surpassed as the world's largest natural gas deposit.

Prorating the medium global estimates of recoverable hydrocarbons over roughly 70×10^6 km^2 of known and prospective sedimentary basins gives a median energy storage density of just 360 MJ/m^2. Many small hydrocarbon fields store less than 1 GJ/m^2; extensive giants such as Hugoton–Panhandle in Texas or Alberta's Pembina store less than 10 GJ/m^2. The richest fields contain between several tens to many hundreds of gigajoules per square meter. Alaskan Prudhoe Bay rates about 25 GJ/m^2; Saudi al-Ghawar, the world's largest oilfield, contains close to 100 GJ/m^2; California's Ventura–Rincon, 300 GJ/m^2; Kuwaiti al-Burqan, with the world's second largest reserves, about 1000 GJ/m^2.

The Green River formation, the world's largest concentration of oil interspersed in the shales of Colorado, Utah, and Wyoming, has total energy content rivaling al-Burqan's density, but its deposits of richer shales (yielding at least 100 and as much as 400 L/t of rock) prorate to no more than 185 GJ/m^2 and reserves recoverable with established techniques to a mere 12 GJ/m^2 (Dinneen and Cook 1974). Canada's oil sands are poorer, containing a total of 25 GJ/m^2, of which 3 GJ/m^2 are recoverable by surface mining.

Assuming eventual doubling of existing extractable coal reserves, the total ultimate potentially recoverable reserves of conventional fossil fuels would be 6.75×10^{22} J, with about 60% in coals and 20% each in oils and gases. As of January 1988 the totals of proved recoverable reserves of the three principal fossil fuels added up to about 3.1×10^{22} J, with coals accounting for two-thirds of the total and oils taking just over half the remainder. Addition of oil shales and oil sands could as much as double the hydrocarbon total, but estimates of eventual recoverability of these deposits are clearly premature.

Continental division shows Asia's great supremacy (nearly 50% of all fossil fuel reserves) and Africa's surprising lack of mineral fuel endowment (less than 10% of global coal and hydrocarbon reserves). Estimates for the world's most populous nations convey the limitations arising from large populations: in spite of its enviable fossil fuel patrimony China's per capita reserves are an order of magnitude smaller than those of the United States and the Soviet Union. On the other hand, as Chinese–Japanese or Soviet–French comparisons indicate, shortages or near absense of domestic fuel resources stimulate efficient use of energy, leading to superior economic performance.

9.2 FROM EXTRACTION TO COMBUSTION

Coal burning was only a marginal source of energy in ancient or medieval societies and it had no influence on determining their technical development (Nef 1957). The first country accomplishing the shift from phytomass fuels to coal was England, and this conversion preceded the often described onset of the late eighteenth-century industrial "revolution" by some 200 years (Nef 1932; Harris 1974). Between 1540 and 1640 almost all English coalfields were opened up for exploitation; by 1650 the country's annual coal output passed 2 Mt, 3 Mt annually were extracted in the early eighteenth century, and over 10 Mt were extracted annually by its end. Coal became not just a new fuel supplanting wood, but its large-scale use required solution of many technical and organizational problems connected with its mining, transport, and combustion.

Rising coal demand led gradually to larger and deeper mines. The deepest shafts did not go below 50 m in the mid-1600s; they surpassed 100 m in the early 1700s, reaching 200 m by 1765 and 300 m by the 1830s. Extractive efforts were initially energized solely by human labor. Hewers, working with picks, wedges, and mallets, were assisted by marrows (putters) filling baskets, loading them on wooden sledges, and dragging them to pit bottom. There onsetters hung the baskets on ropes, windsmen hauled them up, and banksmen carried them to storage heaps. Boys as young as six or eight years started to do lighter tasks, but in some collieries the heaviest work fell on women or teenage girls as they carried coal to the surface by ascending a series of ladders with heavy baskets supported by bent backs and forehead straps.

Available descriptions of ascending 35 m to the surface with up to 75 kg of coal translate, when assuming body weight of 60 kg and speeds around 0.2 m/s, to exertions of close to 300 W, clearly near the bioenergetic limit, especially when considering the precarious nature of the task (a heavy back load on a steeply set ladder). This may be the most painful illustration of the inevitability of subsidizing the introduction of a new source of primary energy by liberal application of the dominant energy, in this case the exertion of female muscles.

Where human muscles were inadequate, horses became indispensable: for powering the whims or treadmills used for pumping water from deep shafts and for hoisting coal from larger pits. After 1667 horses, as well as donkeys were also used underground for hauling. Waterwheels and windmills did some of the pumping and hoisting.

Transportation of coal in heavy horse-drawn wagons was feasible only over very short distances; for more remote destinations boats or ships were indispensable, the necessity leading to extensive development of canals, whose era ended only with the expansion of railways.

Successfully produced coke was used in malt drying during the 1640s but its use in ironmaking came only half a century later (1709); it was generally adopted only by the end of the eighteenth century. Ironworks, whose output defined the mechanical civilization of nineteenth-century Europe, were thus dependent on phytomass fuel longer than other advancing industries. By the late eighteenth century coal mining was gaining importance in many European countries (most notably in Belgium, Bohemia, and parts of Germany and France) and by the early nineteenth century this extended to the United States.

At that time the British coal industry was still producing more than 80% of worldwide coal output and it continued to dominate global extraction until the late 1870s. Increasingly reliable statistics show the country's extraction share at no less than 60% by 1850, about 53% in 1870, 45% a decade later, and 36% in 1890, when it was still about 30% ahead of the rapidly rising U.S. output, which became the world's largest just at the century's turn (Fig. 9.5).

Since then the global output has nearly quadrupled with the first doubling before World War II coming overwhelmingly from the expansion of manual labor and the second one from rapid diffusion of mechanized extraction favoring economies of scale and resulting in sharp labor productivity increases. In the United Kingdom mechanical extraction accounted for 2% of the national total in 1950 but by 1975, with 95%, it was carried as far as practicable. In the United States average performance of about 10 t/worker·day is far ahead of European means (2–4 t/worker·day), while the mid-1980s Chinese mean was just below 1 t. Extraction productivities thus span a broad range, from no more than a few hundred kilograms (2–6 GJ) in primitive small rural Chinese mines to over 30 t/worker·day (about 600–800 GJ) in the largest American surface mines.

Power densities of the most productive surface extraction are typically 10–20 kW/m^2 (Germany's Fortuna has about 13 kW/m^2, Victoria's Latrobe Valley up to 28 kW/m^2); averages for large rich coal basins with open-cast mining are 2–10 kW/m^2. In underground operations the highest densities, 1–2 kW/m^2, are achieved by long-wall mining (Peng and Chiang 1984). This technique commonly recovers 90% of coal in seams suitable for its deployment. In comparison, the traditional room-and-pillar technique leaves in place at least

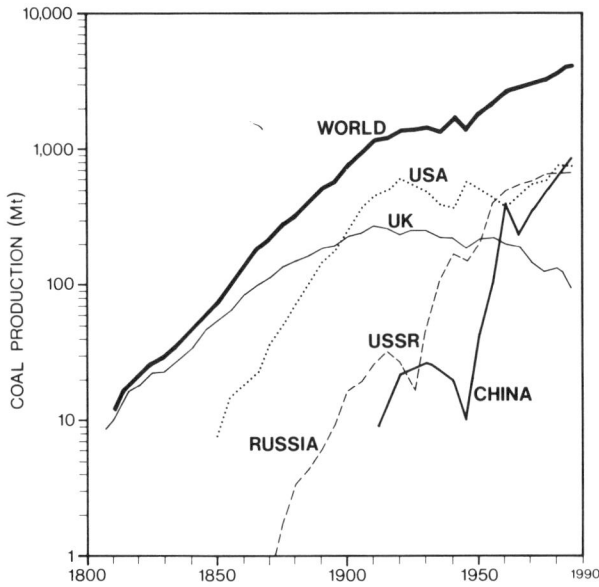

Figure 9.5 Nearly two centuries of global coal production. The United Kingdom dominated the field until the closing decades of the nineteenth century. The Soviet Union and China now rival the gross U.S. output but higher U.S. quality still translates into the largest energy equivalent. Plotted from UNO data (1956, 1976, 1980–).

50% of coal and its recovery densities in smaller pits with thin seams may be as low as 100–200 W/m^2.

Surface mining can achieve virtually total recoveries in flat seams and exponential growth of mining machinery capacities (the largest overburden shovel dippers grew from just 2 m^3 in 1900 to over 100 m^3 by the 1970s) opened the way for extraction of seams under more than 200 m of overburden in progressively larger mines. The largest surface mines in the Soviet Union, Europe, and North America have been designed to extract 15–50 Mt of coal a year, amounts equaling total annual fossil fuel production of many smaller nations.

Maximum overburden/coal ratios have grown from just 1–2 before World War II to 5–6, and ratios of up to 10 are in planning. Lower costs, greater safety, and virtual elimination of health risks associated with underground mining have been additional factors responsible for the growing adoption of surface mining; in 1990 it produced about 60% of U.S. output and 40% of Soviet output but less than 5% of Chinese extraction. The Chinese are also the only major coal-producing nation using more than 80% of its large coal output as it comes from the mines; all other major producers have invested in coal-cleaning facilities which process 60–90% of the fuel. Washing, based on the difference of specific gravities between lighter coal and heavier incombustible waste, is the standard practice and it may be preceded or followed by crushing and sorting to uniform sizes to meet the requirements of different markets.

Long-distance transportation of coal remains dominated by railways. The development of unit trains represents the best solution of an inherently difficult task of moving bulky and dirty solids. Compared to traditional railroad cars of 10–40 t and trains of up to 5000 t, a typical large unit train carries 10,000 t in 100 permanently coupled low-tare aluminum cars of 100 t pulled by three 2.2 MW diesel locomotives on runs of $10^2 - 10^3$ km. Slurry pipelines have not fulfilled their initial promise while belt conveyors and off-road trucks are restricted to shorter runs. Overseas transportation of coal in midsize bulk carriers is a modest affair compared to shipments of crude oil.

In contrast to centuries of coal use preceding the era of industrialization, utilization of hydrocarbons as fuels is of very recent origin. Notable exceptions included the burning of bitumens in Constantinople's *thermae* during the late Roman Empire and the Chinese use of natural gas in evaporation of brines in Sichuan (Forbes 1964). The latter practice, going back to at least 200 B.C., was especially remarkable as the gas came from boreholes up to 1400 m deep drilled with a simple percussion technique: heavy iron bits were suspended by bamboo cables from a derrick and were raised by men jumping rhythmically on a lever.

Two millennia later Colonel E. L. Drake used essentially the same technique at Oil Creek in Pennsylvania. His secondhand percussion drilling rig powered by a small steam engine penetrated 10 m of rock to complete a 21 m deep well on August 27, 1859, the day generally considered to mark the beginning of the oil era (Brantly 1971). Soon the exploitation of Romanian (Ploesti) and Caspian Sea (Baku) deposits started to compete with U.S. production, and the first refineries, pipelines, and tankers were creating the beginnings of a new, complex industry. Total consumption of refined oil products— mostly for household lighting and as lubricants and waxes—grew slowly during the decades of rapid coal-mining expansion before 1900.

Only the invention of internal combustion engines, improved refining methods (high-temperature, high-pressure cracking in 1913– 1915, catalytical cracking in 1937), and rising Western affluence

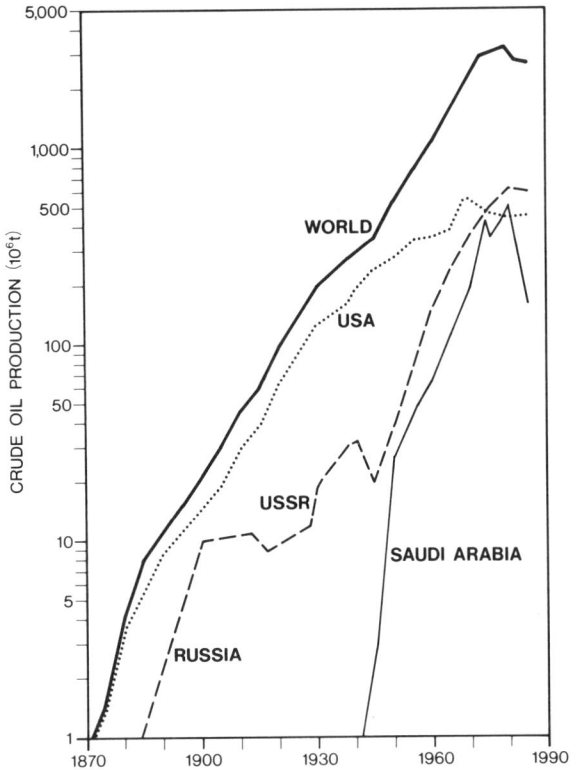

Figure 9.6 Global production of crude oil rose from 1 Mt in 1870 to nearly 3 Gt during the 1980s. Overwhelming U.S. domination ended during the 1950s with the Soviet ascent; the most notable development of the 1980s was the Saudi fall. Based on the same statistics as Fig. 9.5.

spurred oil consumption, first in the United States and later in Europe and Japan. Global production went from 200 Mt in 1930 to 1 Gt by 1960 (Fig. 9.6); the energy content of crude oil extraction surpassed that of coal at about the same time. Although the bulk of the global natural gas reserves is associated with crude oil, the early decades of the oil and gas industry saw most of the gas flared. Only after World War II, with the extension of large-diameter pipelines, expansion of petrochemical industries, and rising household demand for clean fuel, did the production of natural gas skyrocket (Fig. 9.7).

But the rapid growth of oil extraction in many remote and un-developed regions with little or no market for gas in their proximity led to further absolute increases in gas flaring. By 1975 about 50 Mm^3/y was flared, 10 times the amount wasted in 1930 and an

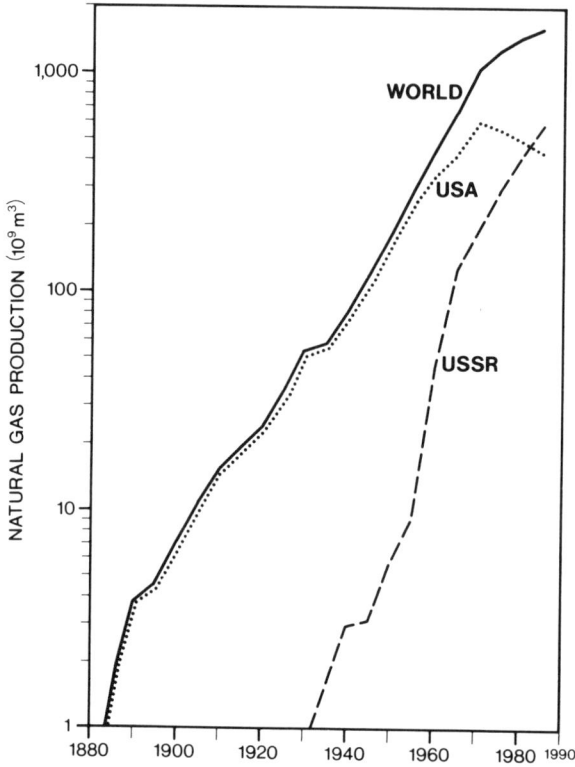

Figure 9.7 Global production of natural gas. (Huge volumes that were flared everywhere during the early years of the industry are not included.) Based on the same statistics as Fig. 9.5.

equivalent of about 3% of global natural gas production. Major sites of these flarings—the Persian Gulf, Libya, Algeria, Nigeria—were spectacularly evident on the nighttime satellite images as the brightest spots on Earth, their intensities dwarfing glares of the largest cities (Croft 1976).

Discoveries of the world's six largest supergiant oilfields between 1927 and 1958 (Kirkuk 1927, Burqan 1938, Ghawar 1948, Safaniya-Khafji 1951, Rumaila 1953, Ahwaz 1958) turned the Middle East into the most concentrated region of high-quality energy supplies. Easy worldwide availability of this inexpensive fuel, whose combustion was essential in energizing the unprecedented spell of global economic growth between 1945 and 1973, was made possible by rapid growth of oil tankers (Ratcliffe 1985). After going up from just over 2000 to over 20,000 dwt between 1884 and 1921, capacities of the largest tankers stagnated for over a generation, took off again after 1945, and reached a plateau in excess of 500,000 dwt by the early 1980s.

Physical limits encountered by such huge ships (depths of ports and channels, distances to stop), safety of navigation, risks of large accidental oil spills and consequent high costs of insurance, and a weakened oil market have been the major factors limiting their further growth. Shipments of natural gas in liquefied natural gas (LNG) tankers are much more expensive and hence much less important. With the exception of the United States, where long-distance pipelines have been in place since the 1870s, continental transportation of hydrocarbons also expanded dramatically only after World War II.

Pipelines are superior to any other form of land transportation: their compactness (a line 1 m in diameter can carry 50 Mt of oil a year), cleanliness, reliability, and safety (hence excellent environmental acceptability) translate into unmatched economies. Large American lines from the Gulf to the east coast have been eclipsed by the Soviet pipelines carrying Tyumen gas from western Siberia to western Europe, a distance of nearly 6500 km bridged by pipes 2.4 and 1.4 m in diameter.

Proliferation of exploratory and production drilling since 1945 has been helped by advances in geophysical prospecting and improvements in rotary drilling, which now accounts for more than 95% of completed wells. This technique, whose commercial success was first proved with drilling the famous Spindletop gusher (Beaumont, Texas) in 1901 and whose performance was greatly improved by the introduction of the rolling cutter rock bit by Howard Hughes in 1909, employs a rotating drill bit attached to the end of a threaded sectional pipe through which the pumps also force a drilling fluid, removing the cuttings and cooling the bit.

Steam-driven cable-tool rigs could drill no deeper than 2500 m by the 1920s but the deepest productive wells drilled by rotaries surpassed 3000 m during the following decade, 5000 m during the 1950s, and 9000 m by 1980. But most oil and gas wells completed worldwide are still less than 2 km deep. Notable innovations of the 1980s have included the diffusion of downhole drilling motors and expansion of exploratory directional drilling at angles up to 70–80° from vertical and horizontal drilling of production wells, which may not only increase production rate but also improve the overall recovery (Petzet 1988).

More than 20% of recoverable oil reserves is offshore and bold advances followed the completion of the first well drilled out of land sight in 1947 off Louisiana (first offshore drilling from piers started in California in 1897). New exploratory rigs—initially mostly jack-ups, later various semisubmersibles—extended the search to waters up to 2000 m deep.

While reserve densities of commercially exploited coal and hydrocarbon deposits overlap, power densities of oil and gas extraction are necessarily much lower when prorated over the total reservoir area. Coal mining will remove at least half and often all of the fuel in accessible seams while oil and gas production exhausts the reservoir gradually, usually over a period of many decades, and the best unaided recovery rates do not surpass 40% of the original oil content. Even the richest fields do not have extraction densities surpassing 200 W/m^2 of their total area: al-Burqan's production since 1938 prorates to about 180 W/m^2, Ventura–Rincon has been producing some 130 W/m^2 since 1916 and al-Ghawar's cumulative output since 1948 has been around 10 W/m^2, all values much below the typical coal extraction densities of 1–20 kW/m^2.

But these power density comparisons seem quite different when looking at land actually claimed by the extraction. In the world's richest oilfields individual wells will produce at least 1 PJ of crude annually; the Middle Eastern mean during the time of peak outputs was nearly 12 PJ per well and al-Ghawar can flow at up to 40 PJ per well. Usually less than 1% of these oilfields is actually taken up by surface structures or reserved for the right-of-ways of gathering pipelines. Actual extraction power densities of these fields are thus at least between 10 and 20 kW/m^2.

In contrast, annual production average for about 650,000 U.S. wells operating in the mid-1980s was just 34 TJ per well. Modal value for the dominant stripper wells was below 25 TJ per well and the best estimates available for U.S. oilfields fall within 1–3 kW/m^2. Typical power densities of natural gas extraction are as high as 10–15 kW/m^2, but this rate is reduced by up to an order of magnitude once the rights-of-way for gathering pipelines and field facilities for gas processing are added to well enclosures.

Potential for additional crude oil recovery remains large. Even for the lightest oils primary recoveries are rarely more than 40% of the fuel in the parental rock. For heavy oils the share is below 10%, with usual means in most oilfields between 25 and 35%. Standard secondary recovery pressurizing reservoirs with water or natural gas lifts 5–10% more. Several techniques of enhanced oil recovery bring still better results. Steam drive can recover up to 30–60% of the remaining oil-in-place, carbon dioxide flooding 20–30%, surfactant flooding 15–

40%, and in situ combustion 15–25%. All of these techniques are expensive and their eventual diffusion and degree of performance will be determined by the cost of alternatives available to replace liquid hydrocarbons. But even a uniform 30% recovery increase would mean that the global mean of crude oil extraction would still fall short of 50%.

Surface coal mines will claim more land than just the area over-lying the worked seam for temporary displacement of overburden before eventual reclamation, and also for on-site transportation and mine buildings. Extraction power densities may be as low as a few hundred watts or as high as several kilowatts per square meter. Published estimates for various U.S. operations imply a range from about 300 W to 2.2 kW/m². An acceptable generalization would be to conclude that fossil fuels are produced mostly with power densities ranging from several hundred to several thousand watts, with 400–4000 W/m² being a good conservative range including the majority of coal mining and oil and gas extraction rates.

Removal of hydrogen sulfide, moisture, and other undesirable ingredients from natural gas can be accomplished with minimal land requirements; throughput power densities may be as high as 70 kW/m² and rarely are below 50 kW/m². Coal preparation and cleaning are more space-intensive, with throughput densities typically between 8 and 10 kW/m². Crude oil refining, producing a wide variety of highly flammable gases and liquids, requires safety precautions for siting or processing and storage facilities, prevention of spills, and fire fighting. Minimum spacing of 60–75 m is mandatory to separate many parts within a refinery.

Typical operational space requirements of modern refineries trans-late to throughput power densities of 3–4 kW/m²; extreme power densities of American operations range from 3.3 to 23.5 kW/m². Inclusion of the land for safety buffer zones (and also for future expansion) reduces these rates to 0.6–2.4 kW/m² (Gary and Hand-werk 1984). An average refinery processes daily about 100,00 bbl of crude oil (7.5 GW), and the largest facilities (Pernis, Netherlands; Ras Tanura, Saudi Arabia; Falcon, Venezuela) have capacities of 25–45 GW.

Typical corridors claimed by railways are 20–30 m wide; where cuts and fills are needed the right-of-ways may extend to more than 100 m. Pipelines usually need 25–30 m for construction; afterwards only access strips of up to 10 m may be necessary, Compressor stations take up to 20,000 m² at 80–120 km intervals; pumping stations take up to 10 times as much every 130–160 km. Tanker terminals and LNG facilities need relatively small areas but require extensive security zones. Exclusive unit train railroads built to move coal (500–1500 km) to large power plants have annual throughput densities of 100–400 W/m^2. U.S. data, and assumption of 7–10 m rights-of-way for operating lines, prorate to average throughputs of 200–300 W/m^2 for natural gas and 350–480 W/m^2 for crude oil. Crude oil tanker and LNG terminals have throughput densities of 30–60 kW/m^2 and underground oil storages can accomodate up to 5 kW/m^2.

Chemical energy in coals and hydrocarbons is converted to heat (and also light) by combustion. This process of rapid oxidation of carbon and hydrogen (reaction times are between 0.1 ms for gases and 1 second for pulverized coals) is the single most important anthropogenic energy conversion of industrial civilization. The range of combustion temperatures is limited at the lower end by the lowest temperature supporting a stable flame (just short of 1000°C) and at the upper end by practical difficulties of containing the hot flame within solid walls; the hottest firebox flame in large boilers is about 1600°C.

Complete combustion of 1 g of carbon requires 2.66 g of O_2 (11.53 g of air), producing 3.66 g of CO_2 and releasing 33 kJ/g; corresponding figures for complete combustion of hydrogen are 7.94 g of O_2 (34.34 g of air), 8.94 g of water, and 121 kJ/g. Oxidation of sulfur or hydrogen sulfide represents a negligible contribution. Combustion devices have been designed in enormous variety to use the released thermal energy directly in space heating and in numerous industrial (and some agricultural) processes or for conversion to kinetic energy in mechanical prime movers.

9.3 MECHANICAL PRIME MOVERS

Evolution of the first practical inanimate prime mover started with Papin's 1690 experiments with a small (7.5 cm in diameter) atmospheric engine. Papin's tiny device was followed in 1698 by Savery's only partially successful steam-operated pump (maximum 750 W) and in 1712 by Newcomen's engine working at atmospheric pressure with the highest effective output of 3.75 kW. Condensing steam on the underside of the piston made Newcomen's engine very inefficient (0.5–0.7%).

James Watt's revolutionary contribution in his appropriately titled 1769 patent, *A new Method of Lessening the Consumption of Steam*

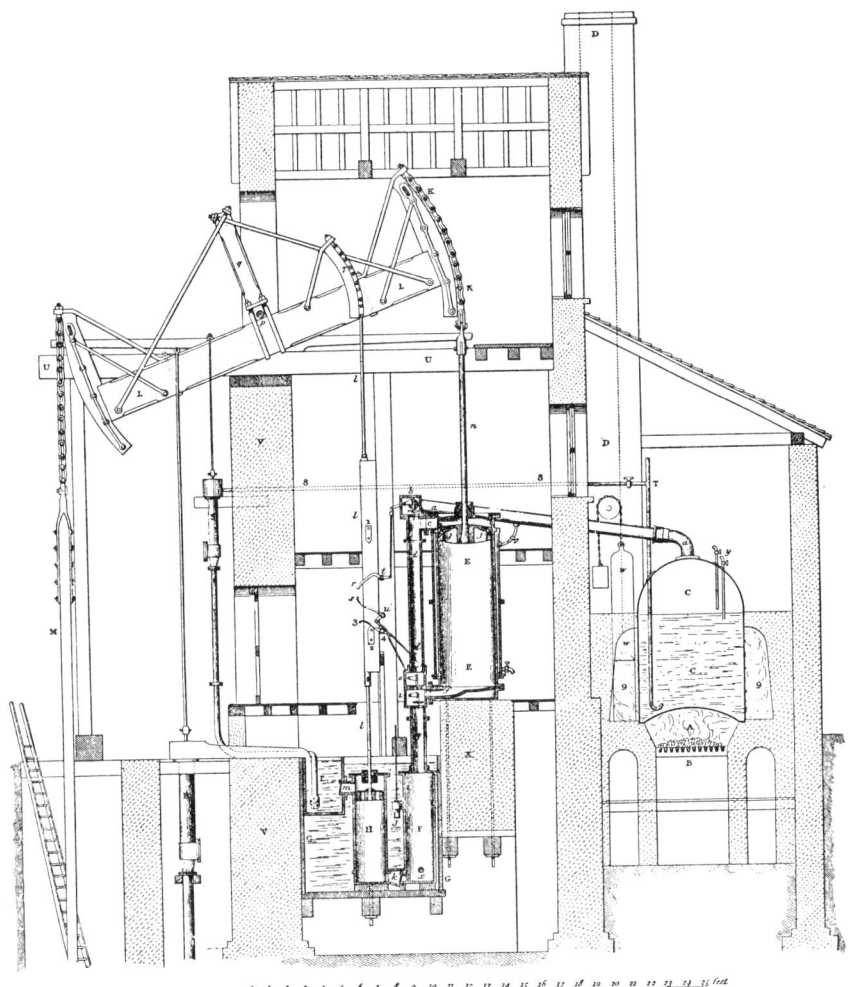

Figure 9.8 Watt's single-acting steam engine built in 1788 for mine drainage. Boiler (C) is located in an outhouse, cylinder (E) is kept in an insulated steam jacket, and there is a separate condenser (F) connected to an air pump (H) to maintain vacuum. (From Farey 1827.)

and Fuel in Fire Engines, was the introduction of a separate con- denser. An insulated steam jacket around the cylinder and stuffing box, an air pump to maintain vacuum in the condenser, sun-and- planet gearing, and, later, a double-acting engine (steam moving the piston also on the downstroke) and centrifugal governor to maintain constant speeds with varying loads were other notable innovations making Watt's engine a rapid commercial success (Fig. 9.8). By

1800, when the 25 year extension of the original patent expired, Watt and Matthew Boulton, his financing partner, built about 500 engines (60% rotative) rated mostly at 8–16 kW; the largest were just over 100 kW.

The average rating of 20 kW was much higher than the means for eighteenth-century watermills (3.7 kW) or windmills (7.5 kW) but lower than the power of many large waterwheels built to serve the expanding manufactures. But the engine allowed an unprecedented freedom of industrial location. An intense period of innovation during the first half of the nineteenth century—between Trevithick's and Evan's introduction of the high-pressure boiler in 1802 and Corliss's invention of the valve mechanism in 1849—made the steam engine more efficient and much more versatile. Its portability, adaptability, dependability, and durability ensured its dominance as *the* prime mover of nineteenth-century industrialization (Dickinson 1939; von Tunzelmann 1978).

Its stationary uses ranged from sawing wood and stones to powering belt drives in factories, from compressing air to driving the first electricity generators; its kinetic uses revolutionized both land and waterborne transport with railways and steamships. Large-scale manufacture of steam engines led to improved machining and integrated design practices that served as foundations of twentieth-century engineering advances. Availability of such concentrated power brought changes in organization of industrial production and in transportation. Both of these transformations were soon reflected in extensive urbanization, migration, growth of trade, and shifts in international relations.

In a way, the steam engine became a victim of its own success: as its performance grew, more was demanded of it than it could deliver. Its improvements and adaptations during the nineteenth century were admirable; maximum ratings went up from about 100 kW to 3 MW (Fig. 9.9), largely owing to a hundredfold increase of operating pressures (from 14 kPa to 1.4 MPa) and resulting in best efficiencies climbing from just 2.5 to 25%. But the engines had their inherent weaknesses: they were massive and hence impractical both for very large stationary applications and light mobile use, and they were relatively inefficient. They could not fill two important emerging needs: to power generation of electricity, a technique that demanded unprecedented prime mover capacities; and to provide a convenient energizer for road transport, which required lightweight, compact power plants.

Steam turbines and internal combustion engines filled these needs

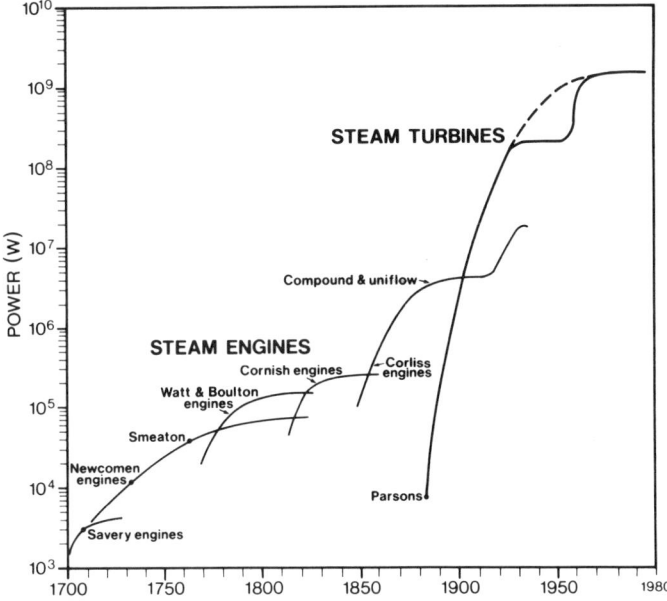

Figure 9.9 Nearly three centuries of power ratings of steam-driven prime movers: first the slow rise of steam engine, then the rapid ascent of steam turbine. Data from Dickinson (1939), von Tunzelmann (1978), Haywood (1980), and Bannister and Silvestri (1989).

in superior ways. Only in railway transport did steam engines retain their global indispensability until after World War II. The superiority of steam turbines for delivery of rotational power is obvious: steam engine rotations rarely surpassed 100 rpm, whereas modern turbines have up to 3600 rpm and are working under pressures of 14–34 MPa and temperatures up to 600°C. Turbines can be built in capacities from a few kilowatts to over 1 GW using only a fraction of the materials needed for steam engines (1–3 g/W for the largest sizes, compared to 250–500 g/W for large steam engines) and achieving efficiencies up to 40–43%. This makes them an excellent source of power for electric generators, compressors, centrifugal pumps, and ship propellers.

There was only a brief period between the first successful experimental design and major commercial applications of steam turbines. Charles Parsons followed his first patented 1884 reaction turbine design by a 75 kW public power station in Newcastle in 1888, the first condensing turbine of 100 kW in 1891, and the first 1 MW unit for the Elberfeld station in 1900. The exponential rise of turbine ratings was interrupted only during the late 1920s, but it resumed in the mid-1950s and reached a plateau in the early 1970s (Fig. 9.9).

The first working prototype of an internal combustion engine, running on an explosive mixture of gas and air, was built in 1859 by Etienne Lenoir. Shortly afterward Beau de Rochas outlined the working of a four-stroke cycle, a precept turned into commercial reality by Otto's 1878 horizontal engine running on coal gas. Decisive commercial breakthroughs came only with introduction of gasoline and diesel engines. Gottlieb Daimler's 1885 patent for a small, light, high-speed, gasoline-powered, single-cylinder vertical engine came in the same year, when Karl Benz was building the world's first car powered by a much slower horizontal gasoline engine. The combination of Daimler's engine, Benz's electrical ignition, and Maybach's carburetor (patented in 1893) provided the key ingredients of modern car engine.

Subsequent development of Otto cycle engines has been surprisingly conservative. Major historical shifts included a steady rise of compression ratios, increase of average power, and decline of weight/power ratio. In the United States typical compression ratios rose from about 4 in the early 1920s to about 10 by the 1960s, then declined slightly, averaging 8–9 by the mid-1980s. The first mass-produced car, Ransom Olds's Curved Dash, had a single-cylinder 5.2 kW engine and Ford's Model T (16 million produced between 1908 and 1927) had a four-cylinder 15 kW engine; most of the passenger cars built during the 1980s are powered by engines rated at 50–120 kW; for example, the Honda Civic GL has 63 kW, the Lincoln Town Car 112 kW (Adler et al. 1986).

Otto's engine needed nearly 270 g/W in 1880, Daimler and Maybach's radical redesign brought the weight down to 40 g/W by 1890, before the end of the century the best ratios approached 5 g/W, and since then it has gradually declined to as low as 3 g/W for trucks and to just around 1 g/W for passenger cars. Decline of weight/power ratios was much faster for aircraft engines (Gunston 1986). A 4-cylinder engine powering the Wright brothers' first flight on December 17, 1903, needed 9.1 g/W, the Wright Cyclone R 3350 28-cylinder engine powering B 29 bombers during World War II weighed 0.67 g/W, the ratio only slightly surpassed in the 1960s by another Wright (R-1820-82A) with 0.59 g/W (Fig. 9.10).

Even these excellent achievements could do little to boost the relatively low efficiency of gasoline engines. Their actual peak performances are no higher than about 25%, inferior to diesel engines,

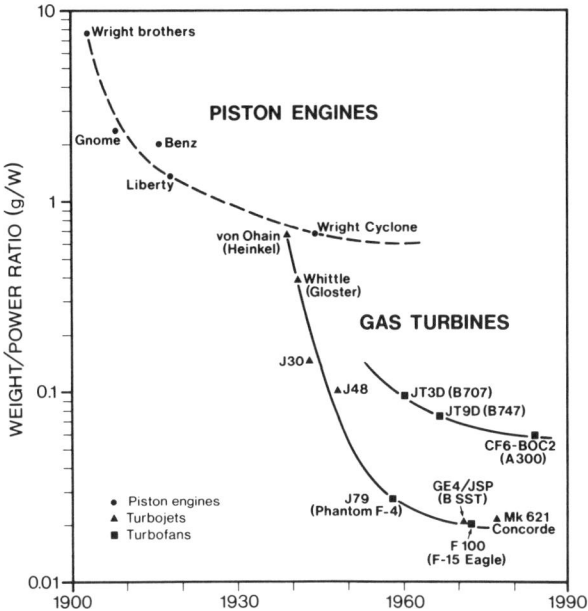

Figure 9.10 A light yet powerful prime mover was imperative for the success of heavier-than-air flying machines. A plot of the weight/power (in g/W) ratio shows the success in producing first lighter reciprocating motors and then better jet engines: the F-15 engine weighs less than 1/400 per watt than Wright's home-made motor. Data from Constant (1981), Gunston (1986) and Taylor (1986).

whose operation can come very close to theoretical maxima of that distinct power cycle. The high compression ratio (15–24) of the air charge to spontaneously ignite the fuel injected into the cylinder distinguishes Rudolf Diesel's internal combustion engine, introduced in 1892. The disadvantage of a heavier (initial weights commonly 40–60 g/W, maxima up to 120 g/W), low-speed (about 300 rpm) engine was more than compensated for by superior thermal efficiency (at least 20% higher than that for Otto engines) and by the possibility of using a variety of cheaper, yet more energy-dense, liquid fuels.

The first niche conquered by the engine was in marine propulsion where its weight was of little consequence. Submarine engines (up to 400 kW by the beginning of World War I) were followed by increasingly larger ship plants, eventually reaching up to 30–36 MW (10–50 g/W), and by the 1950s diesel engines dominated all forms of marine transportation. On land the replacement of steam and gasoline engines by diesels came first during the 1920s among heavy, self-propelled earth-moving and farming machines and switching-yard locomotives, followed by the conquest of the truck and bus market (where the engines of up to 350 kW, weighing 3–9 g/W, can log up to 600,000 km without overhaul) and domination of railway traction (engines up to 3.5 MW weighing 5–10 g/W).

Modifications for higher engine speeds (up to 2000 rpm) and lowering of engine weights (to 2–5 g/W of shaft power) opened the way for automobile applications during the 1950s. Very large stationary diesels (up to around 30 MW) have been used for electricity generation in remote locations as well as for standby capacities. The best operational efficiencies have surpassed 42% and performance above 30% should be attainable with most well-maintained engines. It is only in flight that diesels could not succeed.

The third type of revolutionary prime mover did not succeed until the late 1930s when Whittle in England and von Ohain in Germany designed the first acceptable gas turbines to power the new military jet aircraft (Constant 1981). Post–World War II perfection of this powerful prime mover first displaced the reciprocating engines in fighter-bombers and, starting in 1958, also on transoceanic passenger flights. This transformation reached a temporary plateau with the introduction of hundreds of "jumbo jets" during the 1970s (the first flight of a Boeing 747 was in 1969): all of them are powered by large turbofan engines capable of more than 200 kN of thrust and delivering up to 60 MW. Low weight/power ratio and high ratio of thrust per frontal area have characterized the evolution of these increasingly powerful aircraft gas turbines. As with piston engines, the early rapid gains were followed by more modest improvements (Fig. 9.10).

Other notable niches conquered by gas turbines are in driving centrifugal compressors along natural gas pipelines (starting in the late 1940s, in sizes of 1.5–15 MW), in oil fields and oil refineries, in chemical syntheses (most notably in ammonia production since the early 1960s), and in steel mills; powering fast trains, hydrofoils, and military and cargo ships with engines of up to 15 MW; and driving electric generators used for emergencies, peaking service as well as base load (units mostly 15–80 and up to 150 MW). The best efficiencies in these applications range from 25 to 35% and the weight/power ratios of the largest stationary gas turbines are around 2 g/W.

The only prime movers highly outperforming gas turbines in terms of the ratio of power delivered per unit of weight are liquid- or solid-propellant rocket engines powering missiles and space vehicles. These large-thrust jet propulsion engines are used to accelerate loads to high velocities, often in stages, in short periods of time. Founders of modern rocket science—be it Konstantin Tsiolkovsky or Hermann Oberth—correctly envisaged rapid advances (von Braun and Ordway 1975). In 1942 the 13.8 m long, ethanol-powered German V-2 missile had a maximum range of 340 km and a destructive payload of 1 t; its 931 kg engine had a sea-level thrust of 249 kN, imparting the maximum speed of 1.7 km/s during its 68 second burn. This translates to a maximum power rating of about 6.2 MW and an engine weight/power ratio of 0.15 g/W.

In contrast, during their 150 second firing the liquid fuel (kerosene and hydrogen) engines of the 109 m tall Saturn C 5 rocket, which sent *Apollo 11* on its journey to the Moon on July 16, 1969, had to impart the escape velocity (11.1 km/s) to the mass of 43 t by providing a combined thrust of nearly 36 MN, or an equivalent of about 2.6 GW. Even when including all the fuel in the weight of the 3 booster rockets with 11 engines, their weight/power ratio would be just 0.001 g/W. The thrust/weight ratio of individual rocket engines is as high as 150, an order of magnitude above the best military jets.

Rapid transformation of chemical energy in fuels to kinetic energy of rising rockets has been perhaps the most dramatic demonstration of powerful conversions mastered since the end of World War II—but the most important, although certainly much less spectacular, energy transformation that has matured during the twentieth century has been the large-scale generation of electricity.

9.4 ELECTRICITY

Rise of electricity represents a technical revolution unlike any other in industrial energetics. Previous inventions filled specific needs and could be readily inserted into existing systems. Newcomen's engine displaced horses, Watt's invention severed the dependence on water-wheels, and Fourneyron's turbine prevented the demise of water power. Only subsequent diffusions of these innovations, their greater reliability, lower cost, and higher performance, led to a gradual overall transformation of systems into which they were introduced. Not so with electricity. Here the whole system had to be put in place before the idea became viable: the current had to be generated on a scale sufficiently large to allow for distribution to numerous, more or less distant, consumers who had to be provided with electricity-powered converters.

One man's vision was central to this endeavor:

> Edison was a holistic conceptualizer and determined solver of the problems associated with the growth of systems.... Edison's genius lay in his ability to direct a process involving problem identification, solution as idea, research and development, and introduction into use.... Edison is most widely known for his invention of the incandescent lamp, but it was only one component in his electric lighting system and was no more critical to its effective functioning than the Edison Jumbo generator, the Edison main feeder, or the parallel-distribution systems. (Hughes 1983)

The rapidity with which Edison and his many co-workers put the system in place is astonishing. On October 21, 1879, after months of repeated failures, a cotton sewing thread was carefully carbonized, inserted into a glass globe evacuated by suction pump, and connected to an electricity supply from a dynamo. In Edison's words, the light bulb "burned like an evening star." Less than three years later, on January 12, 1882, the first power plant built by Edison Electric Light Company of London at Holborn Viaduct started to generate electricity. The first American station followed in September 4, 1882. Located in New York's financial district at 257 Pearl Street; it lit 400 lamps installed in restaurants, shops, and offices within an area covering about 0.8 km^2.

Lighting improvements brought by incandescent bulbs were truly revolutionary. Whereas candles delivered 0.1 lm/W and gas mantles could not do better than 1–2 lm/W, modern 60 W successors of Edisonian bulbs emit about 20 lm/W, giving the irradiance of 1200 cd/cm^2, three orders of magnitude greater than a candle! The great lighting revolution of the late nineteenth century was certainly as important an ingredient of industrial modernization as the rise of steam and internal combustion.

Electrification was accelerated by the victory of alternating current in the "battle of the systems." In this dispute Edison stood on the losing side: George Westinghouse and Sebastian Ferranti were the winners. In North America the conflict was basically over in 1893 with the Niagara Falls project designed to deliver high-voltage alternating current. Edison's Pearl Street station used noncondensing steam engines with a capacity of 93.2 kW and a weight/power ratio of about 32 g/W. Parsons's reaction turbine soon made such heavy installations obsolete. Soon afterward power plants called for unit capacities an order of magnitude higher, too large to be served by steam engines.

Transmission of high-voltage alternating current and its distribution to customers at low voltages was made possible by the introduction of the first practical transformers by William Stanley in 1885–1886. Without these devices, which convert high current and low voltage into low current and high voltage and vice versa, electricity distribution distances would have to be minimized and the inevitable decentralization of power generation would have precluded taking the advantage of enormous economies of scale associated with larger sizes of turbines and plants.

The largest demand for electricity lay in the conversion of clumsy, noisy, and inefficient steam-powered shaft-and-belt drives used to run machines of industrial enterprises. This huge market was unlocked by Nikola Tesla's design of a practical alternating-current induction motor in 1888. The accomplishments of Edison, Westinghouse, Parsons, Stanley, and Tesla during the 1880s are still indispensable a century later. Principles remain but particulars have changed.

Between 1885 and 1985 the largest turbogenerator sizes increased by five orders of magnitude, reaching the range of 1.2–1.5 GW (Fig. 9.9). Working steam pressure rose from just around 1 MN/m^2 for the first commercial units to as high as 35 MN/m^2 for supercritical turbines introduced in the 1960s; however, 31 MN/m^2 appears to be the optimum for greatest efficiency gains. Steam temperatures rose from 180°C for the first units up to 650°C by 1960, with typical optima at 560–600°C. Cross-compound units, with steam going sequentially through two turbines, account for most of the largest sizes. Mechanical stokers feeding small lumps of coal were dominant until the 1920s, when they were rapidly replaced by pulverized coal firing (particle sizes below 75 μm) in large multistory units now standard in all large stations.

But this efficient mode of combustion is inevitably a concentrated source of numerous air pollutants. Pulverized coal burning proceeds with excess air at flame temperatures of 1600–1800°C, ideal conditions for producing large volumes of nitrogen oxides and for oxidizing virtually all of the present sulfur to sulfur dioxide; incombustibles generate hard, slagging fly ash and heavy metal emissions. Fly ash emissions have been gradually eliminated by installation of electrostatic precipitators (patented in 1909, universally introduced only after 1950), and various flue gas desulfurization (FGD) techniques, commercialized only during the 1970s, can remove most sulfur dioxide.

Handling difficulties with fly ash, disposal problems with both the abrasive solids and wet sulfate sludge from FGD, and maintenance requirements for desulfurization systems have considerably lowered

coal's attractiveness as a power plant fuel. The most promising technique to overcome these problems and to make coal a more attractive
power plant fuel is fluidized bed combustion (Howard 1983). First
patented for gas generators by Fritz Winkler in 1921, the technique
entered the stage of commercial power plant applications only during
the early 1980s with installation of units ranging from 20 to 160 MW.

A fluidized bed is a layer of small noncombustible particles (usually limestone) kept aloft in riotous motion by air forced through
perforations in a base plate; coal (or other pulverized fuel) is introduced into the bed in quantities smaller than 5% of total load and
superior mixing rates make it possible to burn it at just 700–900°C.
These low combustion temperatures preclude fixation of atmospheric
nitrogen and hence a voluminous generation of nitrogen oxides,
virtually all sulfur in the fuel is removed by reacting with limestone,
releases of heavy metals are minimized, and dry, powdery ash is
nonslagging and easier to handle.

Thermal efficiency gains resulting from higher operating pressures
and temperatures brought the best power plant performance from
5% for the first stations built around 1900 to over 40% by the 1970s;
improvements in national averages paralleled this trend with a lag of
7–10%. Between 1900 and the late 1980s the power of largest transformers increased 500 times and their voltage increased about 15
times while their weight/power ratio declined by an order of magnitude and their efficiency reached practical limits at over 99%
(Coltman 1988). Distribution of electricity kept pace with generating
advances. In the United States the transmission started with 4.6 kV
and it has been carried through a series of leaps (6.9, 23, 69, 115, 230,
345, 500 kV) to 765 kV.

Direct-current transmission made a useful return in high-voltage
links by undersea cables and in long-distance connections between
hydrogenerating stations and major load centers. In 1954 the
pioneering Sweden–Gotland cable carried 20 MW at 100 kV over
96 km, the first English Channel crossing in 1961 had a capacity of
160 MW at ±100 kV, and New Zealand's two islands were connected
in 1965 by a ±250 kV tie carrying 600 MW. On land the West Coast
Pacific Intertie (1440 MW at ±400 kV over 1330 km) in 1970 was
followed in 1972–1977 by the Nelson River–Winnipeg link (1620 MW
at ±450 kV over 890 km) and in 1976 by Zaire's Inga–Shaba line
(560 MW at ±500 kV over 1700 km). These ties were surpassed in the
late 1980s by an 800 km ±600 kV link between Itaipu, the world's
largest hydrostation, and São Paulo.

Combustion of fossil fuels in large boilers proceeds at very high power densities (mostly 2–13 MW/m^2) and modern turbogenerators are also quite compact (Parsons's first 100 kW turbine produced about 26 kW/m^2 while the turbogenerators over 1 GW put out up to 10 MW/m^2). The heart of any power plant—boilers and machine halls—is thus only a small part of the site. Switchyards take up much space at every site (typically 50–75 kW/m^2) but overall densities are determined above all by the plant's size, its fuel storage, air pollution controls, and water-cooling facilities.

Whereas on-site hydrocarbon storages of plants supplied by gas pipelines or fuel oil barges may be quite small, coal piles required for 60–90 days of operation are necessarily extensive. With heights of 5–12 m these piles have storage densities of 25–100 kW/m^2, similar to those of switchyards. Electrostatic precipitators take up about as much space as boilers (owing to large volumes of hot flue gas) and desulfurization units can be even more compact (up to 400 kW/m^2), but the disposal of captured fly ash and sulfate sludge requires much land.

Combustion of typical steam coals with about 22 MJ/kg and 10% ash will generate annually about 250 g fly and bottom ash per installed watt. After dry transport or hydraulic sluicing the ash is deposited in fills or ponds to a depth of 5–10 m requiring 20–40 m^2/MW annually (assuming specific density of 1.3 g/cm^3). During its lifetime of 35–50 years a coal-fired power plant will need anywhere between 700 and 2000 m^2 of ash disposal space per installed megawatt. Higher ash contents can raise these rates by up to 50%. Flue gas desulfurization facilities have been taking up 400–600 m^2/MW and the lifetime disposal of sludge requires mostly 200–600 m^2 per installed megawatt.

Once-through cooling drawing water from streams or ocean has minimal spatial claims but spray ponds need about 400 m^2/MW and ordinary cooling ponds between 4500 and 5000 m^2/MW (their area is independent of the pond's depth). Cooling towers cut this huge demand while allowing for a greater density of water's reuse on major streams. Counterflow wet cooling towers with natural draft have been the most frequent choice, but more efficient mechanical draft units as well as dry cooling towers have been diffusing steadily. Both wet and dry natural draft towers require 30–60 m^2/MW; mechanical draft units need as little as 10 m^2/MW.

These and numerous other environmental effects of fossil fuel conversion have gradually become as prominent in considering the future expansion of modern energy systems as is the capital cost of new facilities. Chapter 12 is devoted to these environmental implications and complications, but this must follow the discussion of major trends in production and conversion of fossil fuels, contributions of renewable sources to global energy supply, and energy costs of energy and principal manufactured products, as well as the energy cost of modern farming with its critical dependence on fossil fuels and electricity.

10

FOSSIL-FUELED CIVILIZATION: PATTERNS, TRENDS, COSTS

Wind power, water power, and wood fuel are parts of the year-to-year revenue of sunshine ... But when coal became king, the sunlight of a hundred million years added itself to that of today and by it was built a civilization such as the world had never seen.

—Frederick Soddy
Cartesian Economics

Principal attributes distinguishing mature fossil-fueled civilization from its predecessors are easy to list. Most notably, there is a secure, abundant, and varied supply of food far in excess of any imaginable needs (persistence of significant intake disparities and even of hunger within rich societies is a matter of distributional, that is, social, inequalities). This reflects the high output density of modern agriculture, which is subsidized by numerous direct and indirect inputs of energy. Most conspicuously, there is a high level of material affluence reflecting large amounts of energy invested in production of commodities as well as in provision of energy-intensive services. Mechanization of agricultural and industrial labor has changed the human role in productive tasks primarily to that of a controller and manager of fossil fuel flows and allowed more time for leisure than is needed for employment.

A particularly notable component of this affluence is the high mobility of products and people. High mobility was made possible by the introduction of progressively more powerful, lighter, and more efficient prime movers and it became a mass reality with increasing affluence accompanied by greater amounts of leisure time. Systems producing, converting, distributing, and consuming energies maintaining this affluent civilization are enormously complex. These complexities cannot be appreciated without looking at essential attributes of fossil energy industries, at the contribution of nonfossil conversions integrated in the predominantly fossil-fueled systems, at grand patterns of energy utilization, at energy costs of energy, materials, products, and services revealed by detailed process analyses, and at considerable energy conservation opportunities

10.1 ENERGY PRODUCTION AND DISTRIBUTION

Long-term exponential growth of fuel extraction and electricity generation can be traced since the onset of the fossil-fueled era on levels ranging from global to local, its smooth progression interrupted only by wars or severe economic setbacks. This aggregate growth has been driven by increased unit sizes, be it for individual techniques or complete production systems, a trend resulting in gradual concentration of outputs accompanied by emergence of extensive transportation and distribution systems creating strong intranational and international dependencies and, in the case of easily transportable crude oils, evolving into a truly global market.

Qualitative changes have been no less obvious. Production of oils and gases has grown much more rapidly than mining of coals and an increasingly higher proportion of fuels has been converted to electricity, the most convenient and the most flexible of all energies. Improvements in performance, already detailed for various prime movers, extend to higher extraction, conversion, and distribution efficiencies of whole systems. All of these general trends, outlooks, and limits are assessed in this chapter in appropriate historical, spatial, and thermodynamic perspectives.

Global production of fossil fuels can be charted with acceptable accuracy almost from its very beginnings (Mitchell 1975; U.S. Bureau of the Census 1975). Inevitable uncertainties and omissions are less important than cumulative errors inherent in converting the raw fuel data to energy equivalents. The best available reconstruction of worldwide fossil fuel production spanning the years 1860–1953 (United Nations Organization 1956) probably overestimates the total pre-World War II output, but the error is no greater than 5–10%. Better figures are available for post-1945 production (United Nations Organization 1976, 1980–).

A semilogarithmic plot of global fossil fuel production shows a hundredfold rise from about 2.6 EJ (82.5 GW) in 1850 to 265 EJ (8.4 TW) in 1985 broken into four distinct periods (Fig. 10.1). A nearly perfect exponential rise for three generations averaging 4.3% a year until the beginning of World War I was followed by a period of slow increases until 1932, a resumption of vigorous growth (4.5% a year) lasting until 1974, and most recently a near-stationary state with a production growth rate of 0.7% a year between 1975 and 1985.

The British production lead ended in the early 1890s and the United States has been dominant ever since; the Soviet Union has been the number two producer since the late 1940s and China surpassed Saudi Arabia's falling output by 1981. Recent distribution of global output is somewhat less uneven than in the past (mainly owing to China's ascendance) but the skewness is still considerable: in 1988 the five largest producers—the Soviet Union, the United States, China, Canada, and Saudi Arabia—with one-third of the world's population produced two-thirds of all fossil fuels and the two superpowers, with just one-tenth of mankind, extracted half of all coals, oils, and gases.

Aggregate growth has been accompanied by a universal qualitative substitution of hydrocarbons for coal: they surpassed coal's contribution by 1960 and since 1970 they have been providing about two-thirds of all fossil energies. Energy transitions follow fairly smooth curves when the fractions of fossil fuel supply (F) are plotted

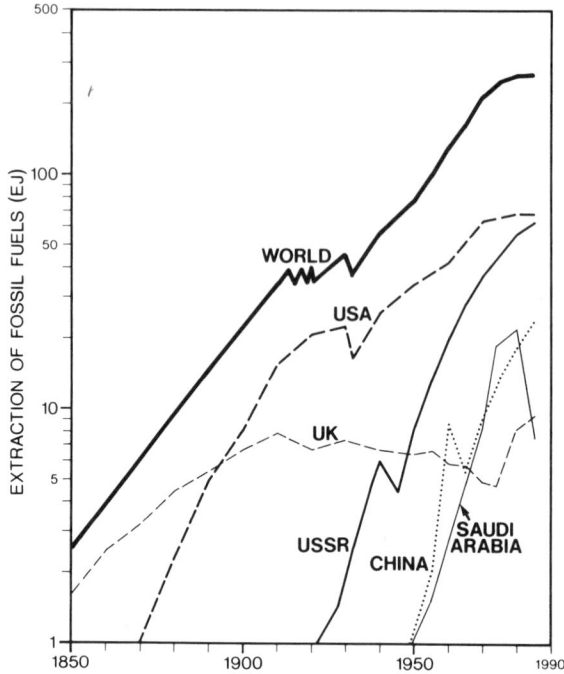

Figure 10.1 Global extraction of fossil fuels, 1850–1985. The dominance of the United States, the rise of the Soviet Union and China, and the rise and fall of Saudi Arabia were the most notable trends of the twentieth century. Based on data from the United Nations Organization (1956, 1976, 1980–), U.S. Bureau of the Census (1975), and the Central Statistical Ofice (1987).

together with the commercial fuelwood supply; when the production figures are expressed as log $F/(1 - F)$ the pattern becomes even more orderly (Fig. 10.2). Substitutions are slow, with about a century needed to reach half of the market, and surprisingly regular: in spite of many possible perturbations the penetration rates remain constant over long periods of time. "It is as though *the system had a schedule, a will, and a clock*" (Marchetti and Nakicenovic 1979).

This conclusion was based on the global data set ending in 1970. Since then there has been a leveling off of coal and wood share, a decline of crude oil, and a slowdown in natural gas penetration (Fig. 10.2). These shifts are almost cetainly more than temporary aberrations. Closer looks at energy transitions on a national level show much less orderly substitution curves and the marked regularity of the past experience on the global level is no guarantee of continuation. It seems highly unlikely that natural gas will become the world's

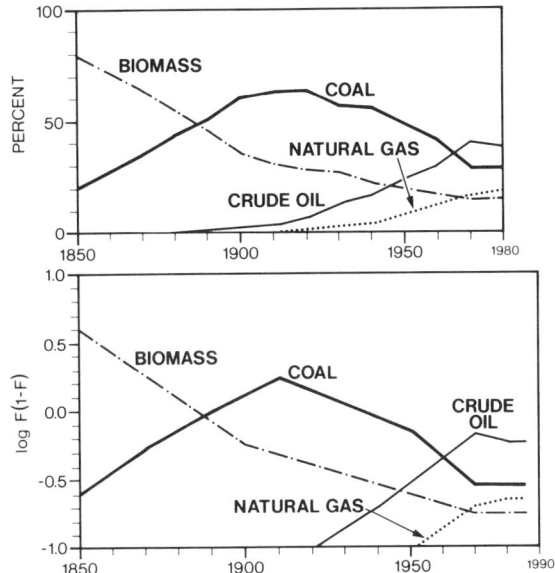

Figure 10.2 Global primary energy transitions have followed a highly regular course resembling normal (Gauss–Laplace) curves (upper graph). Even more orderly worldwide substitution emerges when the shares of individual resources are expressed as $\log F/(1 - F)$. Based on the same data as Fig. 10.1 and on Marchetti and Nakicenovic (1979).

leading fossil fuel during the first decade of the twenty-first century (a system on a schedule would bring it there by about 2010) or that coal will cover no more than 10% of all supply by the year 2000. Transitions will remain slow and orderly, but the slopes of penetration curves are not immutable and the peaks to be reached are not preordained.

The other qualitative shift has been away from the direct use of fuels, coal above all, through rapidly expanding generation of electricity. Early strategies for expanding electricity generation, especially evident during the 1920s and 1930s, have included several universal components (Hughes 1983): pursuit of economies of scale, concentration of fossil-fueled capacities in or near large load centers, development of high-voltage links to transmit electricity from remote hydrostations, promotion of mass consumption (charging of differential rates), and interconnections of smaller systems resulting in greater supply security, lower installed and reserve capacities, and automated central controls.

After World War II most of these thrusts intensified. The only

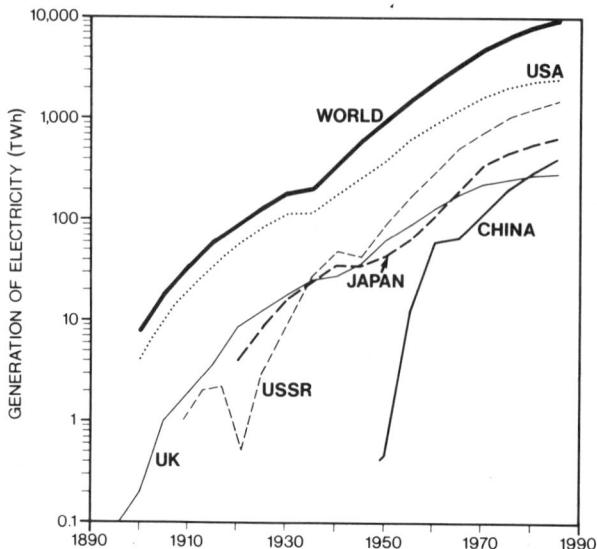

Figure 10.3 Global generation of electricity, 1900–1985. As with fossil fuel extraction, the most notable features are the U.S. dominance and the Soviet rise. Based on the same sources as Fig. 10.1.

major departure from earlier trends was the location of fossil-fueled power plants away from the cities, the shift brought about largely by strengthening air pollution regulations and by scale economies of huge mine-mouth stations. Global generation has grown smoothly, with just a short break in the early 1930s. Between 1900 and 1930 the average annual expansion rate was approximately 10.5%, and between 1935 and 1970 it was just over 9%. Subsequent growth has been roughly halved, to 4.5% for the years 1970–1985, a reflection of declining demand throughout the rich world (Fig. 10.3).

Although many countries have always produced most of their electricity from falling water, fossil-fueled generation has been globally dominant, accounting for just short of 60% in the 1920s, 67% of the total in 1950, 75% in 1970, and 64% in 1985 (post-1970 relative decline has been caused largely by rising nuclear generation). When assuming average e_1 (first-law efficiency: see the Appendix) of 33% the 1985 global fossil-fueled electricity generation would have consumed 25% of worldwide production of fuels compared to less than 10% just after the end of World War II.

Fossil-fueled electricity generation is even more unequally distributed than fuel extraction: in 1985 the United States produced almost exactly 30% and the six largest producers (the United States, the Soviet Union, Japan, China, West Germany, and the United Kingdom) accounted for nearly 70% of the world total of 6.2 PWh. Opinions about the future of fossil-fueled generation range from expectations of a relatively early return of higher growth to arguments about the desirability of actual consumption declines in the rich countries.

Another perfect illustration of future inapplicability of past trends concerns the growth of unit sizes and production systems and continuing concentration of outputs; examples from every sector show that past expansions are not replicable. Small, shallow pits that ushered the coal-mining era had outputs of a few hundred kilowatts and the combined production of Europe's richest coal basins in the early decades of the nineteenth century added up to less than 1 GW per coalfield. In contrast, by 1990 many surface mines extracted over 1 GW and several coal basins deliver in excess of 50 GW. Leaving aside the exceptional Middle East fields whose huge outputs (surpassing 50 GW per field for the top dozen at full flow) are primarily a matter of resource abundance rather than superior extractive techniques, there are thousands of hydrocarbon fields producing more than 100 MW.

During the first century of fossil-fueled electricity generation installed capacities of the largest power plants rose from a few hundred kilowatts to 4–5 GW, individual utilities have grown from enterprises serving a handful of city blocks to regional and national systems with capacities up to 10^{10}–10^{11} W serving areas of 10^4–10^6 km^2, and interconnections in Europe and North America enable trade of electricity at a multigigawatt level on semicontinental scales. But electricity trade, amounting to just over 2% of total generation, is very small compared to international fuel sales: in 1985 these included 10% of all hard coal, nearly 40% of crude oil, and 14% of natural gas output, altogether about 80 EJ (2.5 TW). The oil trade is a truly global undertaking and several principal export streams (Persian Gulf–Japan, Middle East–Europe) deliver 10^{11} W over distances up to 12,000 km.

A study of growth curves shows that most of the techniques underpinning these impressive systems either have ceased growing or are increasing at much slower rates than the historical trends would indicate. This change has become perhaps most conspicuous in the electric industry where the typical sizes of new generating units have been declining, but it is also easily discernible in extractive, processing, and transportation endeavors—indeed, we may never see a larger overburden-removing shovel, a bigger refinery, or a more

Table 10.1 Extraction, Processing, and Transportation Efficiencies of Fossil Fuel Energetics[a]

Process	Underground Coal	Surface Coal	Crude Oil	Natural Gas
Extraction	50	85	30	75
	(40–60)	(80–95)	(20–40)	(70–90)
Processing	96	96	88	96
	(92–98)	(92–98)	(86–92)	(94–98)
Transportation	98	98	98	98
Storage	98	98	99	99
Combustion	95	95	96	96
	(85–98)	(85–98)	(88–98)	(88–98)
Electricity generation	38	38	38	38
	(34–42)	(34–42)	(34–42)	(34–42)

[a] Single values show typical rates; common ranges are listed in parentheses.

voluminous supertanker. Reasons for this end of growth have not been primarily technical but rather environmental, economic, and social.

There is also no possibility of replicating the past advances of the top extraction and conversion efficiencies, although there is still much room for overall system improvements. Extractive endeavors recover as little as 20% of liquids in place in some heavy-oil fields—and in excess of 95% of coal in the best surface seams. Losses during coal processing vary mostly between 2 and 5%, crude oil refining leaves behind on the average about 12% of the initial energy content in nonenergy products (lubricants, waxes, asphalts), and processing of natural gases before pipeline transportation lowers their original heat content by 2–4% (Table 10.1). At the point of conversion the heat content of fossil fuels thus may represent as little as 17 or as much as 80% of the original energy.

Transportation losses are already minimal; a slight decline of gross energy content during processing actually improves particular utilities of fuels (cleaned and pulverized coal for power plants, specialized refined products, higher homologs separated from raw natural gas) and most of the removed energy is recovered (coal briquettes, liquefied petroleum gases) or goes into important nonenergy products (lubricants, paving materials). But major gains await in higher extraction efficiencies, above all through increased shares of surface coal mining and through wider application of secondary and tertiary oil-recovery techniques. These increasingly demanding options will face greater competition from a variety of nonfossil conversions, which are augmenting the fossil fuel base and which will eventually have to take over the provision of energies for global civilization.

10.2 NONFOSSIL CONTRIBUTIONS

The importance of these contributions is mixed, ranging from critical but inefficient and environmentally ruinous inputs (fuelwood and crop residues throughout the poor world) to mere curiosities (tidal electricity generation), from well-established, universally adopted large-scale conversions (hydroelectric generation) to promising but still only marginal new small-scale techniques (photovoltaics), and from systems that could conceivably totally displace fossil fuels (nuclear fission) to options that will always remain only of local or regional interest (geothermal electricity generation).

Biomass is by far the most important source of nonfossil energy. Since 1945 millions of pages have been written about commercial applications of nuclear energy but by 1990 fission reactors generated over 6 EJ of electricity—while roughly five times as much energy was consumed worldwide, largely in poor countries, in biomass fuels. Even when considering electricity's superior final-use efficiencies (in excess of 90% vs. no more than 20% for biomass fuels), fuelwood, crop residues, and dung still provided the world with roughly as much useful energy in 1990 as did nuclear power plants.

Unfortunately, it is impossible to offer an accurate account of biomass consumption as most of the harvesting and utilization of such fuels is done by more than 400 million families in the poor world. Many approximations are available on a national level, showing the dependence on biomass energies to be in excess of 80% for most of black Africa, about 50% for Indonesia, 40% for India, and 30% for China and Brazil. Most reliable are the still rare national rural energy surveys. China's extensive survey indicated the actual availability of less than 15 MJ of effective energy a day per family, an average shortfall of over 20% in comparison with the estimated minimal need; seasonal and regional shortages (especially in northern and eastern China) are often more than twice that share (Smil 1988).

Dimensions of China's rural energy shortage are thus enormous. About 60% of all peasants, or some 500 million people, have a serious fuel deficit for at least three to five months—and similar existential limitations prevail in many other regions. Most acutely affected are Africa's Sahelian zone and Namibia, Swaziland, Lesotho, and parts of Botswana in the south, in Asia the Nepali hills, large parts of India, Bangladesh, Pakistan, Afghanistan, and Thailand, and much of Central America (Food and Agriculture Organization 1980).

Wood, straw, and dung availability, climatic differences, and cooking and heating habits are responsible for large consumption variation but most surveys and estimates indicate flows equivalent to 0.5–2.5 m^3 of air dry wood per capita a year, or about 5–25 GJ. Published rates (not easily comparable) for countries whose rural population is still almost totally dependent on wood (including charcoal) range from 8–10 GJ for Zambia, Zimbabwe, Madagascar, Kenya, and Ethiopia to around 15 GJ for Angola, Ghana, Cameroon, Sudan, Nigeria, and Thailand and up to 34 GJ for equatorial Gabon, still with extensive tropical rain forest (Hall et al. 1982; Smil 1983).

For heavily deforested, densely populated nations whose peasants burn any available combination of forest fuels (stemwood, branches, stumps), grasses, shrubs, crop residues, and dried dung the rates are obviously much lower, going from 1.9–3 GJ for Banglandesh to 7.8 GJ for China and 5.9–7.7 for India. The best conclusion would be that those traditional rural societies still dependent on biomass fuels for all of their household thermal energy needs consume annually as little as 5–9 GJ per capita of crop residues and dung just for simple cooking, and they are reasonably well off with 15–30 GJ in warm climates. Potential savings from introduction of better stoves are enormous but practical success in displacing the traditional designs has been limited.

In a series of 1979 Chinese tests stoves burning cornstalks had just 8–14% efficiency and common firewood stoves averaged just 15%. The reasons: their combustion chambers and stoking inlets were too large, causing unnecessary heat loss and allowing excessive cold air intake, there were no grates, and there were even no chimneys. New designs of the early 1980s—with proper fire grates and chimneys, appropriately sized combustion chambers and fuel inlets, and 1.5–2 volumes of excess air for optimum burn-up have thermal efficiencies at least 25 and up to 44% (Smil 1988).

A conservative estimate for the poor world's annual phytomass consumption in the mid-1980s would be at least 15 EJ of woody phytomass, 3 EJ of crop residues, and less than 1 EJ of dried dung. A total close to 25 EJ would seem to be a more likely value, with

China, India, Brazil, and Indonesia being the expected largest consumers. To this should be added fuelwood consumed in the rich
countries, mostly for household heating and by forestry and pulp and
paper industries in North America, northern Europe, and the Soviet
Union. World Energy Conference (1986) statistics put this total at
just over 5 EJ, with more than half consumed in the United States.

The grand total of global biomass energy consumption thus would
be somewhere between 25 and 30 EJ, and equivalent of roughly 10%
of the mid-1980s worldwide fossil fuel combustion. Nearly all of this
energy is used directly as space, cooking, and processing heat and
only a minuscule part is converted to electricity or to gaseous or
liquid fuels. For example, the world's largest alcohol conversion program, Brazil's PROALCOOL, produces annually about 200 PJ of
ethanol for motor vehicles—while the country burns at least 2 EJ of
fuelwood (World Energy Conference 1986).

Power densities of biomass energetics are very low, reflecting
the inherently low efficiency of photosynthesis. Woody phytomass
burned throughout the poor world is gathered with densities ranging
from just 0.02 W/m^2 for twigs, branches, and leaves in arid environments to about 0.15 W/m^2 for sustainable selective stem cutting in
moist forests. Plantations of fast-growing trees—willows, poplars,
eucalyptuses, leucaenas, pines—can yield anywhere between 0.1 W/m^2
in dry northern climates to 1 W/m^2 in the best stands in humid
regions or with irrigation, but values around 0.5 W/m^2 would be more
typical upper rates.

Logging residues from timber clearcutting of rich natural stands
can provide a nonrecurrent yield of 1–4 W/m^2 but most of them may
not be usable (too remote, too dirty) and will be burned on the site.
Crop residues used for household combustion are harvested with
densities ranging from 0.01 W/m^2 from low-yielding cereals (assuming
straw yield of 1 t/ha and one-third of it used for fuel) to 0.4 W/m^2 for
sugarcane bagasse (cane yield of 100 t/ha, all of the bagasse burned).

Conversion of water's potential energy to electricity in hydropower
plants is the second most important nonfossil energy input. Hydroelectric generation started in the same year as thermal production
(in Wisconsin in 1882) and major advances were achieved in construction of increasingly higher dams before 1900 in Alpine countries,
Scandinavia, and the United States. Between the world wars came
the start of state-supported development of large hydro projects in
the United States (most notably the Tennessee Valley Authority) and
in the Soviet Union. By 1939 there were hundreds of plants with
installed capacities of 10^7–10^8 W but only two American projects—

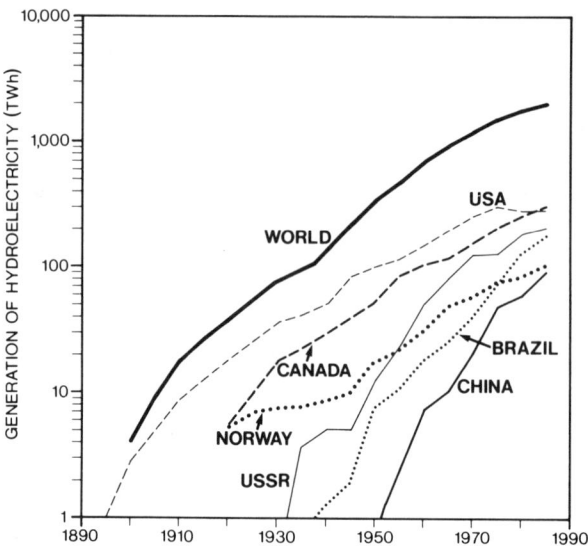

Figure 10.4 Global generation of hydroelectricity. Long-term U.S. dominance, recently surpassed by Canada, and the rapid rise of Brazilian and Chinese contributions are seen. Based on the same sources as Fig. 10.1.

Hoover Dam on the Colorado and Bonneville on the Columbia—had in excess of 1 GW.

Giant Grand Coulee on the Columbia (6.494 GW) was completed in 1942 and between 1945 and 1985 nearly 120 stations with capacities in excess of 1 GW were put into operation in more than 30 countries. The United States has the largest number of such plants, including the world's second largest station (Grand Coulee, whose capacity eventually will reach 10.83 GW); Itaipu on the Paraná between Brazil and Paraguay is the largest, with 7.4 GW and with ultimate power of 12.6 GW. The United States also has the largest installed capacity (almost 70 GW in 1988), followed by Canada and the Soviet Union. These three countries generated 40% of global hydroelectricity in 1985 (Fig. 10.4) when all hydrostations accounted for roughly 23% of total generating capability and when they generated over 20% of global electricity.

But hydrogeneration is even more important for many relatively small producers throughout the poor world: its shares are over 90% of all electricity in many African countries and 80–90% in South America. Most of these countries have a considerable potential to develop small-scale hydro resources, the sites with capacities below 2 MW. China has led these efforts with about 9 GW in small hydrostations in 1985. The post-1973 renaissance of interest in small-scale hydroelectricity has extended also to those rich countries that have run out of practical options for large sites (nearly all of Western Europe, the northeastern United States, Japan).

Peak technical achievements in large dam construction include the height of 325 m (Rogun on the Vakhsh in Tadzhikistan, completed in 1985), dam volumes in excess of 200 Mm^3, reservoir capacity of nearly 170 Gm^3 (Bratsk on the Yenisey), and the dam crest of 82 km for Yacyreta–Apipe on the Paraná between Paraguay and Argentina to be completed in 1996 (Mermel 1988). The largest turbines are Francis machines with more than 500 MW (second stage of Grand Coulee, 700 MW), and the largest proposed station (Turukhansk on the Lower Tunguska) would have a capacity of 20 GW.

Hydrogeneration is obviously highly space-demanding and its power densities span three orders of magnitude. Plants on the lower courses of rivers operating with relatively small gradients and huge water volumes in big reservoirs—that is, most of the world's largest hydrostations—occupy extensive areas, commonly surpassing 10^9 m^2. Areas of the seven largest reservoirs add up to the size of the Netherlands; the two largest, Akosombo on the Volta in Ghana (8730 km^2) and Kuybyshev on the Volga (6500 km^2), approach the size of such small countries as Lebanon and Cyprus. These huge projects generate electricity with power densities below 1 W/m^2. Power densities of 0.5–1.5 W/m^2 are common for midsize (50–200 MW) projects.

In contrast, large plants on the middle and upper courses have high gradients, smaller reservoirs, and power densities on the order of 10^1 W/m^2: Itaipu rates nearly 9 W/m^2, the final Grand Coulee almost 34 W/m^2, and some Alpine stations surpass 100 W/m^2. All of these density values are for installed capacities. Load factors of most hydro stations are commonly below 50%; many projects are also built largely for peaking power and will have even shorter operating hours. Effective power densities of most hydro stations are then merely one-third to one-half of the theoretical rates.

An insurmountable dilemma meets every attempt to assess the contribution of hydrogeneration (and other nonfossil electricity) to the primary energy supply. Primary electricity can be easily converted in terms of its heat value, that is, 1 kWh = 3.6 MJ, but this approach obviously undervalues hydroelectricity's contribution in comparison with fossil fuel generation, which, depending on its efficiency, generally requires 9–14 MJ of fuel for each kilowatt-hour.

Consequently, it has been customary to convert primary electricity at the rate of the prevailing average thermal generation efficiency.

This option almost certainly overvalues hydroenergy's contribution, since few countries would generate as much electricity if they were forced to do this by burning fossil fuels. International statistics use a straight thermal equivalent, which results in relatively lower overall primary energy production and consumption levels in countries with a very large hydro component in their total supply. When converted at a straight thermal equivalent rate the mid-1980s hydrogeneration contributed a little less than 3% of the global primary energy.

An increasingly important component of hydroelectric generation is pumped storage, a process using nighttime, weekend, or low-demand season power (i.e., mostly thermal electricity) to pump water to an elevated reservoir to be released during peak-demand hours through turbines into a lower reservoir. In Europe the pumped storages frequently have been used to enhance system stability. With their high heads (up to 1400 m, commonly 200–500 m) and limited load factors, power densities of pumped storage generation are quite high, mostly 1–3 kW/m^2. Reversible pump–turbine units serving these projects have reached 300–400 MW, and the largest pumped storage plants—the 2.1 GW Bath County in Virginia and 1.8 GW Dinorwic in Wales—will be surpassed by several projects designed to deliver up to 3.6 GW. Worldwide installed capacity of pumped storages was about 65 GW in 1988 (Mermel 1988).

Nuclear generation has been rapidly approaching hydrogeneration's contribution but it faces a much more uncertain prospect. Its rise has been spectacular, with the first commercial reactor coming just 14 years after the first sustained chain reaction on December 2, 1942. The British Calder Hall (202 MW) was followed in 1957 by the U.S. Shippingport (141 MW); during the next decade a rising wave of orders pushed the average size of new nuclear power plants to over 1 GW by 1970. At the time of OPEC's oil embargo in October 1973 there were more than 400 reactors in operation, under construction, or in various planning stages in 13 industrial and 7 industrializing countries. The industry appeared to have a very bright future.

In reality, the years after 1973 became a time of nuclear retreat. Reasons for this development include complications accompanying early stages of every new technique (important but not a decisive complication), escalation of construction cost (a crippling but not inevitable reality), public perception of intolerable risks involved in routine operation (a critical concern naturally heightened by the 1979

Three Mile Island accident and the 1986 Chernobyl disaster), and the unsolved problem of long-term disposal of highly radioactive wastes. But the retreat has not been universal: France, the Soviet Union, and Japan have gone ahead with extensive programs.

As the reactors ordered during the 1970s have been entering operation the share of nuclear generation in global electricity production has gone from just 1.5% in 1970 to 8% in 1980 to 17% in 1990. France has the highest dependence (about 70%), with Belgium close behind, in the United States, the country with one-third of all working reactors, the share was about 17% in 1990. When converted at the straight thermal equivalent the worldwide nuclear generation represented only about 2% of global primary energy production in 1990. Most of the 345 units in service (totalling about 300 GW) or under construction were pressurized water reactors (PWR) where the coolant circulates through the core in a closed loop and transfers its heat in a steam generator.

These reactors have grown to nearly 1.5 GW (the two largest are the Soviet Ignalina at 1450 MW and the French Chooz at 1457 MW) and their e_1 is about 32%. A boiling water reactor (BWR) generates steam by passing water directly between the elements of nuclear fuel; the efficiency of BWRs is also just over 30%. The largest nuclear power plant, Japan's Fukushima with 9096 MW in 10 units, was actually the world's largest electricity-generating facility of the late 1980s. In both PWR and BWR ordinary water acts both as a coolant and as a moderator needed to slow the neutrons and to improve the efficiency of conversion. In pressurized heavy water reactors (PHWR, the base of Canada's national program with 32% efficiency), heavy water (D_2O) is a moderator and organic compounds, gas, water, or heavy water is the working fluid. Gas-cooled reactors (GCR, the foundation of British nuclear development) use graphite as moderator and helium or carbon dioxide as coolant.

Frequently reported problems with nuclear generation obscured an outstanding performance of the best units and systems. Many reactors achieve impressive annual load factors of over 99% (Japanese Ikata had 100% in 1986), lifetime records are over 85% (Canada's Pickering and Bruce), and the best national lifetime averages are close to 80%. Power densities of nuclear fission are high: the rates for nuclear reactors are mostly 50–150 MW/m^2, values an order of magnitude higher than those of fossil-fueled boilers. Overall generation densities are also high, typically 2–3 kW/m^2. Addition of space requirements for the complete fuel cycle—mining and processing of uranium ores, uranium enrichment, production of fuel elements, reprocessing of spent fuel, and storage of radioactive wastes—lowers these values by no more than 20% (Mielke 1977).

Other modes of nonfossil electricity production are globally much less significant: geothermal output is two orders of magnitude, photovoltaic generation three orders, and tidal and wind production four orders of magnitude smaller. Utilization of geothermal energy for electricity generation started in Italy just after 1900 at the dry steam field of Lardarello. In 1960 Wairakei (New Zealand) and the Geysers (California) went on stream; the three plants added up to just 368 MW of installed capacity. Two decades later the global total was up to nearly 2.5 GW; in 1985 it reached 4.72 GW with California (2 GW), the Philippines (894 MW), and Mexico (645 MW) in the lead (Cataldi 1986).

Power densities of geothermal generation—mostly between 20 and 50 W/m^2—are comparable to those of Alpine hydrostations. Best developmental prospects are clearly for relatively small stations (up to 100 MW) in high-gradient regions in many poor countries where such capacities could make a major addition to regional supply. Direct use of geothermal heat has been most extensive in Iceland, largely for space heating and hot water in Reykjavik but also in greenhouses and in industry. In 1985 these uses added up to about 1.1 GW, or 20% of the global total of 5.6 GW. Hungary and the Soviet Union were close behind, followed by France, China, and the United States. This ranking, compiled by World Energy Conference (1986), excludes balneological applications; their inclusion puts Japan first (with 4.8 GW) and raises the global total to about 12 GW.

There is no shortage of windy places, and machines located at these sites (average wind speed at least 7 m/s) would intercept wind energy at rates higher than 200 W/m^2 of vertical area swept by rotating, horizontal-axis propellers. The best sites—northern Texas, western Oklahoma and Kansas, and coastal California—have mean annual vertical wind power densities over 500 W/m^2 at 50 m above the ground. Translating these rates into surface power densities is far from straightforward. Full development of a particular site will have to allow for sufficient spacing between the units needed for elimination of upstream wakes and for vertical and horizontal replenishment of kinetic energy. Spacing equal to five rotor diameters will avoid excessive wake interference but the space allowance for wind energy replenishment in the lowest 100 m of the atmosphere may require separation as large as 35 rotor diameters.

If the spacing of five rotor diameters were sufficient, a wind farm with machines 15 m in diameter, each rated in 7 m/s at about 35 kW, would have 177 units/km^2 and intercept the total of some 6 MW or 6 W/m^2. However, no more than 59.3% of the wind's kinetic energy can be extracted by a rotating horizontal-axis generator (Betz limit) and the actual performance will be no higher than 80% of this maximum, prorating to output power density of 2.8 W/m^2. Even in America's windiest regions such units could work only about 60% of the time and the actual output power density would be about 1.7 W/m^2. Wilshire and Prose (1987) give detailed figures for developments in California where the land disturbed by construction amounts to about 0.4 ha/unit and the total occupied land amounts to 1.2 ha/machine, or almost exactly 2 W/m^2 for the rated performance.

After decades of small-scale generation for isolated households on windy plains of Americas and Australia the first major conversions of wind to electricity came only in the late 1970s. By 1985 about half of the worldwide generation of some 750 GWh (0.012% of the fossil fuel–generated total) came from the Altamont Pass area in California (Smith 1987). A year later the wind farm's 6700 wind turbines with installed capacity of 630 MW generated about 550 GWh, implying a very low load factor of 10% and effective power density of just 0.84 W/m^2. Several sites in California account for over 90% of global commercial wind-powered generation. Danish turbines provide about 5%, and the rest is widely scattered. There are also still nearly 300 MW of small family-size generators in Argentina, about 50 MW in South Africa, and 20 MW in Uruguay.

Photovoltaic generation of electricity is perhaps the most appealing renewable conversion owing to its unattended and silent operation, absence of moving parts resulting in low maintenance requirements, inherent modularity facilitating construction, and minimal environmental impact (Zweibel and Hersch 1984; Hamakawa 1987; Hubbard 1989). The technique has well-established niches in powering space vehicles and satellites as well as a variety of Earth-based devices ranging from remote relay stations to portable calculators and wrist watches. Virtually all of these applications have been based on crystalline silicon cells. These are basically large-area diodes sensitive to about 44% of incident radiation, with response region stretching from 350 nm into the invisible infrared (just short of 1.1 μm).

Heat, reflection, and absorption losses limit the best conceivable performance of crystalline cells to about 28% and efficiencies of 14–22% can be achieved in everyday applications. Advances in crystalline–silicon cell production lowered the cost of installed power by an order of magnitude between the mid-1970s and the mid-1980s.

The best prospects for further substantial lowering of costs is with thin film materials made of amorphous silicon, copper indium diselenide and cadmium telluride. Another promising approach is a two-junction cell made of gallium arsenide atop a base of silicon: it's efficiency, under concentrated sunlight, surpassed 30% in 1988. Average conversion rate around 15% in large arrays could provide a competitive alternative for utility-sized peaking-power systems.

With sustained 15–20% efficiency power densities of photovoltaic cells would be at best 13–35 W/m^2 but in large central stations the actual plant area would be at least double the cell surface, reducing the effective conversion densities to 6–18 W/m^2. Cumulative global shipments of photovoltaic cells (in terms of peak power) rose from less than 100 kW in 1974 to nearly 100 MW by 1987, and the largest operating power plants—led by 6.5 MW Carissa Plains installation in California—reached the level where they could start making a difference to a utility's peak demand. Fulfillment of the high hopes for cheap, mass-produced, efficient photovoltaics before the end of the century would mean a fundamental shift in electricity generation in sunny parts of the world, but a subsequent massive decentralization of electricity generation seems unlikely.

While the prospects of a photon route to solar electricity generation are steadily improving, the thermal route—a field of heliostats reflecting the sunlight to a central receiver generating high-pressure steam—was demonstrated in 1982 on commercial scale in the Solar One, a pilot plant near Barstow, California. The station, with 1818 heliostats of 39 m^2, is rated at 10 MW for 4 hours on the winter solstice and for nearly 8 hours on the summer solstice (Kreith and Meyer 1983). With best efficiencies (about 22%) and at the sunniest sites such central stations could produce electricity at about 60 W/m^2, in less sunny locations the densities would be 20–40 W/m^2, and for less efficient systems they would be no more than 15 W/m^2. The Solar Electricity Generating System (SEGS) operating near Daggett (Mojave Desert) since 1984 is using tracking parabolic trough collectors and oil as a working medium to generate steam. Rated at 43.8 peak MW the facility has 165,000 m^2 of collection surface and occupies about 360 ha, for a peak power generation density of just over 12 W/m^2.

Flat plate collectors for space and water heating in houses, institutions, and industries have made much greater practical contributions. Stationary black flat plates can transfer up to 60% of incident radiation to water at useful temperatures of 40–50°C; everyday efficiencies of well-designed systems are 35–40%. This means that the power production densities of flat plate collectors would average mostly between 30 and 100 W/m^2 while the peak noontime rates under clear skies could be as high as 500 W/m^2. Globally, the number of these installations reached several million by the mid-1980s but it is impossible to offer a meaningful estimate of their contribution.

Tidal generation remains a singular curiosity: the Rance power plant near Saint-Malo has had 240 MW installed in reversible turbines since 1966. The Soviet Kislaya Guba (400 kW; 1968) and Canadian Annapolis (17.8 MW; 1984) are experimental setups, as are a few Chinese ministations. Rance's power generation density is about 14 W/m^2 while that of the two most promising potential sites—the Bay of Fundy in Nova Scotia and the Severn estuary in the U.K.—would be, respectively, 15–16 and nearly 13 W/m^2.

Undoubtedly, the share of nonfossil contributions to the global energy supply will continue its gradual rise but the bulk of consumption growth will come, at least during the next two generations, from the same source as the past spectacular expansions: from combustion of fossil fuels.

10.3 CONSUMPTION PATTERNS

Perhaps the single most revealing marker of fossil-fueled civilization has been the exponential rise of per capita energy consumption. Typical annual wood and charcoal consumption in richer preindustrial societies ranged between 20 and 40 GJ per capita. In forest-rich, thinly populated North America of the eighteenth and nineteenth centuries it peaked at atypically high rates of 70–100 GJ per capita. If the e_1 of fireplaces, stoves, and simple furnaces had averaged 10%, such consumption rates would have prorated to no more than 2–4, and exceptionally to 7–10 GJ of useful energy per capita.

During the nineteenth century countries with rich coal resources increased their rates of gross energy consumption rapidly: from just 20 GJ in 1800 to 116 GJ in 1900 in Britain; from a negligible amount to 105 GJ during the same time in the United States. Per capita consumption of coal in countries that had to import most of the fuel—be it in warmer France or colder Sweden—remained relatively low, surging only after World War II with large imports of oils, as it did in Japan (Fig. 10.5). In 1985 rich countries—in North America and Europe, as well as the Soviet Union, Japan, Australia, and New Zealand—averaged just over 150 GJ per capita; with overall mean conversion efficiency of about 40% this prorates to about 60 GJ of

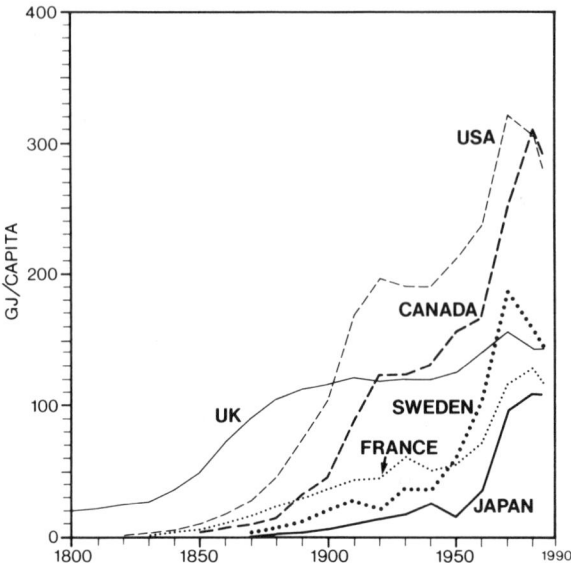

Figure 10.5 Long-term trends of per capita energy consumption in half a dozen leading rich economies. A huge post-1900 gap between North America and western Europe and Japan is the most notable feature of these rates. Calculated from sources listed in Fig. 10.1 and from Mitchell (1975).

useful energy, an order of magnitude higher flux than the averages during the last decades of the biomass era. Moreover, this quantitative jump was accomplished through deliveries of qualitatively superior forms of energy.

These gains have not been universal and the fairly reliable gross consumption figures available since 1950 for virtually all countries of the world show no lessening of the huge gap between the rich and poor consumers, as well as surprisingly large differences among the countries within these distinct groups. The absolute range (all values refer to the late 1980s) is from less than 1 GJ per capita for the poorest African countries to 290 GJ per capita for Canada (the United States is second with 280 GJ). All of these measures are just statistical abstracts. As the poorest people in the poorest countries do not consume directly any fossil energies or primary electricity, the difference in modern energy consumption between a subsistence pastoralist in the Sahel and an average Canadian easily may be larger than a thousandfold!

In 1950 industrialized countries consumed about 93% of all commercial primary energy. Subsequent development of poor countries reduced this share: in 1985 the rich world containing a quarter of the global population consumed just over four-fifths (82%) of all primary energy. The United States alone used about 27% of the total. In contrast, the poorest quarter of mankind—some 15 sub-Saharan African countries, Nepal, Bangladesh, the countries of Indochina, and most of rural India—consumed just 2.5% of the worldwide flow of fossil fuels and primary electricity. No other comparison illuminates more starkly the chasm separating the two worlds.

One common trend has been the greater use of better fuels. Higher heat content, easier and safer extraction, inexpensive seaborne and virtually invisible continental transportation, cleaner and much more convenient combustion, and a great flexibility of utilization account for the ascent of hydrocarbons and relegation of coal to a few specialized markets. Shifting patterns of final uses mark distinct eras of fossil-fueled civilization. When coal overtook wood in the United States in 1885 some 40% of it was used by railways, about 14% was converted to coke, and the rest was about split among industrial boilers and residential heating.

By 1985 there was no trace of the railway market (it disappeared by the early 1960s), the coking share fell to about 8% of the total, but electricity generation took more than 80% of all coal. Crude oil's first market was as a substitute for whale oil in lighting and only the diffusion of internal combustion engines created the first mass demand for the fuel. National consumption peculiarities are clearly discernible in the shares of light and heavy oil products: while the United States consumes 50% of all liquid fuels as gasoline, Japan's share is less than 20%—but residual fuel oil accounts for nearly 33% of Japanese use while it is just 8% of the U.S. total.

Given the advantages, the steadily rising shares of electricity in final consumption of rich nations have been unavoidable. Only electricity offers the following combination: instant, effortless consumer access; ability to step into every consuming niche and be converted into motion, heat, light, and chemical potential with unmatchable control, precision, and speed; silent, clean (at the point of final conversion), and extremely reliable individualized delivery; and capacity for easy accommodation of growing or changing uses. And this energy can be produced from a wide variety of (often inferior) fuels, its conversion to heat can be accomplished with nearly perfect efficiency, it can provide temperatures higher than combustion of any fossil fuel, and its utilization requires no inventory.

Not surprisingly, electricity became the energy of choice for residential use in affluent societies. But electricity's ascendance first revolutionized industrial production when it became the dominant source of motive power. As revolutionary as was the replacement of waterwheels by steam engines in nineteenth-century factories, this step did not change the mode of distributing mechanical energy needed for countless productive tasks: factory ceilings were clogged by complex arrangements of iron or steel line shafts connected by pulleys and belts to parallel countershafts which were belted to individual machines. If the prime mover (be it a waterwheel or a steam engine) broke down, if the line shaft cracked, if a belt slipped—the whole assembly had to stop; conversely, if most of the machines did not need to work, the whole system was still running.

Electric motors changed this rapidly, first just driving relatively short shafts for groups of machines and, since the first decade of twentieth century, increasingly as unit drives. Studies by Devine (1983) and Schurr (1984) document the rapidity of this critical transition in the United States: while the total installed mechanical power in manufacturing grew roughly 400% between 1899 and 1929, electric power capacities grew nearly 6000% and reached over 82% of the total compared to less than 5% at the end of the nineteenth century. Since then the share of electric power has changed little; the substitution was practically complete just three decades after it began.

Individualized power supply translated into superior productive efficiencies opened the way for flexible plant design and easy expansion, enabled precise machine control and highly focused power applications, did away with the overhead clutter, noise, and health risks, freeing the ceilings for installation of better illumination and ventilation—and resulted in higher labor and capital productivity.

While aggregate national consumption figures for individual fuels and electricity have been available for decades, systematic analytical studies of energy flows through national economies are of surprisingly recent origin. In the United States it was only in 1971 when a Stanford Research Institute (1971) report outlined the movements of fuels and electricity and allowed a closer look at the pattern of consumption. After 1973 it became obvious that sensible management of energy consumption cannot be accomplished without first appreciating the complexities of actual flows, and this realization has led to a proliferation of sectoral use studies. Still, this information remains limited and not highly accurate, requiring frequent approximations.

Longitudinal comparisons of sectoral uses in rich countries show three general trends: gradually declining shares of industrial consumption and slowly increasing residential and transportation demand reflecting the rising affluence of consuming societies. During the 1980s industrial energy use was below 50% of the total in all rich Western countries where the transportation sector took 10% (in small European nations) to nearly 30% (in North America) of all primary energy. The share of residential consumption is a highly reliable measure of a country's affluence—certainly a more revealing indicator than GNP per capita. In the mid-1980s it stood at just about 15% in the Soviet Union, 20% in Japan, and about 30% in the United States.

Buildings are either the largest or second largest consumers of energy (behind industrial conversions) in all rich societies: in the United States they took about 40% of all fuels and 75% of electricity in the mid-1980s. Space heating (and cooling) is the dominant demand in every rich nation (50–80% of all residential consumption), followed by lights and electric appliances and water heating. High efficiency of final conversions, ease of operation, near absence of maintenance, and cleanliness have favored natural gas and electricity as the preferred energies for indoor use. Pre-1973 performance of both houses and commercial buildings was astonishingly poor in North America and better, although far from satisfactory, in Europe.

Since then many improvements have taken place but further large savings remain possible (Baird et al. 1984; Schipper et al. 1985; Rosenfeld and Hafemeister 1988). For private houses standardized comparisons of energy requirements are usually done per unit area per heating degree day (in °C to the base of 18). Average for the U.S. housing stock was about 160 $kJ/m^2 \cdot$ degree-day in 1980, new buildings put up during the 1980s rate between 100 and 120, super-insulated houses go down to 30–50, and the most efficient designs can do with as little as 15–20. In terms of actual consumption, national means of total primary energy consumed in housing in the early 1980s prorated (all values are rounded) to just 20 W/m^2 in Japan, 40 in the United States and Germany, 50 in Sweden, and 70 in Canada.

These power densities reflect more than the climatic stress (Canada's average of about 4600 degree-days compared to Japan's 1800; the U.S. mean is 2600, German 3200, Swedish 4000). Different life styles and affluence levels set different consumption baselines. Japanese and British tolerate much lower indoor temperatures (below 15°C) than Americans and Canadians (typically above 18°C) and commonly heat only some rooms in the house or, in the Japanese case, just parts of some rooms. Canadians and Swedes heat on the average 2.5–3 times as much water as most Europeans or Japanese.

Refrigerator ownership is nearly universal throughout the rich world but clothes dryers, dishwashers, and freezers have widely differing saturation rates.

Efficiency of principal energy converters differs widely from 50% for solid-fueled furnaces, to 55–65% for well-adjusted traditional oil and gas furnaces with standing pilot light, and up to 91–96% for condensing nonvented gas-fired units (Macriss 1983). And even greater differences separate heat losses of walls, windows, doors, and roofs in poorly built pre-1973 dwellings and in new superinsulated houses with very low rates of air infiltration. Total energy requirements of such structures are no more than 30 W/m^2 even on Canadian prairies where houses built during the 1950s needed more than 150 W/m^2.

A similar trend has been evident for commerical buildings. Most of the multistoried glassy structures put up in North America between the early 1950s and the early 1970s averaged 110–140 W/m^2 of floor area. This means that a typical 20 story glass building of those years required 2–2.5 kW/m^2 of its foundation—and that New York's 110 story World Trade Center uses about 12 kW/m^2 for a total of 80 MW. By the mid-1980s the primary energy required by new office buildings was below 50 W/m^2 and many all-electric buildings have been designed for just around 10 W/m^2, or less than 30 W of primary energy per square meter of occupied area (or, for a 50 story structure, no more than 1.5 kW/m^2 of foundations).

Climate, height, concentration, and specific performance of commercial buildings combine to produce an order of magnitude range for consumption power densities of urban central business districts (CBDs). Low-profile (buildings mostly less than 10 stories high), low-density (at least two-thirds of the area taken by roads, parking, and greenery) CBDs in mild, equable climates will have annual consumption averages below 50 W/m^2 of the CBD area. High-rise (average over 20 stories), high-density (two-thirds of area built up) CBDs in colder midlatitudes will rate over 700 W/m^2. Most of the heating needs in properly designed high-rises can be met from lights and electric motors.

Lights, indoor and out, have also become much more efficient during the first century of their diffusion. In 1882 efficacy (the quotient of the total luminous flux emitted by the rated power, measured in lumens per watt) of carbonized fibers in Edison's 100 W lamps was 1.4 lm/W, eventually improving to 2.5–3 lm/W. Osmium, introduced in 1898, delivered 4 lm/W, tungsten filaments in vacuum (first in

1912) 10 lm/W, and tungsten in inert gas-filled bulbs 20 lm/W. The most efficient incandescent light of the 1980s, a 10 kW source for film studios, puts out 33.6 lm/W (Weast 1988).

But even the best incandescent lamps are rather inefficient. Using 1.47 mW/lm as the standard mechanical equivalent of light makes the 10 kW lamp no more than 5% efficient in converting electricity into light and the rate for the common 100 W bulb is only 2.6%. New techniques boosted performance: a standard 40 W warm white fluorescent tube has nearly 12% efficiency and lasts at least 25 times longer (20,000+ hours) than a 100 W incandescent bulb. The best performers are high-intensity discharge metal halide (up to 110 lm/W) and high-pressure sodium lamps (up to 140 lm/W) with efficiencies near or over 20%.

There are only a few exceptions to oil fuels' dominance of transportation market, such as largely coal-fired railways of China and Brazilian cars running on ethanol from sugarcane. In the late 1980s nearly 30% of global refinery output was motor gasoline, just over 20% diesel oil for land vehicles and ships, and jet fuels accounted for about 3%. Waterborne transport powered by diesels and gas turbines is both relatively efficient and incontestably the cheapest way to move goods or people; pipelines are best for bulk movement of liquids and gases but trains, powered also by diesels or by electricity, are the best performers on land.

By far the greatest changes in the energetics of transportation thus can be affected by improving motor vehicles, above all the performance of passenger cars. The automobile was a European invention, but affluence and mass production combined to give the United States more than 90% of the world's cars, trucks, and buses before World War II—and still 60% by 1960 (Fig. 10.6). But by 1983 Europe matched the U.S. total and is now the world's largest market for new vehicles.

Everyday performance means are very low. The best practical e_1 of the Otto cycle is around 32%; frictional losses bring this down to 26% and partial load factors, inevitable during the urban driving which constitutes most car travel, reduce this to 19–20%; accessory loss and automatic transmission may nearly halve the total so that the effective e_1 is no more than 10–12% and often can be as low as 7–8%. Peculiarities of American automotive preferences, decades of declining oil prices, and a tradition of heavy Detroit design meant that the performance of the U.S.-dominated fleet was actually getting worse: by 1974 American cars needed roughly 15% more energy per

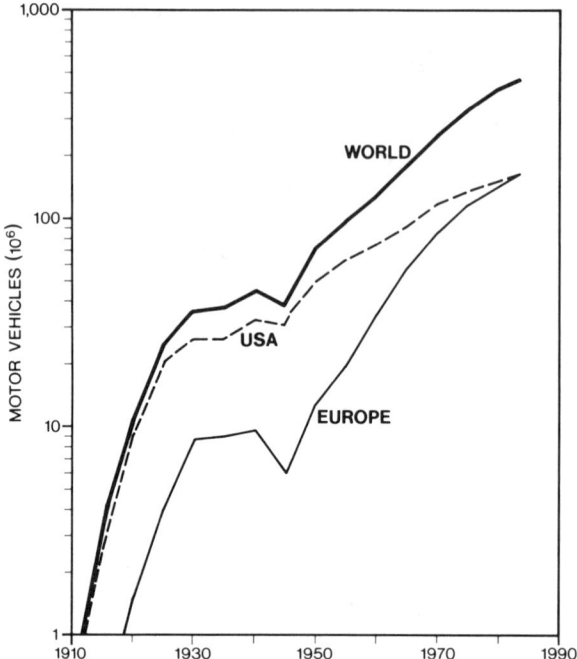

Figure 10.6 Global registration of motor vehicles (passenger cars, trucks, and buses) shows the long-lasting U.S. dominance and rapid European increases. Based on data in Motor Vehicle Manufacturers Association (1988).

kilometer than their counterparts of the 1930s. Sharp turnaround was brought about by OPEC's oil price rises and by the growing importance of European and Japanese car imports.

Between 1974 and 1988 the fuel consumption mean for the U.S. car fleet fell by almost exactly 50% to 3.1 MJ/km. Impressive as this belated decline was, the new rate is still over 8 L/100 km at a time when there is no shortage of cars averaging 6–8 L/100 km even in city traffic and when the best performers among gasoline-fueled vehicles need just 4.3–4.8 L/100 km (1.7–1.5 MJ/km) when running at 90 km/h.

Energy requirements of buildings and transportation combine to reach their highest densities in cities and industrial conurbations. Surprisingly, minima in such poor hot-weather cities as Calcutta or Bombay have been 10–15 W/m^2, the same as in warm but affluent Los Angeles. In the former instance the low per capita energy use and the near absence of private cars is counterbalanced by high dwelling densities and the presence of energy-intensive industries in residential areas; in the latter case the extraordinary urban sprawl dilutes the effect of very high demand for transportation fuels and air conditioning.

Similarly surprising equality applies in the case of poor, densely packed, and heavily industrialized Shanghai (about 35 W/m^2) and most large European and North American cities with high per capita energy demand but also with much lower densities of dwellings and industries (typical range of 30–60 W/m^2). All of these rates have been annual means: during the coldest winter months Canadian, Swedish, and Russian cities can go up to 150 W/m^2 and their downtowns well over 500 W/m^2.

Thus far I have looked at the performance of energy converters or at the power densities of various energy-consuming facilities. A very different, yet no less revealing and no less fundamental way of appraising energy use in modern societies is by the analysis of energy costs directly and indirectly incorporated in the production of various goods. (The same approach can be extended to services, although there the usually high share of skilled human labor makes the accounts less revealing than in the case of extracted or manufactured products).

10.4 ENERGY COSTS

To identify the best options for efficiency improvements and to formulate long-range investment strategies it is necessary to go beyond sectoral appraisals. Industries with relatively high shares of energy costs in their total operating expenditures have always closely monitored their fuel and electricity consumption but other enterprises could rarely answer the simple question, How much energy does it take to make this product? This neglect was rapidly remedied following OPEC's first round of oil price increases with the birth and diffusion of energy analysis, a new branch of energy studies tracing the fuel and electricity costs of individual products and services (Chapman 1974; International Federation of Institutes for Advanced Study 1974; Verbraeck 1976; Long 1978; Thomas 1979).

Process energy analysis, rather than an approach based on input–output tables, is the preferred tool. It is basically a variant of standard econometric analysis using the square sectoral matrix of national economic activity to extract the values of direct and indirect energy inputs. But limitations of input–output approach are obvious: sectoral input aggregates do not distinguish among the kinds of fuels used, monetary values must be converted to physical energy equivalents (necessarily imprecise where inputs are mixes of fuels), and the analysis reaches only heterogeneous categories, not particular products.

In contrast, process energy analysis must first identify the sequence of physical operations required to produce a particular item, then account for all significant material and energy inputs into the process, and finally assign the best available energy equivalents to direct fuel and electricity inputs and to energy costs of raw materials. The first two parts of the exercise are not only of great heuristic value, but they are also indispensable for any successful management improvements and conservation efforts. The choice of system boundaries is the most important determinant of the utility of the whole exercise.

Limiting the analysis to direct energy inputs to the final process stage may be fairly satisfactory in the case of simple smelting of metallic ores—but it is insufficient in analyzing the energy cost of car manufacturing. To account for energies embodied in rolled steel or plastic bottles the analysis has to move to the next level, embracing the inputs needed to produce the materials used in the manufacturing process; this level also includes direct energies needed to produce and distribute the direct fuel and electricity inputs—and it could also approximate the energy costs of equipment used in manufacturing or assembly.

The third level would trace the energy costs of capital equipment used in the production of material inputs and in the provision of direct processing energies. Higher levels of regression are rarely necessary. The most sensible recommendation is to carry the exercise only to the level whose contributions are larger than the combination of inevitable measurement errors and analytical uncertainties encountered during the preceding stages. Energy analysis through the first two levels will commonly include 90–95% of all energy inputs through the fourth level.

In practice there is a need for much quasi-detective search for varied information and for many approximations. Perhaps the greatest difficulty with energy costs published since 1973 for hundreds of products is the frequent lack of clearly specified analytical boundaries, which is also the major source of often large discrepancies in overall unit values. Differences in published energy costs for apparently identical products are also caused by tracing different production processes, comparing identical processes in plants of different age and maintenance practices, and including or excluding the energy costs of salable by-products.

Boustead and Hancock (1979) gathered more than 900 energy cost

Table 10.2 Typical Energy Costs of Common Materials[a]

Material	Energy Cost (MJ/kg)	Remarks
Aluminum	227–342	Metal from bauxite
Bricks	2–5	From clay
Cement	5–9	From raw materials
Copper	60–125	Metal from ore
Glass	18–35	From sand, etc.
Hydrogen	9–26	By-product
Iron	20–25	From iron ore
Limestone	0.07–0.1	From quarries
Nickel	230–70	From ore concentrate
Oxygen	6–14	From air
Paper	25–50	From standing timber
Polyethylene	87–115	From crude oil
Polystyrene	62–108	From crude oil
Polyvinylchloride	85–107	From crude oil
Sand	0.08–0.1	Excavated
Silicon	230–235	From silica
Steam	3–4	From water
Steel	20–50	Finished from ore
Sulfuric acid	2–3	From sulfur
Titanium	900–940	From ore concentrate
Water	0.001–0.01	From streams and reservoirs
Wood	3–7	From standing timber

[a] Compiled from Verbraeck (1976), Long (1978), Boustead and Hancock (1979), and Baird et al. (1984).

values for nearly 500 different materials and Table 10.2 lists typical ranges for over 20 commodities which are either needed in the largest quantities or have come to play critical roles in the functioning of high-energy societies. After a brief review of energy costs of energy, I will take a closer look at energy costs of several commodities which have been most important in shapping the industrial civilization—iron and steel, building materials, and paper.

The large variety of fuels and of their extraction, processing, and transportation modes precludes any simple generalizations. Coals, especially if used as extracted, should have the lowest energy cost, whereas oil fuels, ready for the market only after fairly energy-intensive processing, should be most costly. A score of published figures for energy cost of U.S. and British coal shows expenditures ranging from less than 100 kJ/kg to about 4 MJ/kg, values translating to delivered efficiencies as high as 99.75% and no lower than about 83% with the U.S. mean around 98% and the British around 95%.

For the richest Middle Eastern oilfields energy invested in exploration (less than 1 kJ/kg) and production (0.5–5 kJ/kg) is negligible in comparison with energies used in shipping (1–3.5 MJ/kg including construction of tankers and oil terminals with storages) and refining (an equivalent of anywhere between 4 and 10% of crude oil input, or 1.7–4.4 MJ/kg). Chapman and Hemming's (1976) calculations for two North Sea oilfields, Auk and Forties, show the net energy requirements for exploration, appraisal, and field development at, respectively, about 840 and 230 kJ/kg, two to three orders of magnitude above the lowest Middle Eastern needs. Yet even in these demanding cases the overall energy costs of oil delivered to refinery —about 1 MJ and 500 kJ/kg—were repaid in oil produced in less than three months.

Crude oil is thus delivered to refineries with efficiencies surpassing 97 or even 99.5% and the overall efficiency for individual refined product delivered to customers may be as low as 80 and as high as about 93%; values of 83–88% may be typical for most of the products used in the 1980s. For natural gas the main postproduction cost is energy consumed in driving pipeline compressors; overall field flaring and pipeline distribution losses may surpass 5% of produced fuel and the delivered energies then may be anywhere between 85 and 98% (typically just over 90%) of the extracted gas. For secondary fuels derived from coal the costs will range between 0.70–0.88 for coke and 0.65–0.81 for manufactured gas.

Overall costs of fossil-fueled electricity generation span a rather wide range determined by the cost of fuel used (including highly variable costs of transportation), conversion efficiency (mostly 34–42% but as low as 25%), internal consumption by the plants (as low as 2%, as high as 8% with efficient fly-ash removal and FGD facilities), and transmission losses (typically 7–9%). Minimal fractions of original energy in the delivered electricity may be no higher than 0.2, and maxima can go up to 0.33. Inclusion of construction costs will decrease these fractions by less than 5%.

Energy costs of nuclear generation are clearly higher even when the analysis must remain incomplete. Production of enriched uranium fuel in gaseous diffusion plants is highly energy-intensive, equaling about 5% of the station's gross annual output during 25 years of operation, and construction costs may represent a further loss of

6% (Chapman et al. 1974). With an equivalent of 1–2% going into uranium mining and up to 15% taken by self-consumption and distribution losses, even the best nuclear stations would have an overall efficiency no higher than 0.26. But as it is difficult to factor in the energy cost of decommissioning and as there is no way to include the ultimate energy cost of long-term (10^4 years) disposal of radioactive wastes, any arguments about energetic desirability of fission generation rest on grossly incomplete assumptions.

Similar accounts of hydroelectric stations are complicated by the multipurpose function of dams and reservoirs: is it appropriate to charge the high construction and relocation costs just against electricity generation when flood control, irrigation, drinking water supply, and recreation may be equally important? In any case, consumption of energies sustaining modern civilization represents either very large (for primary fuels), large (secondary fuels), or quite satisfactory (fuel-generated electricity—especially when considering the unmatched quality and flexibility of this energy) energy gains.

Energy analyses of common materials are somewhat easier. Concentrating at iron and steel as the leading concern is obvious. Fashionable misperceptions may see contemporary civilization dependent more on information flows and low-energy electronic networks rather than on bulky material inputs, but iron is still by far the most important metal extracted from the crust: worldwide iron ore production surpassed 500 Mt in 1960 and reached 900 Mt in 1979; the 1987 total was 917 Mt.

Steel (global output in excess of 700 Mt since 1978, at 723 Mt in 1988) provides a large part of the fundamental physical infrastructure of modern civilization (railways, pipelines, buildings, bridges, roads) as well as a high-quality, relatively inexpensive, and durable raw material for manufacturing an enormous variety of products, including all of the essential machinery for energy industries. Ironmaking was always one of a few energy-conscious industries and the rapid diffusion of coke-based iron ore smelting after 1750 was accompanied by impressive improvements of energy efficiencies (Bell 1884; King 1948; Heal 1975; Hyde 1977).

Small (less than 20 m^3) late eighteenth-century blast furnaces used up to 300 MJ/kg of metal. The key subsequent innovations were the diffusion of more powerful steam-driven bellows (1780–1820), Nielsen's hot-blast process in 1829 (supplanting the use of cold air), and radical changes to blast furnace shape and size (starting with Lothian Bell in the 1840s). Later improvements included much more efficient production of higher quality coke in by-product recovery

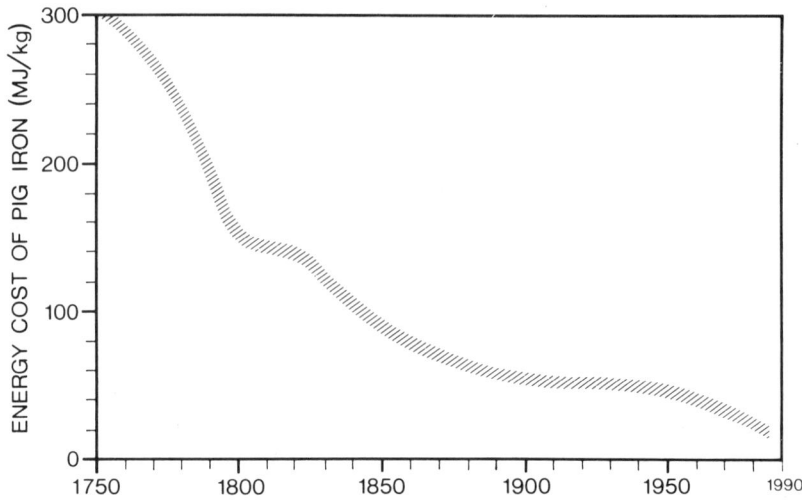

Figure 10.7 Long-term decline of energy cost of pig iron production, 1750–1985. Based on historical series of British and U.S. rates in Bell (1884), King (1948), and Heal (1975) and on data in *Iron Age*.

ovens (displacing the inefficient beehives), enrichment of charged ores, progressively higher blast temperatures, and further changes in furnace design, above all the introduction of large-hearth units after 1950.

Combination of these developments lowered typical energy costs of pig iron to less than 25 MJ/kg by the early 1970 (Fig. 10.7). The world's largest blast furnace—a 5070 m^3 unit blown-in in 1976 at the Kyushu Oita Works—producing 12,000 t of pig iron a day consumes just 14.5 MJ/kg of hot metal; its power rating of 2 GW and power density of 11.9 MW/m^2 of hearth area make it an equivalent of a large electricity-generating station. A new alternative to blast furnace iron is a variety of direct reduction processes producing solid sponge iron by chemical reduction of high-quality ores using natural gas, coal, or coke-oven gas (Davis et al. 1982). By 1985 global capacity of such plants was almost 5% of the worldwide output of pig iron.

Modern steelmaking is dominated by basic oxygen furnaces in which molten pig and scrap iron reacts with supersonic oxygen jets to lose its impurities. These furnaces need just 0.5–1.5 MJ/kg of liquid steel; old open hearths (Siemens–Martin furnaces) typically consume 4–6 MJ/kg and electric arcs consume 6.4–13 MJ/kg. In all integrated plants liquid steel is then either cast into oblong ingots or made into slabs, blooms, or billets. Energy-efficient continuous casting needs only 1.4–2 MJ/kg, traditional casting and slabbing mostly 2.5–6 MJ/kg. Overall cost of semifinished steel made from the blast furnace pig iron ranges from as little as 15 to as much as 30 MJ/kg. Similarly, direct reduction methods coupled with electric furnace steelmaking need 16–26 MJ/kg of hot metal.

The mid-1980s national averages were nearly 25 MJ/kg in the United States and about 20 MJ/kg in Japan; the global average was no less than 23 MJ/kg. Naturally, steelmaking from scrap iron in electric arc furnaces (charges may be 100% scrap) and basic oxygen furnaces (the charge is limited to about 30% scrap) is much less energy-intensive and the typical energy costs for all steel would be about 20% lower than the figures for metal made from ore, that is, mostly 15–20 MJ/kg. With a global average close to 19 MJ/kg the worldwide 1980s production of about 700 Mt of crude steel would have required about 13.5 EJ annually, an equivalent of about 5% of global primary energy production.

Reheating of ingots (1.3–3 MJ/kg), alloying, secondary rolling of billets, blooms, and slabs, precision casting, and forging of a huge variety of steel products can easily double the energy cost of finished steel: energy analyses published for finished steel products show costs of 35–65 MJ/kg. Steel products have traditionally made up by far the largest proportion of cars' mass. During the 1970s and 1980s substitution of aluminum, plastics, ceramics, and composite materials for steel and iron lightened the total weight of most vehicles (Flink 1985), but the use of these energy-intensive materials has not lowered the specific energy cost of making cars.

By the mid-1980s it still cost between 110 and 150 GJ to produce most cars, with specific rates as high as 150 MJ/kg for some small European models. A medium-sized five-passenger car weighing 1.1 t and requiring at least 120 GJ to produce would consume about 50 GJ of fuel and oil annually, so its production cost would be less than 20% of its lifetime energy cost of some 700 GJ (assuming 10 year use with some 80 GJ going toward spares, repairs, garaging, and road maintenance). With annual output at about 30 million units, the global energy cost of passenger car making in the late 1980s was close to 4 EJ.

Modern buildings contain a large assortment of finished materials resulting in a large variation of their embodied energies. North American and Japanese family houses use lumber, costing as little as 4 MJ/kg, as their main structural material; in contrast, high-rise buildings have steel skeletons (more than 35 MJ/kg) and aluminum cladding (more than 250 MJ/kg). The energy cost of baked clay bricks ranges between 4 and 8 MJ/kg, that of cement between 3 (for clinker ground with blast furnace slag and gypsum) and 9 MJ/kg (for portland cement for which the clinker is ground with gypsum); national averages show the lowest value for the United Kingdom at 6 MJ/kg, with the United States at nearly 9.5 MJ/kg.

Ready-mixed concrete, made from cement, gravel, and water, costs less than 1 MJ/kg; reinforced concrete, including 100 kg of steel bars per cubic meter, costs at least 2.5 MJ/kg. American drywalls (gypsum between paperboard) cost up to 6 MJ/kg, insulating materials generally 15–25 MJ/kg, plate glass 20–35 MJ/kg. These costs translate to gross energy requirements of 2–30 GJ/m^2 of floor area. Energy cost of residential space varies from 3 to 8 GJ/m^2 with floors and roofs being usually the largest items (Baird et al. 1984). U.S. data show that warehouses come cheap (5–7 GJ/m^2), high-rise apartments take 8–9 GJ/m^2, stores, restaurants, hotels, and industrial buildings need 10–13 GJ/m^2, and hospitals and office buildings 18–20 GJ/m^2.

These values translate to totals of about 500 GJ of energy embodied in an average three-bedroom North American bungalow, 25 TJ in a 5 story apartment building with 75 flats, and nearly 1 PJ in a 100 story skyscraper with 1000 m^2/floor. Ratios of construction to operation energy requirements for residential buildings may be as low as 3.1 years for a typical English semidetached house and as high as 9.8 for a New Zealand timber frame structure; a U.S. single-family house rates between 6 and 9, high-rises up to 14 years.

A paragraph on paper, the carrier of civilization's memories and messages, ends this section. The electronic age was to usher paper's gradual demise, but heavier advertising, demand for computer printouts, copying and packaging have helped to boost global demand for paper and paperboard from 170 Mt in 1980 to nearly 240 Mt in 1990. The cheapest unbleached papers for packaging cost less than 20 MJ/kg but good-quality writing or book paper costs over 30 MJ/kg. Most printing costs are in the vicinity of 2 MJ/m^2.

All energy costs have one thing in common: they are still substantially above the fuel and electricity inputs achievable with the best available conversion and production techniques: no survey of high-energy civilization would be complete without stressing the enormous conservation potential in every sector.

10.5 CONSERVATION

As Rose (1986) noted, the term "rational and effective energy use" is preferable to "conservation," yet I will sweep under that imprecise but established cover all appropriate approaches whose opportunities and benefits have been generously extolled (Ford et al. 1975; Lovins 1977; Socolow 1977; Gibbons and Chandler 1981; Hu 1983; Rose 1986; Gibbons and Blair 1989). The opportunities may be divided among a trio of broad strategies: doing without; maximizing conversion efficiencies; and reducing the use of energy-intensive products through better design and extensive recycling of materials.

Doing without is a much underappreciated option in affluent societies habituated to the idea of growth, but its impact is obvious when comparing the per capita energy use in the Western world of the early 1960s with that of the late 1980s: was life with 10 (the U.S. case), 30 (in Canada or West Germany), or even 40% less energy (nearly as much in France) so unbearable? Voluntary frugality, however, has been in short supply throughout the rich Western world and in the absence of acute social crises, democracies are averse to adopting proscriptive measures.

Maximization of conversion efficiencies runs into some technical limits: thermodynamic limits have been approached in various energy-intensive chemical syntheses, the best large electric motors are near-perfect converters of electricity to rotary power, and many boilers and furnaces are, in e_1 terms, more than 90% efficient. But this still leaves ubiquitous opportunities for efficiency improvements ranging from relatively modest gains whose multiplication by 10^5–10^7 units translates into huge savings to surprisingly large jumps, such as the energy savings of superinsulated houses where halving the total needs is not exceptional. Further savings can come from structural changes: in comparison with a three-bedroom, single-story house an equally sized two-story building is 15% more energy-efficient (Burchell and Listokin 1982).

Small improvements in automotive efficiency can bring massive gains as a result of gradual adjustments and widespread applications of existing techniques (Reitz 1985). Cars have been unnecessarily too powerful. Unless the driving requires unusually rapid acceleration, travel on uncommonly steep roads, or heavy towing, there is no reason to have any standard passenger cars rated over 35 kW; even the Volkswagen Golf is overrated (40 kW) and the Honda Civic is nearly twice as powerful as necessary (63 kW). Weight reductions reduce power requirements: front-wheel drive with transversely mounted engine and replacement of steel by lighter materials are the key routes to take.

Reductions of aerodynamic drag still have far to go before they become nonfunctional. Lean-burn low-friction engines, continuously variable transmissions, and greater diffusion of diesels are other ingredients of the strategy which could see national car fleets averaging below 1.75 MJ/km (5 L/100 km) before the end of the twentieth century. This would correspond to drive train efficiencies around 20%. Only such improvements would bring some sanity into North America's incredible waste of automotive fuels. In 1985 U.S. gasoline consumption of 12.76 EJ was almost exactly equal to the *total* primary energy use in Japan (13.08 EJ)—a comparison that goes a long way toward explaining the disparate economic fortunes of the two countries. In global terms, America's 1985 gasoline use accounted for nearly 5% of the total fossil fuel conversion.

Airlines are a minor fuel consumer compared to cars and trucks (even in United States jet fuel is equivalent to just about one-sixth of gasoline use) but the high share of energy in their total operation costs has led to very successful conservation efforts. Turbofan engines mounted on wide-bodied jets have brought the greatest efficiency gains (Sampl and Shank 1985), and there are considerable further opportunities to continue this commendable trend: advanced turbofans and turboprops should cut the 1985 specific fuel consumption another 25–35% by the year 2010.

Huge combined gains are also achievable with modernization of lighting. Outdoor lighting now relies on high-performance halide and sodium lamps, but substituting fluorescent lights for indoor incandescent bulbs represents a large potential for conservation of electricity. The conservation potential of electric motors—especially the AC-polyphase induction machines rated between 750 W and 100 kW, which dominate industrial applications—spans a wide range of efficiency gains. Large gains for small motors (from 68 to 85% for machines below 1 kW) and small improvements for larger machines translate into huge aggregate gains when multiplied by millions of operating units. Payback periods are just a year or two.

Utilization of waste heat offers an enormous conservation potential in power plants and industrial enterprises requiring high-temperature steam or dry heat. Cogeneration—the use of a single primary heat source to produce simultaneously electricity and thermal energy saving 10–30% of fuel in comparison with separate generation—is hardly a novel concept, but the onset of a rapid diffusion of available techniques dates only to the 1970s (Williams 1978; Hu 1983; Clark 1986).

Qualitative mismatches between sources and final uses abound in any high-energy society. The most frequently cited example is the use of fossil fuel–generated electricity for resistance heating. In general, rich societies need 20–40% of their final energy as low-temperature heat (below 100°C) and e_2 (the second-law efficiency: see the Appendix) of delivering this heat would soar with wider application of solar conversions. Repairing this mismatch has been one of the principal objectives of the soft energy path (Lovins 1977), but a simplistic maximization of thermodynamic efficiency should not be the sole guiding reason for restructuring the energy supply.

Reducing the use of energy-intensive materials is an approach applicable equally well to soft-drink cans and jumbo airliners: in 1960 the average aluminum can weighed nearly 22 g, in 1989 it was 17.7; and lighter high-thrust engines have revolutionized mass long-distance flying. But these design improvements yield limited energy savings compared to recycling of most metals and paper. Using scrap in steelmaking saves at least 10 and up to 25 MJ/kg. Typical reductions in recovering nonferrous metals from scrap as opposed to their smelting from ores are at least 250 MJ/kg for aluminum, around 50 for copper and zinc, and 30 for lead (Wilson 1979).

Paper is the only extensively reused nonmetallic material, with the highest national recycling shares (Japan, Netherlands, Germany) at or over 50%. Even higher recycle rates could be possible with generally available special collection arrangements. De-inked pulp can be made with about 60% less energy than pulp from wood (savings of at least 15 MJ/kg) and there are considerably lower environmental impacts. Glass recycling cuts a few megajoules per kilogram—but returning bottles for at least half a dozen refills is a much more rewarding option, saving up to 16 MJ/kg. Benefits from recycling energy-intensive plastics are obvious but the heterogeneity of wastes makes this task difficult. The best outlook is for recycling of mass-produced wastes such as polystyrene cups, fast-food containers, and soft drink bottles.

Combustion of municipal wastes is emerging as a particularly desirable way of material disposal not only because of its considerable energy yield (with heat content ranging from 4 MJ/kg for vegetable refuse to just over 44 MJ/kg for polyethylene) but also because of greatly reduced demand for increasingly scarce landfill sites. Precombustion sorting and flue-gas controls increase the costs but there is

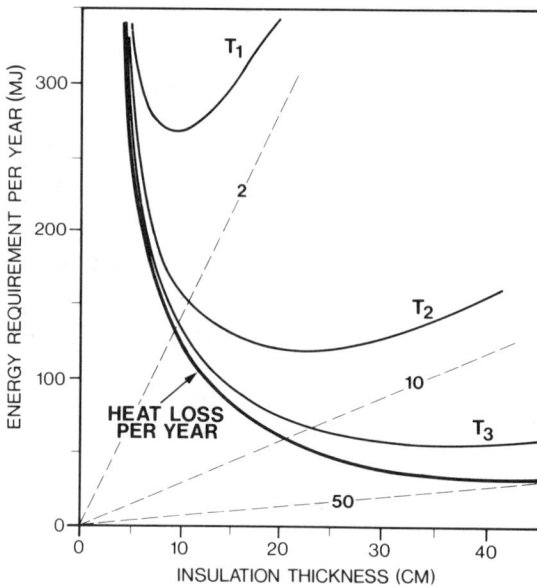

Figure 10.8 Longer amortization time justifies higher energy expenditures in conservation effort. This fundamental relationship is illustrated by a comparison of three variants of house insulation prepared by Spreng (1978). T_1–T_3 curves trace total energy requirements (for heating and for the insulation), while the dashed lines show energy invested in insulation divided by the number of years (2, 10, and 50) of energy amortization time.

no better long-term choice for managing these wastes with energy profits and environmental benefits.

Conservation springs from what Socolow (1977) labeled "inverted emphasis" in facing the energy dilemma. Instead of traditional concentration on enlarging energy supply, the inverted approach focuses on deliveries of particular energy needs and inevitably discovers huge opportunities for savings. But there is no equivalent of the spectacular giant oilfield discoveries, no dramatic single-item short-cut. No particular conservation efforts can reduce national energy use by large margins. But sustained accumulation of numerous improvements can easily add up to one-tenth, and eventually one-fifth, or even one third, of the total.

We use energy to save time but we can also use information to save energy. To take an extreme case, the profound realization of the futility and vanity of human existence leading to the contemplative life of Taoist sages, Christian and Buddhist monks, or Hindu swamis substitutes ultimate information for countless energy demands of the secular world. Less dramatically, information used by a well-educated and highly motivated population can lead to substantial energy savings through better construction, rationalized transportation, and purchases of more durable goods. Successful conservation efforts thus require prodigious numbers of informed individual decisions and long-term commitment: both can be a problem in many rich societies.

The importance of time can be impressively illustrated by a comparison of energy optima in house insulation (Fig. 10.8): longer amortization spans justify higher energy inputs and generate greater life-cycle savings. Energy conservation thus demands important socioeconomic adjustments in return for its long-term benefits. But its effects go beyond energy savings: as it reduces environmental impacts it offers rewards that cannot be matched by increased supply.

And energy conservation opportunities abound also in modern food production, which uses—directly as fuels and electricity, indirectly as machinery and agricultural chemicals—increasing amounts of fossil energies. There will be no fundamental alternatives to this dependence in the foreseeable future. Consequently, the rational management of global food production requires a solid understanding of our reliance on fossil fuel subsidies in farming.

11

ENERGETICS OF FOOD:
MODERN FARMING AND
FOOD CONSUMPTION

The great conceit of industrial man imagined that his progress in agricultural yields was due to new know-how in the use of the sun . . . and that higher efficiencies in using the energy of the sun had arrived. This is a sad hoax, for industrial man no longer eats potatoes made from solar energy; now he eats potatoes partly made of oil.

—Howard T. Odum
Environment, Power, and Society

We know of no alternative to managed photosynthetic conversion of solar energy to crop phytomass to feed large populations. Until the last decades of the nineteenth century our management consisted merely of recycling part of the captured solar flows through animate labor and the return of organic matter. These practices could be highly productive, but they required the participation of 80–95% of the adult population supplemented by child labor—and they could not remove vulnerability to famines. Only during the early decades of the twentieth century, with production of the first practical tractors and synthesis of ammonia, has Western agriculture entered an era of fundamental transformation. This was later intensified by the introduction of pesticides and high-yielding cultivars. Since the 1950s the same wave has been sweeping the poor world.

Systematic study of the energy cost of modern food production started only in the early 1970s (Heichel 1973; Pimentel et al. 1973; Leach 1975) but the subsequent wave of interest produced a large number of comprehensive reviews or detailed case studies (FAO 1977; Stout 1979; Pimentel 1980; Fluck and Baird 1980; Organization for Economic Cooperation and Development 1982a; Smil et al. 1983; Stanhill 1984; Singh 1986; Helsel 1987).

The following sections examine the basics, the evolution and the extent of energy subsidies in food production, concentrating first on field crops and animal foods and then on the integrated energy cost and output levels and efficiencies of modern farming. Processing and preparation of food cannot be neglected, for in industrial societies these activities consume more energy than farming. The chapter closes with an assessment of our surprisingly inadequate knowledge of food availability and consumption.

11.1 CROPS

Only the invention of the internal combustion engine, a powerful but light prime mover, made it possible to mechanize field farming. The first tractor manufacture was established in 1905; subsequent fundamental innovations (all American) included power takeoff in 1919, mounted-type implements in 1924, power lift in 1930, low-pressure rubber tires two years later, and hydraulic lift in 1935. Diesel tractors were introduced in 1931, liquefied petroleum gas (LPG) as a fuel a decade later (Dieffenbach and Gray 1960). In the United States the capacity of gasoline tractors surpassed that of horses before 1930; in Europe the rapid adoption came only after World War; the process is still under way throughout the poor world, which in 1985 had only about as many tractors as the United States alone, that is, roughly 20% of the global total of nearly 25 million units.

Compared to early machines (up to 450–500 g/W), modern tractors are much lighter (just 70–80 g/W for sizes over 100 kW) and larger. In North America average maximum belt power rose to 50 kW in 1985 (the largest machines rate 300 kW). In contrast, in 1985 some 40% of Chinese tractors were small two-wheel units rated at about 3.5 kW. Average annual fuel consumption of American gasoline tractors (in L/h) can be approximated by multiplying maximum power (in kW) by 0.305; diesel tractors use about 0.75 and LPG tractors 1.2 times as much. Consumption for specific field operations is a function of soil, crop, and rolling resistance, and working speed.

With diesel-fueled machines typical ranges for half a dozen common field tasks are as follows (all values in MJ/ha): moldboard plowing 600–1200, disking 200–4900, planting 80–160, ammonia application 150–300, cultivating 100–200, and grain harvesting 250–500. For gasoline-powered machines the rates are about a third higher. The trend toward reduced, or conservation, tillage resulted in a major decline of fuel needs (Unger and McCalla 1980; Gebhardt et al. 1985). The main advantages are reduced soil erosion, improved water retention, and greater flexibility of land use. Compared to conventional practices disk-and-plant tillage may need 66% less fuel, slot-planting 75% less (Phillips et al. 1980).

Disadvantages include the necessity to rely on larger applications of herbicides, increased opportunities for pest damage, and lower soil temperatures shortening the growing season. Conservation tillage is not spreading as fast as initially envisaged. Its extreme form, no-till farming, would obviously bring the greatest energy savings, but the practice is also the one most circumscribed by environmental factors (Gersmehl 1978).

To find the energy subsidies into field machinery the average weights of machines and implements used for a particular cropping cycle are prorated per hectare over a period of expected service (10–20 years). With energy cost of tractors and major implements at 70–120 MJ/kg and with most staple crops requiring 10–30 kg/ha, these indirect energy subsidies amount to 700 MJ to 4 GJ/ha. Mark-ups of 5–15% cover energy costs of maintenance and repairs. With nearly 25 million tractors averaging 30 kW and 500 hours of work, global mid-1980s fuel consumption was about 3.5 EJ, and energy cost of machinery production (10 t for tractor and associated implements, 90 MJ/kg, 15 year span) would total 1.5 EJ.

Water requirements range from 350 mm for peas to 2200 mm for bananas, with cereal grains, oil crops, legumes, and vegetables generally needing 500–800 mm during their 90–150 day growing period (Doorenbos et al. 1979). Modernizing agricultures first use simple mechanical pumps with inefficient ridge-and-furrow irrigation; later come more energy-intensive solid-set, big-gun and center-pivot sprinklers, which are now the dominant setup in the United States (Splinter 1976). Well-maintained pumps should last 20–25 years and their efficiencies may fall with age from 75% to below 40% (Paul 1983). Efficiencies of electric motors are high (89–97%), resulting in the best overall pumping performance between 65 and 70%. Internal combustion and diesel engines will deliver best overall efficiencies of, respectively, 15–18 and 20–25%.

Total volume of water to be delivered will depend on net irrigation needs (evapotranspiration plus leaching minus soil water stores plus precipitation) and on application efficiencies, which may be as high as 95% for center pivots; good field practices should average 65–75% while furrow irrigation may be only 30–40% efficient. Energies required to produce and install irrigation systems may be a significant fraction of overall lifetime costs (Batty and Keller 1980).

The extremes would be a surface irrigation without runoff return (less than 500 MJ/ha) and a solid-set sprinkler at about 6 GJ/ha; a center pivot would cost about 1 GJ/ha. Global dependence on irrigation has trebled since 1945 to over 220×10^6 ha in 1985, with 60% in Asia and 20% in China. With half of the land mechanically irrigated at an average cost of 2 GJ/ha global operating costs in the mid-1980s were about 220 PJ, and about 45 PJ went into installing and replacing the irrigation equipment annually.

Chemical fertilizers represent the largest indirect energy subsidy in nonirrigated farming. No other innovation has raised the yields more than the availability of nitrogen, phosphorus, and potassium in inorganic compounds (Engelsted 1986). The treatment of phosphate rocks by diluted sulfuric acid, producing ordinary superphosphate, started to diffuse by the 1870s. An abundant supply of synthetic nitrogen fertilizers became a possibility only in 1913 with Haber–Bosch synthesis of ammonia. The onset of large-scale fertilization was postponed until the early 1950s, when the increased planting of new cultivars could take full advantage of intensive fertilization and the use of all three macronutrients grew exponentially (Fig. 11.1).

Average global N, P, and K applications in the mid-1980s were, respectively, 45, 10, and 15 kg/ha; the rich world's means were nearly 60, 15, and 25 kg/ha; and on the most intensively farmed land they were

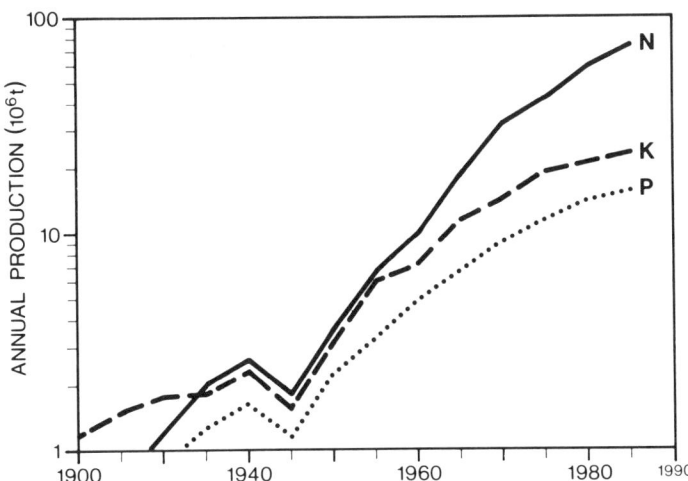

Figure 11.1 Exponential rise of global production of inorganic fertilizers. Haber-Bosch synthesis of nitrogenous compounds, mining of phosphate rock, and extraction of potassium compounds have formed an irreplaceable foundation of modern agricultural productivity. Plotted from data in Emsley and Hall (1976) and FAO's *Fertilizer Yearbook*.

as high as 550, 70, and 110 kg/ha. Energy costs of fertilizers vary widely (Blouin and Davis 1975; Mudahar and Hignett 1981; Dovring 1984; Helsel 1987). Potassium production involves either conventional shaft mining of KCl, costing as little as 4–5 MJ/kg, or solution extraction, which needs 15–20 MJ/kg. A liberal average of 10 MJ/kg would mean that K applications in the mid-1980s required about 250 MJ/ha of the rich world's farmland; poor countries use little of it.

Mining phosphate rock also costs no more than 4–5 MJ/kg, but the subsequent syntheses raise the rate to 18–20 MJ/kg for ordinary superphosphate (8–9% P) and 28–33 MJ/kg for diammonium phosphate (up to 23% P). At no less than 20 MJ/kg the global application of 10 kg P/ha would translate to about 200 MJ/ha of arable land, while the rich world's use of 15 kg at 27 MJ would average about 400 MJ/ha.

The Haber–Bosch process remains the core of ammonia production. Natural gas is now both the leading feedstock and the principal energizer of the synthesis, and technical advances (above all the introduction in the mid-1960s of centrifugal air compressors, which saved 95% of electricity used to run the reciprocating machines) have lowered the energy intensity of the best ammonia plants to around 35 MJ/kg N. Low energy cost and high nutrient content (82%) of ammonia have made it the dominant nitrogenous fertilizer in North America while urea (70–110 MJ/kg N, 45% N), requiring no special equipment to store, handle, and apply, is favored throughout the poor world.

During the mid-1980s the average global application represented an investment of 3.6 GJ/ha. Intensive fertilization in the Corn Belt or in Honshu paddies thus would cost as much as 20 GJ/ha; in Dutch fields it would cost up to 45 GJ/ha. Worldwide cost of synthesizing 75 Tg of N would be about 5.5 EJ; P and K would add, respectively, only about 350 and 250 PJ, for a grand total of about 6.1 EJ.

Pesticides are produced by energy-intensive processes from petro-chemical feedstocks but their low application rates translate to minor subsidies. Synthesis of commonly used active ingredients requires 100–200 MJ/kg and the total energy costs, including formulating, packaging, and marketing, generally are 200–300 MJ/kg; means of 220, 270, and 275 MJ/kg were offered as representative for U.S. applications (Pimentel 1980; Dovring 1984; Helsel 1987). As the usual recommended applications are 1–2 kg/ha, even the use of the most energy-intensive compounds will translate to subsidies no higher than 1 GJ/ha. Global energy investment into pesticide production was about 500 PJ in the mid-1980s.

Energy analyses of crops have much wider error margins than those of industrial goods. Field extremes are marked by wheat crop-ping, costing as little as 6–10 GJ/ha—and by establishing a vineyard for table grapes, which requires 200 GJ/ha for years before the first harvest. Typical rates are 8–15 GJ/ha for dryland cereals, 40–65 for rice, 40–90 for potatoes, 25–100 for vegetables, and 50–150 for tree crops. These rates are dwarfed by winter heating of greenhouses, requiring several terajoules per hectare. Global energy subsidies in crop farming sum up to about 12 EJ—machinery 5 EJ, fertilizers 6.1 EJ, pesticides 500 PJ, and irrigation almost 300 EJ—or roughly 8 GJ/ha of cropland, with means of 10.5 GJ/ha for the rich countries and 7.5 for the poor.

11.2 ANIMAL FOODS

High metabolic requirements of heterotrophs mean that energy sub-sidies of animal foods will have to be considerably higher than in plant agriculture. Beef production is the most energy-intensive form of animal husbandry. In North America it is a combination of inten-sive feeding and extensive use of western grazing land. Detailed analyses (Trenkle and Willham 1977; Yorks et al. 1980) showed that even beef ranching can be fairly energy-intensive. Recorded extremes are 18–130 MJ/kg of gain by weaner calves. Managing feedlots costs 3–11 MJ/kg of gain with expected economies of scale.

These subsidies are dwarfed by the cost of feed. Actual subsidies depend on the length of confined feeding (60–150 days) and the share of concentrates in the ration (30–40%). Young animals are more efficient converters of feed into protein; steers need about 13% less energy than heifers to gain 1 kg of body weight. Net energy needs during feedlot finishing of animals (400–500 kg of slaughter weight) will be 50–65 MJ/kg of daily gain, corresponding to 100–125 MJ/kg of metabolizable energy (NRC 1984). These gains would require an equivalent of 110–140 MJ/kg of gross feed energy inputs.

When produced by fairly extensive cropping, 1 kg of feed may need just 3 MJ and the feed would cost as little as 30 MJ/kg of beef. Case studies taking into account the needs of breeding herds have come up with total energy subsidies of 60–150 MJ/kg of feedlot beef; typical U.S. Midwest values are 80–110 MJ/kg of dressed meat. Throughout the rich world pork is now produced by a virtually identical method of intensive feeding in confinement, taking only six months to slaughter. Soybean meal is the dominant supplier of protein, grain corn of carbohydrates. American pigs need 75–90 MJ (5–6 kg) of concentrate feed to gain 1 kg of liveweight, an equivalent of 15–60 MJ of energy subsidies. All other requirements add up to about 10 MJ/kg of liveweight for the total of 25–70 MJ/kg.

Broilers are the most efficient converters of feed into lean meat—and also the only species whose efficiency of feed conversion has been improved appreciably (in the United States, about halving the feeding units per kilogram of gain between 1930 and 1980) with breeding (Fig. 11.2). Total costs per kilogram of liveweight broilers were calculated at 30 MJ for northern and 27 MJ for southern states of the United States and 33 MJ in England (Ostrander 1980). Egg production starts with rearing pullets to the point-of-lay, an investment worth 110–140 MJ/bird. A laying hen requires 2.5–3.5 kg feed/kg eggs (annual output is 200–250 eggs). Feed costs are at least around 230 MJ/y, with overall minima about 450 MJ/y. The U.S. mean was estimated at 525 MJ, the U.K. average at nearly 600 MJ/y.

Milk is the least energy-intensive animal food: only about 1 kg of concentrate feed is needed to produce 1 kg. Mixtures of grazing and grain feeding can translate into very different energy subsidies. Other inputs, dominated by those for milking and water heating, are usually no more than one-third of the subsidy totals. The U.S. mean is about

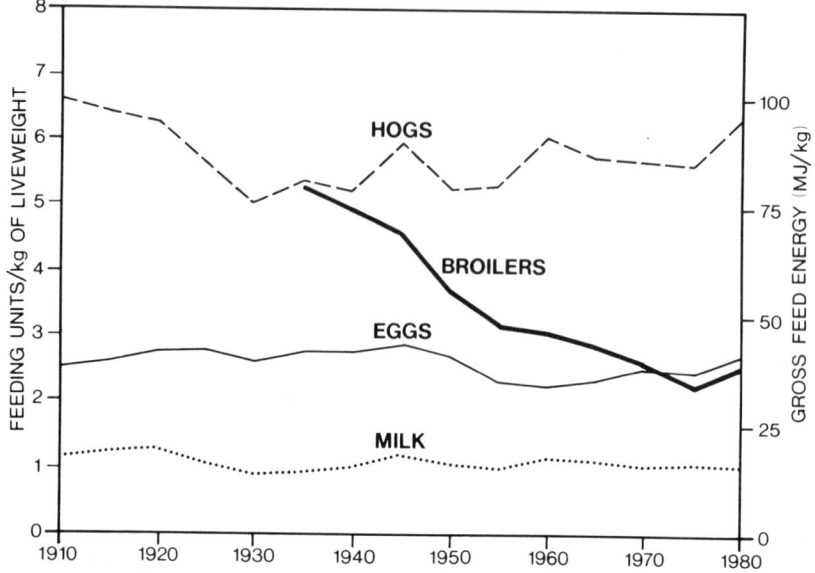

Figure 11.2 Long-term U.S. trends in conversion efficiency of feed to animal products. Broilers have been the only case of a major decline. Plotted from annual means in USDA's *Agricultural Statistics*.

6.8 MJ/kg of milk, the U.K. value 7.6 MJ/kg (Oltenacu and Allen 1980).

No single measure can serve as a clear yardstick of efficiency. In rich societies no animal food is eaten primarily for its energy content; in fact, the fat, the most energy-dense component of these foods, is frequently avoided or discarded. Complete protein is the most desirable nutrient in animal foods and in this respect young, lean birds have an obvious advantage as feed converters. But the ruminants are able to digest cellulosic phytomass in crop by-products and forages.

Gross energy in feed is converted to dressed meat with efficiencies of about 5% in cattle and roughly 10–15% in pork and poultry; these respective shares drop to less than 2% and 5% and to up to 7% for cooked edible meat. Feed energy is converted to eggs with about 12–17% and to milk with 14–18% efficiency. These differences are not perfectly mirrored in the rates of energy subsidies going into protein production because the relatively high electricity and fuel use in broiler and battery egg operations narrow the gap. While 1 g of beef protein needs at least 600 kJ of energy subsidies, pork protein can be grown with 400–500 kJ/g, chicken and egg protein with 300–

Table 11.1 Evolution of Global Population Densities, Harvests, and Energy Subsidies, 1900–1985

	1900	1925	1950	1975	1985
Population (10^9)	1.7	2.0	2.5	4.0	4.8
Harvested area (10^9 ha)[a]	1.10	1.20	1.25	1.46	1.48
Population density (people/ha)	1.50	1.70	2.00	2.75	3.25
Annual harvest (EJ)	6	9	12	25	35
Annual yield (GJ/ha)[a]	5.5	7.5	9.6	17.1	23.6
Annual subsidies (EJ)	0.1	0.5	1.5	8.0	12.0
Subsidy density (GJ/ha)	0.1	0.4	1.2	5.5	8.1

[a] The Food and Agriculture Organization and the League of Nations statistical yearbooks were used to compile the series of harvested areas and yields.

350 kJ/g, and milk protein needs just around 200 kJ/g. Energetic advantages of lactoovovegetarianism and the market price edge of chicken are obvious.

Much of the ocean fish protein costs as much as does beef or pork protein. Modern fishing is totally dependent on liquid fuels and the British skippers' rule of thumb in the 1980s—it takes 1 t of fuel to get 1 t of fish—can be a useful initial approximation: it translates into an energy subsidy of about 500 kJ/g of protein, roughly split between fuel for powering the ship and for refrigerating the catch. In contrast, Rochereau (1980) found the northeastern U.S. fishery much more efficient, averaging just about 75 kJ/g of protein.

Explanation of this excellent performance lies in the overwhelmingly small size of the fishing vessels and natural richness of the area. Rawitscher and Mayer (1977) calculated the energy subsidies for a dozen species of U.S. seafood ranging from 15 kJ/g of protein for sardines to 1.8 MJ/g for fresh shrimp. The best efficiencies in aquaculture come with polycultures based on herbivorous species, a pattern well tested by East Asian experience with different carps: as little as 150–200 kJ/g may be needed to produce 1 g of carp protein.

11.3   MODERN FARMING: GAINS, COSTS, EFFICIENCIES

The most important gain from intensifying energy subsidies has been a sharp rise in cropping productivities. Approximations in Table 11.1

tell the story. Between 1900 and 1985 the world's cultivated area grew by a third but the harvest of edible crops expanded nearly sixfold, the result of a greater than fourfold rise of average productivity made possible by an eightyfold increase of relative energy subsidies (from about 0.3 to some $25 \, mW/m^2$). In 1900 the global harvest prorated to just 10 MJ/day per capita, offering a slim margin above the average food needs of 9 MJ/day and greatly limiting the extent of animal feeding. In 1985 the daily mean of 20 MJ per capita secured high meat and dairy diets for the rich world and improved animal protein intakes in many poor countries.

Simple division shows that modern farming increased the number of people supported by cultivating 1 ha from about 1.5 in 1900 to 3.25 in 1985. But this is a comparison of two qualitatively different rates: if the overwhelmingly meatless diets of the early 1900s would still prevail, the average hectare of farmland could now support at least six people. The best modern performances are much higher. China of 1990 was feeding 11, and its most populous provinces about 17 people/ha. The average Chinese diet is still meat-poor compared to Western expectations, but its total energy content is less than 10% behind the mean of rich Japan and its protein supply is adequate (Smil 1986). Dutch farming supports about 20 people/ha and U.S. agriculture can provide for at least a 12 people/ha on a generous diet rich in animal protein.

Potential for further yield increases is best seen by comparing average national harvests with localized record yields: for U.S. staple crops the ratios range from 0.15 to 0.40; the yields of Dutch wheat are more than four times and those of Japanese rice are about twice as large as the respective global averages (Fig. 11.3). But many environmental stresses will always limit the gains realizable even with optimum cultivars and high subsidies.

Sharply lower labor requirements of subsidized agriculture have been responsible for a gain no smaller than that ascribed to the higher yields. A tonne of American wheat needed 137 hours of labor in 1800, 56 in 1880, and less than 2 hours in 1980; in just two generations, between 1940 and 1980, labor needs were reduced from 32 to 2 hours. Every rich country has experienced a massive release of farm labor—the U.S. share went from 63.7% in 1850 to 2% in 1985—and the same process is now under way throughout the poor world. The frequently quoted increase in numbers of people supported by one farm worker—from 4.7 in 1850 to 90 in 1985 in the United States—is both an underestimate and an overestimate. It is an underestimate because of considerable surpluses and exports, an

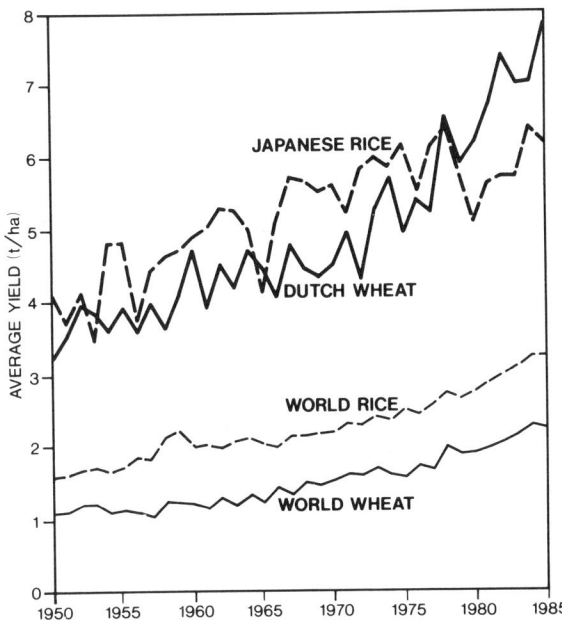

Figure 11.3 Comparison of average global yields of wheat and rice with the best national means, Dutch wheat and Japanese rice harvests. Based on annual statistics in FAO's *Production Yearbook*.

overestimate because a proper account would have to include workers employed in the production of industrial subsidies and in the provision of support services used by farmers. This tally would about double Western "agricultural" employment—but it would still be well below 10%. The universal nature of this process is illustrated by comparing support ratios of countries at different stages of intensification: China at about 4, Brazil at nearly 10, Spain at close to 20.

Comparisons of national subsidy accounts are complicated by non-uniform analytical boundaries, questions of imported feed, choice of energy equivalents for major inputs, and the selection of farmland (Stanhill 1984). Where grassland grazing contributes little food I have calculated average energy subsidy only for the cultivated area; elsewhere I have included the area of managed, improved pastures in the total. The rates range from just 2–3 GJ/ha for Australia and New Zealand to 70–80 for Israel and the Netherlands.

The most noteworthy finding is that some populous poor nations, notably China and Egypt with about 25 GJ/ha, are already subsidizing their farming with intensities matching or approaching those of intensive European agricultures (generally 25–35 GJ/ha). Although their per capita consumption of commercial fuels is still at least an order of magnitude below the rich world's mean, their dependence on nonrenewable energies in farming already rivals the advanced levels.

This counterintuitive reality is understandable in view of high cropping ratios (each hectare yields more than 1.5 crops a year in China, 2 in Egypt), extensive irrigation (50% of all farmland in China, 100% in Egypt), and intensive fertilization (above all, the applications of nitrogen fertilizers). Similarly high subsidy densities prevailed nationwide in the late 1980s in both Koreas and Taiwan and in some Central American nations (El Salvador, Barbados). On a regional scale they were encountered in the best farming areas of Asia and Latin American (Java, Punjab, and southeastern Brazil were the best examples). Other populous poor countries will certainly follow this trend of unsubstitutable dependence on fossil energies in food production.

In relative terms U.S. subsidies of some 10 GJ/ha (USDA 1980) translate into just 3% of all primary energy whereas the share of over 10% is one of China's major final energy uses. For most nations the shares are between 5 and 15% (in Egypt nearly 20%), and the dissimilarities for the countries with very similar energy use and standard of life—United Kingdom about 5%, France at 10%—largely mirror the degree of food self-sufficiency. A rough global estimate of 35 EJ in 1985 equals about one-eighth of worldwide primary energy use (excluding biomass energies). Power densities of these fluxes—mostly 30–100 mW/m^2—are low, reflecting the overwhelmingly intermittent tasks of crop farming spread over large areas.

Unfortunately, there is no suitable measure to evaluate the efficiencies of these energy subsidies and their most plausible future course. Black (1971) introduced the quotient of the harvested food energy and energy invested in the growing process as an efficiency ratio of farming systems. This approach has been used repeatedly to illustrate high returns of traditional cropping (10–30), low and declining energy gains of modern intensive agriculture (below 10), and substantial energy losses incurred by all modern animal production systems (ratios of 0.4–0.05). For critics of modern agriculture such ratios are perfect proof of the dubious course pursued by subsidized farming.

But these ratios are inappropriate, and their use should be avoided (Fluck 1979; Smil et al. 1983). To the uninformed they misleadingly suggest a direct link, a genuine conversion of input energies into food outputs. Any inference that fuel energy is converted to food is, of course, quite erroneous: the relevant energy conversion is photosynthesis and the subsidy inputs are merely removing or moderating factors limiting its potential. If higher energy subsidies result in higher productivities, then the conversion efficiency of a cropping system is clearly increasing. For example, between 1945 and 1985 the subsidy rate in U.S. corn farming tripled—but so did the fraction of solar radiation converted into harvested grain.

The second key error is the simplistic concentration on energy output: farming aims at maximizing productivity of particular cultivars but not the conversion efficiency of sunlight into phytomass. If the latter were the case, we would cultivate only C_4 species—silage corn (where the whole plant is harvested and used) in temperate regions, sugarcane (the highest phytomass producer with year-round growth) in the tropics. Plants are grown not simply for their gross energy content but rather for their unique combinations of nutrients (vitamins and minerals in fruits and vegetables), processing potentials (wheat flour rich in glutens imparting excellent dough-making properties vs. cornmeal), palatability, storability (cereals vs. tubers), and even the presence of undigestible roughage. Energy ratios can never capture these qualitative superiorities and hence they are relevant only to agricultural systems producing fuels (wood, ethanol, methanol).

The third fundamental weakness of energy ratios is their disregard for time–energy and space–energy tradeoffs. Energy subsidies not only produce higher yields but they do so while dramatically cutting all strenuous and tedious labor—and while supporting higher population densities at generally higher nutritional levels. Is it desirable to produce staples with high energy ratios but with heavy exertion involving up to 80% of the labor force, including most of the children—or with lower ratios in systems requiring mostly light work and the participation of just 2–10% of the population?

Writing about fertilizers, White–Stevens (1977) put the choice squarely: "To deny American farmers the use of artificial fertilizers ... would require the removal of 100 million Americans.... It would also become necessary to open up some 250 million acres of new lands ... to return some 20 million nonrural people back to the toil of the soil and stoop labor; to reduce the present standard of living by at least 50% and to retreat to the way of life a century ago." In mechanical terms, matching the 1985 power of American tractors with horses would require building up their stock to at least 250 million, 10 times the record total of the 1910s. Some 300×10^6 ha, *twice* the *total* U.S. arable land, would be needed to feed the animals.

In the short run there is no shortage of conservation adjustments; in the long run only the stabilization, or even decline, of global population can end the rise of energy inputs. At the same time, it must be stressed that energy inputs into farming remain a surprisingly small part of total fuel and electricity use in the food system: processing and preparation dominate the totals in every nation.

Estimates for several rich countries show energies used in food processing and distribution, wholesale and retail, to be at least twice as large as those consumed by field farming and animal husbandry, while food preparation takes 30–50% of all energies used in the country's food system. Scores of food-processing operations have been analyzed for direct and indirect energy inputs, and Singh (1986) presents and extensive review of American values. Cereal milling is the least energy-intensive mode of basic food processing. Wheat flour can be produced with as little as 1.3 MJ/kg, white rice may need no more than 1.5; pasta takes just 4–5 MJ/kg, smoked meats 4–8.

Beer requires only about 1 MJ/kg, wine around 2. Sugar refining needs as much as 35 MJ/kg, cheesemaking 12–20, and production of soft drinks in refillable bottles 20–25 MJ/kg; 20–25 MJ/L is also the range of energy needs for distillation of liquors. Breakfast cereals need in excess of 12 MJ/kg, frozen citrus juice about 20, fruit and vegetable canning takes in excess of 5, oil pressing 10–15 MJ/kg. Delivery to stores takes about a third of the 7.3 MJ/kg needed to produce and retail American white bread; baking itself claims only about a sixth.

Beech (1980) found very similar costs—6.43–7.14 MJ/kg (average 6.99)—in the United Kingdom, with directly used fuel and electricity accounting for nearly 80% and packaging for about 12.5%. He also did a detailed study of home baking, finding a total home cost of 7.8 MJ/kg, a surprisingly small difference from commercial performance. Adding the energy cost of wheat growing (usually 4–7 MJ/kg) and grain milling (at least 2 MJ/kg) would raise the overall cost of bread making to 14–17 MJ/kg.

International comparisons show annual household cooking energy totals ranging from 3.2 GJ in Sweden to 4.9 in the United States and 5.3 in France (Schipper et al. 1985). Efficiencies are no more than 15–30% of consumed electricity or gas and their doubling is possible with better utensil and range designs. Efficiencies are much lower in traditional societies burning phytomass—as little as 2% with open fires, no more than 10–15% with the best traditional stoves—but diffusion of improved stoves has been disappointingly slow (Manibog 1984).

The first refrigerators were marketed by Kelvinator Company in 1914; freezers were introduced in 1940. Refrigeration now consumes 5–10% of electricity in rich nations, and it is rapidly spreading in the cities of poorer countries. Modern retailing is overwhelmingly energized by electricity and refrigeration takes about 60% of all energy in food supermarkets; lighting and space heating/cooling take about 14% each (Doering et al. 1977). Total energy needs vary mostly between 400 and 450 W/m^2 of selling area.

Two simple question concluding the inquiry into food energetics—how much food is available for human consumption and what are the actual daily intakes—are surprisingly difficult to answer. The food and Agriculture Organization prepares annually global sets of food balance sheets offering a comprehensive review of national food supplies. But nearly 70% of all entries are estimated in the FAO's Rome headquarters and cumulative errors can be very large.

Food balance sheets of rich nations at least do not omit any major inputs. In the poor countries, however, they either miss or underestimate important local food supplies, above all the intakes of wild meat and collection of seeds, tubers, fruits, and leaves. But even the best food balance sheets do not inform about actual food consumption: this knowledge can come only from household surveys which show levels markedly below the balance sheet means (in rich countries at least 15% and up to 40%). Differences result from losses during wholesale and retail storage and transfers, kitchen waste, and leftovers thrown away or fed to domestic animals or pets.

Accurate food consumption surveys are rare. Most averages are derived indirectly from income and expenditure surveys and even the consumption studies are based mostly on unreliable 24 hour recalls of past food intakes (Block 1982). Results are often inconsistent and puzzling. Reliable answers to one of the critical questions of human energetics—how much do we eat?—remain elusive. This has a profound effect on the validity of frequent estimates of global or national prevalence of malnutrition and hunger, especially when one recalls our uncertain understanding of nutritional needs.

A combination of dubious recommendations and poor consumption estimates led the first FAO *World Food Survey* in 1946 to conclude that at least 50% of mankind was malnourished, a share lowered to 40% by the early 1960s. By 1981 the FAO estimated 435 million "seriously undernourished" people, an equivalent of about 10% of the world's population. When standard nutritional recommendations are combined with uncertain intake means, the resulting conclusions and policymaking impressions are often misleading. Energetics of human nutrition remains poorly known, making any advocacy of simple intake norms and any citations of supply and consumption estimates highly questionable.

12

ENVIRONMENTAL IMPLICATIONS: NECESSITIES AND CONSEQUENCES

Shall I not have intelligence with the earth?
Am I not partly leaves and vegetable mold myself?
—Henry David Thoreau
Walden

Environmental consequences of the preindustrial quest for energy were far from negligible. Extensive deforestation was the most obvious environmental degradation, with pollution effects limited to generally poor indoor air quality, seasonally high levels of sulfur dioxide and ash in the atmosphere of large coal-burning cities and emerging industrial centers, and discharges of urban wastes into streams.

A century of global diffusion of fossil-fueled industrial societies and subsidized farming changed both the extent and the rates of environmental intervention. During the 1960s, when concern about the state of land, air, waters, and biota emerged as one of the leading preoccupations of industrial civilization, there was no doubt that the rising rates of fossil fuel extraction, distribution, and conversion and the intensifying food production are the key causes of environmental degradation and pollution.

Consequently, much attention has been given to all prominent environmental impacts of industrial, urban, and agricultural energetics including air pollution (Stern 1976–1986), climatic change (National Academy of Sciences 1977; Budyko 1982; Titus 1986), water thermal pollution and oil spills (Esch and McFarlane 1976; Breuel 1981; Pritchard 1987), land degradation (Zachar 1982; Dregne 1983), and nuclear radiation (United Nations Scientific Committee on the Effects of Atomic Radiation 1982; Pochin 1984), as well as general appraisals of energy and the environment (Fowler 1975; National Academy of Sciences 1980a; Rand 1982; Gates 1985).

Instead of systematically surveying these well-documented particularities this chapter concentrates on three general concerns. The first covers fundamental spatial implications of modern energy conversions, the second pertains to the basic metabolism of fossil-fueled civilization, and the last focuses on the most critical environmental consequences of modern energetics, on large-scale impacts interfering with the principal biogeochemical cycles sustaining the functioning of the biosphere. But before addressing these broad topics at least a brief look is needed at an even broader set of links involving space and energy—the complex relationships between power and population densities.

12.1 POWER AND POPULATION DENSITIES

Perceptions, valuations, and utilization of space by preindustrial civilizations were in many ways fundamentally different from attitudes

Table 12.1 Evolution of Intensification of Food Provision[a]

	Energy Input (GJ/ha)	Food Harvest (GJ/ha)	Density (people/km^2)
Foraging	0.001	0.003–0.006	0.01–0.9
Pastoralism	0.01	0.03–0.05	0.8–2.7
Shifting agriculture	0.4–1.5	10–25	10–60
Traditional farming	0.5–2.0	10–35	100–950
Modern agriculture	5.0–60	29–100	800–2000

[a] Based on data in Chapters 6, 7, and 11.

and uses of modern societies. An excellent illustration of this divergence is the treatment of land as a factor of production. Classical economics, born at the beginning of industrial intensification, considered land a critical resource. In contrast, in modern economic thought "land virtually dropped out of the scene, and production was viewed essentially as a synergy of labour and capital" (Slesser 1978). Rising energy inputs, along with space-shrinking prime movers and intensification of agriculture, have been the principal reason of this shift. But nothing—save space migration—can remove constraints imposed on civilization by the Earth's finite area.

In spite of universal intensification of farming, agricultural lands occupy roughly one-third of the Earth's nonglaciated surface and arable land alone about one-tenth. Table 12.1 traces the intensification of land use from foraging societies to modern subsidized agriculture: energy inputs per unit of land rose by four (three to five) orders of magnitude, resulting in equivalent increases of harvests and population densities. These trends will have to change soon. Continuation of this century's growth rate of farming subsidies (about 6% annually from 1900 to 1985) would lead to a thirtyfold rise of energy used in agriculture in just three generations, with the new total surpassing by one-third all of today's global primary energy use.

Reductions in the rate of growth of farming subsidies are forthcoming—and in the absence of commercialized genetic manipulations leading to substantial yield increases such declines would limit further intensification of cropping. And although the world's total of potentially cultivable land is roughly twice as large as the currently cultivated area, unequal distribution means that Africa could expand its cropping nearly fivefold while the reserves are marginal in Asia (Dudal 1987).

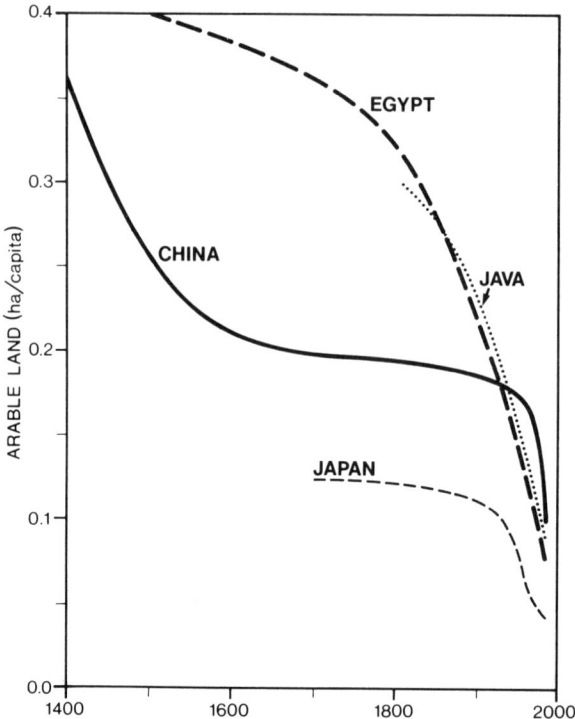

Figure 12.1 Decline of the availability of arable land in East Asia and Egypt. Plotted from data in Perkins (1969), Grigg (1974), and Waterbury (1979).

Higher harvests in the world's most populous continent (nearly three-fifths global total in 1990) thus must come from further intensification—but East Asia already has a very high level of farming subsidies, with average rates above 25 GJ/ha, more than 3 times higher than the global mean and 2.5 times above the U.S. rate. China's controversial one-child policy has its roots in the already well-discernible land constraints. These concerns are made more acute by high rates of farmland losses and by the country's susceptibility to climatic disasters (Smil 1984).

For most poor populous nations the combination of already high energy subsidies, absence of new land reserves, and substantial farmland losses means that they will not be able to support larger populations or higher quality of life within a few generations. For these nations Gamal Hamdan's (1970) succinct appraisal of Egypt's outlook—"Our density is our destiny"—is not hyperbole but encroaching reality (Fig. 12.1). Only as yet unavailable means of genetic manipulation of primary productivity could remove these inexorably approaching limits. In contrast, rich countries have enormous land reserves owing to their high consumption of animal foods: major substitutions of food crops for feed crops could greatly increase population densities supportable per hectare of arable land.

Intensification of energy subsidies in farming has been also the critical factor in the global rise of urbanization and its attendant high residential densities. Concentrated energy in fossil fuels opened the way for large-scale, centralized urban-based manufacturing, but the massive shift of rural labor to cities could get under way only with field machinery and fertilizers displacing traditional animate exertion: by 1800 only some 3% of the global population of 1.2 billion was urban; by 1900 the share was nearly 15% of 1.7 billion people—but by 1975 it topped 40% of 4 billion. This profound socioeconomic transformation has almost run its course in most rich nations but is still accelerating in many poor countries.

Settlement densities brought by this process can be orders of magnitude higher than agricultural rates—although residential densities in some urban areas are only marginally higher than the farmland densities in the most intensively cultivated agricultural areas (2500 people/km^2 in the suburbs of Los Angeles compared to nearly 2000 peasants/km^2 of arable land in Sichuan). But human packing can reach astonishing levels (Hall 1984). Tokyo averages about 12,000, Mongkok, heart of Hong Kong's Old Kowloon, almost 90,000 (Manhattan's peak working population in 1969 was about as dense), and the Wall Street area during working hours close to 250,000 people/km^2. These rates are surpassed only temporarily in East Asian public swimming pools or on Mediterranean beaches with densities over 500,000/km^2.

Maximum residential densities of around 90,000 people/km^2 translate (assuming 40 kg or 400 MJ/person) to anthropomass of 3600 g or 36 MJ/m^2. This is roughly 200 times higher density than that of large herbivorous ungulates in Africa's richest ecocystems and 3–4 times larger biomass than in all bacteria and fungi in rich farming soils. Densities around 10,000 people/km^2, common in crowded multimillion cities, represent anthropomass an order of magnitude higher than total zoomass of rich ecosystems and rival all decomposer biomass in good soils.

There are no preordained reactions to higher population densities. Crowding has no simple relationship with any mechanism or response that has been measured (Freedman 1980). Our territoriality has been elaborated to such a high degree that it is hardly recognized as such (Malmberg 1980). Territoriality of modern urban populations has no energetic foundation (all food comes from outside)—but energetic reasons alone are also insufficient to explain the preindustrial quest for a defensible territory (Lopreato 1984). Even among foraging humans food needs were only one of many determinants of territoriality.

Extreme crowding of Kowloon, Shanghai, or Cairo is not the urban norm, and given the fact that high residential densities do not necessarily produce an inordinate amount of social decay (Tokyo vs. New York) there is no way to argue that, globally, we are nearing the physical limits of urban packing. Urban densities, thanks to human behavioral plasticity, can reach locally awesome levels but large-scale maxima are, or soon will be, limited either by farmland densities or by intolerable metabolic consequences of energy conversions.

12.2 SPACE CONSTRAINTS OF MODERN ENERGY CONVERSIONS

An effective way to deal with these wide-ranging limits is to focus first on the disparities between the utilizable power densities of natural and anthropogenic energy conversions. Most of many impressive natural energy flows remain completely unusable (lightning, volcanic eruptions, avalanches, landslides) while even the best harnessing techniques can capture only small fractions of the remaining fluxes with power production densities well below 100 W/m^2; in contrast, extraction of fossil fuels and thermal generation of electricity proceed with power densities ranging mostly between 1 and 10 kW/m^2.

On the utilization side the power densities range from 20–100 W/m^2 for individual houses to peaks of over 10 MW/m^2 in the largest blast furnaces. Graphic comparison of power densities of various energy supply modes with those of final uses reveals a wealth of spatial implications (Fig. 12.2). The most important conclusions concern the spatial consequences of the eventual transition from fossil fuels to a solar civilization. These contrasts bypass the dubious cost comparisons to demonstrate the fundamental physical difficulties of such a transformation.

A nearly perfect power density overlap between small-scale solar conversions and household needs means that mature, reliable flat plate and photovoltaic techniques could cover significant portions of domestic energy needs in warmer, sunny regions without any fundamental changes of current land use. On a sunny day enough radiation could be captured even in January at 50°N to heat a

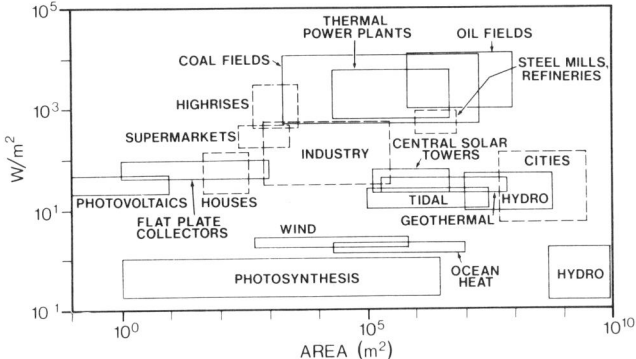

Figure 12.2 Comparison of power densities for major energy sources and uses highlights the differences—often of several orders of magnitude—and overlaps between harnessing of renewable flows and extraction of fossil fuels on one side and the final consumption uses on the other. Since consumption has been based largely on the supply of high power density fossil fuels, the future transition to arrangements dominated by low power density biospheric fluxes will require a major socioeconomic transformation.

superinsulated single-story house by an array of rooftop collectors. But heating the current stock of family dwellings solely with solar energy would be impossible without interseasonal energy storage and without major modifications of neighborhoods and energy needs.

Even ideally oriented roofs would not be sufficient to cover peak daily needs or to carry heating loads during spells of cold but overcast days; two- and three-story family houses would be especially disadvantaged, as would be many old houses with insufficient insulation. In reality, unsuitably oriented roofs limiting the top collection densities are common, as is partial or total shading by surrounding trees and buildings.

Clearly, the most efficient solar housing must be planned and built *de novo*. On the other hand, many large-roofed, single-story buildings with low energy needs (most warehouses, many offices and factories) could install flat plate collectors or photovoltaic cells to generate surplus heat or electricity for neighboring structures. But it would be impractical to have a supermarket or an energy-intensive factory solarized and impossible to attempt such a solution for highrises or a densely built-up downtown. Moreover, efficiency of active solar houses has been far lower than the expected rates: instead of supplying a major portion of winter demand, many systems have been only minor contributors (Shurcliff 1986).

Satisfying transportation needs for liquid fuels in solar civilization would be spatially even more demanding. Some 400 million cars and 100 million trucks and buses registered worldwide in 1986 consumed about 30 EJ of gasolines and diesel oils annually. Even the most productive alternative, cultivation of tropical sugarcane with bagasse used to fuel the distillation of ethanol with the net gain of about 0.3 W/m^2, would require about 25% of the world's arable land—that is, virtually all productive land in the tropics. A corn-based system (fueling machinery with ethanol, distilling with phytomass heat) would have a gain of 0.04 W/m^2, so that the U.S. motor vehicle fleet of the mid-1980s would have needed roughly five times the total U.S. farmland.

Very low power densities of all biomass energy conversions cannot be made attractive even by large efficiency improvements. Only in a few land-rich tropical countries it would be possible to energize relatively small car fleets with conversions of high-yielding sugarcane or cassava. And power densities would impose no smaller strictures on a hydrogen-dominated system. The element would have to be liberated by electrolysis (as yet there are no commercial thermo-chemical processes available), and with 75% efficiency the splitting requires about 16 MJ/m^3. Generation of the requisite electricity by renewable conversions would have space demands greatly surpassing today's claims of liquid fuels.

Nor it would be easy to supplant fossil fuels in iron and steel production. Charcoal would be the only practical substitute for coke. Even with efficient wood conversion provision of enough charcoal just for the 1980s annual output of roughly 500 Mt of pig iron would need nearly 2.5 Gm^3, or about 25 EJ, of wood. Even if all of this wood were grown in intensively cultivated, short-rotation plantations yielding about 5 t/ha, at least 300 Mha would have to be devoted to such plantings, equivalent to about 15% of the world's forests.

Large infrastructure maintenance needs in the rich countries and enormous developmental tasks throughout the poor world will keep the metal's output high. For a solar civilization the only conceivable option besides charcoal-based smelting would be to rely on as yet totally unproven direct-reduction processes using concentrated solar radiation. And it would be even more difficult to solarize the produc-tion of nitrogenous fertilizer. Haber–Bosch synthesis uses mostly natural gas as both a fuel and a feedstock; oil and coal are more cumbersome choices but no large-scale nonfossil alternatives to this technique exist today. Turning to hydrogen from electrolysis of water and to solar heat and electricity would lead to profound organiz-ational changes, relocations, and vastly increased land needs.

More accurate comparisons of fossil and renewable conversions must take into account the spatial requirements beyond extraction and conversion facilities. The controls of environmental pollution and safety precautions claim much additional land and the distribution networks intensify this demand. This is best illustrated by contrasting representative values for power densities of major components of coal-fired electricity generation, the dominant mode of electricity production worldwide.

In a 1 GW station basic plant structures (10 ha) and switchyard (2 ha) are fairly compact and coal pile and cooling towers can be designed with high throughput power densities (about 5 ha each). But even the minimal requirements for disposal of fly ash and desulfurization sludge during the plant's expected 35–50 years of operation add up to 200 ha, an area order of magnitude larger than that of all direct generation needs. Surface coal mining would claim roughly the same amont of land every year—and rights-of-way for even a relatively short transmission link (500 km, 400 kV line) would take up about 10 times as much space (roughly 1750 ha).

Consequently, some coal-fired generating station will deliver electricity with power densities below 50 W/m^2 and in terms of rated capacities their land claims will be roughly comparable with the total land requirements (including transmission) of solar power plants in such sunny regions as the Southwest of the United States. But in terms of actual generation coal-fired stations will come well ahead (their load factors are obviously much higher) and their advantages will increase with higher shares of land reclamation in surface mining (Pasqualetti and Miller 1984).

Similar differences between high-density production facilities and inherently low-density distribution systems can be illustrated by comparing hydrocarbon extraction with pipelining requirements. With average power densities of, respectively, 2 and 10 kW/m^2, U.S. oil and gas extraction occupied less than 400 km^2 in the mid-1980s while the rights-of-way for oil and gas pipelines claimed nearly 5000 km^2.

Two fundamental problems complicate these comparisons: obvious qualitative differences among the areas occupied by energy infrastructures and the duration of their lifetimes. A cleared strip of remote slopeland under a transmission line represents and intervention incomparably less intrusive than a massive coal pile or a huge sludge pond of a suburban power plant. Simple quantitative comparisons also hide the difference between the land occupied permanently and the land that is a part of the right-of-ways. In the latter case there is temporary damage to a relatively narrow corridor, after which nearly all of the land either can return to its natural state or can be used for grazing, crops, or trees.

Still, the lines do restrict land uses and do take away farmland. Esthetics aside, people cannot live under them and there may be health risks in living even close to them. For farmers the losses of cultivable land are more than twice as large for transmission towers sitting in field headlands than for those on fence rows, and up to four times as large in row-crop cultivation than in hayfields. Similarly, land claimed by surface coal mining can be returned to productive uses—or it can languish for decades before eventual rehabilitation. Coal's land claims are further complicated by the fact that unit trains share the railway right-of-ways with other freight. Is proportional allotment the best accounting solution for coal's land share?

And while nuclear power plants can be more compact than coal-fired stations—PWRs have densities between 150 and 300 MW/m^2 of core, and the fenced plant sites rate up to 2 kW/m^2—desirability of low population zones around the plants (Golding and Kasperson 1988) limits the type of land uses in their immediate vicinity while not physically claiming any of the affected land.

Consequently, any proffered global total of the land claimed by energy conversion and distribution can be only a very rough quantitative estimate hiding a host of qualitative differences. Table 12.2 is an attempt at such an approximation resulting in a total claim close to 200,000 km^2, an area roughly equivalent to half of Japan or all of the farmland in France. A much more accurate, though still far from precise assay can be made for the U.S. energy system. A set of conservative power density averages results in a total of some 40,000 km^2, or an average density of about 55 W/m^2 in terms of primary energy. Fossil fuel processing takes about as much space as thermal electricity generation (400 km^2), and these two activities together claim an area roughly equal to the total land used in extraction.

Transportation (excluding highways) takes up over 6000 km^2, and transmission rights-of-way (14,500 km^2) and hydrostation reservoirs (17,500 km^2) are by far the most space-demanding parts of the infrastructure. There are also some outstanding land debts: about 10,000 km^2 of bituminous coal wasteland were reclaimed between 1930 and 1985, but during the same period extraction disturbed about 20,000 km^2. Counting all of this land still as a part of energy infrastructure would drop the mean density to about 45 W/m^2.

The total fixed land claim of some 40,000 km^2 is about five times larger than the country's area occupied by airports and about 50% of

Table 12.2 Approximate Land Claims of Global Energy Infrastructures in 1985

Activity	Power (GW)	Typical Power Density (kW/m²)	Approximate Area (km²)
Extraction	8,300		4,000
Crude oil	3,700	4	1,000
Natural gas	1,900	7	300
Coal	2,700	1	2,700
Refineries	3,700	3	1,200
Transport	7,200		25,000
Oil pipelines	3,600	0.5	7,200
Gas pipelines	1,800	0.2	9,000
Coal transport	1,800	0.2	9,000
Electricity generation	1,100		100,000
Fossil-fueled	700	1	700
Nuclear	200	2	100
Hydro	200	0.002	100,000
Transmission	1,100	0.02	55,000
Total	8,700	0.05	185,000

the space taken up by highways and roads and 40% of the total taken up by urban areas (Barlowe 1979). These comparisons lead to two interesting conclusions. First, for each square meter of residential, commercial, or industrial areas using energies with power densities between 50 and 100 W/m^2 at least the same area, or as much as twice the area, must be taken up elsewhere by energy facilities. Second, when the areas of all places accounting for the bulk of final uses—settlements, highways, railways, and airports—are added together their total of about 200,000 km^2 is just five times the space taken up by energy infrastructures.

One could thus generalize that for highly industrialized affluent nations in the temperate zone at least one-fifth to as much as two units of space are devoted to energy infrastructures for each unit of space in urban areas and transportation corridors. These ratios would have to change with eventual transformation to a purely solar civilization. Mean power density of an affluent solar civilization dominated by direct solar conversions could not be higher than 20 W/m^2 without distribution—compared to about 55 W/m^2 for the existing arrangements including transmission rights-of-way. Satisfying current energy demands of rich nations by renewables would boost the area claimed by requisite energy infrastructures at least threefold, and securing transportation fuels and energies for metal smelting and nitrogen fertilizer synthesis would be even more demanding.

Conclusions from the study of power densities are clear. In fossil-fueled civilization we are shifting downward, producing fuels and thermal electricity with power densities one to three orders of magnitude higher than the common final-use densities, and distribution rights-of-way greatly surpass extraction and conversion needs. In solar civilization inheriting today's urban and industrial systems we would harness energies at best at the same power densities with which they would be used, but more often we would have to concentrate diffuse flows, bridging power density gaps of two to three orders of magnitude. This would increase both the fixed land requirements and transmission needs. Other important considerations are restricted locations of electricity-generating plants harnessing renewable flows and many locally inevitable conflicts with food production requirements.

But there would also be positive effects, with waste heat burden and generation of acidifying gases, objectionable particulates, and water pollutants on a general decline. In contrast, demand for water and nutrients needed for the cultivation of phytomass fuels would go up. At this time it would be highly premature to guess at the balance of the changes, but there is no doubt that the metabolism of an increasingly solar civilization would be very different from the current fluxes.

12.3   METABOLISM OF FOSSIL-FUELED CIVILIZATION

Metabolic similarities between the functioning of fossil-fueled civilization and heterotrophic life are obvious: they both oxidize carbon and hydrogen in, respectively, fuels and foods; they both require water; their conversions heat the surroundings and release wastes that may have undesirable effects on the environment; and they depend on steady material inflows—structural materials in the first case, essential nutrients in the second.

Complete combustion of 1 kg of coal carbon requires 2.67 kg of oxygen (O_2) and the burning of 1 kg of hydrocarbons needs 4 kg of O_2. Annual global combustion of 265 EJ of fossil fuels in 1985 thus consumed nearly 17 Pg of O_2 and combustion of all fossil fuels from

1850 to 1985 required roughly 500 Pg of O_2. In contrast, human metabolism claims around 15 g O_2/kg a day or roughly 1.3 Pg annually for the world's 5 billion people. But small and microscopic heterotrophs have much higher oxygen demands. Assuming 10 g of heterotrophic mass for each of 1.3×10^{14} m^2 of nonglaciated land area with an average metabolic need of 100 g O_2/kg results in annual consumption of 130 Pg. Land heterotrophs alone thus consume an order of magnitude more oxygen than does fossil-fueled civilization.

Even when adding oxygen consumed by the combustion of biomass harvested for fuel or set afire to clear the land the total annual anthropogenic diminution of biospheric oxygen stores represents just 0.002% of the element's atmospheric mass. Complete combustion of all known fossil fuel reserves would reduce the atmospheric oxygen content by less than 0.3%. And exhaustion of all coal, oil, and gas resources (a most unlikely prospect!) would lower the tropospheric O_2 by no more than 1–2%. While fuel combustion and clearing of tropical rain forests bring small oxygen reductions, mining of phosphate ores and their use as fertilizers act in the opposite way.

In any case, levels of atmospheric oxygen content have shown no significant shift from 20.95%. With a better understanding of the complexity of feedbacks controlling the oxygen content of the atmosphere (Holland 1985) we can be certain that the marginal depletion of this vital element is in no way dangerous. In contrast, decline of stratospheric ozone, largely the result of chlorofluorocarbon releases, deserves close attention.

Water's peculiar properties have inestimable consequences for planetary energetics (van Hylckama 1979). If water behaved like other fluids of a similar molecular weight, it would boil at −91°C and freeze at −100°C and its liquid range would be too far below the optima (30–40°C) for life's enzymatic energy conversions. Water's high specific heat, at 4.18 J/g·°C 67% higher than that of ethanol, contributes to heat regulation of all living organisms and especially to the maintenance of homeothermy in higher animals.

Given its strong molecular coherence water has retained high mobility: its low viscosity, 1 mPa·s, means that living organisms need not spend much additional energy in pumping their blood and that waterborne transportation has the lowest energy intensity. And water's high dielectric constant (78.5 at 25°C) makes for easy transport of nutrients in animate bodies (as ions remain separate) and releases more of the metabolized energy for growth and reproduction.

Photosynthetic water needs are very high: even the most

efficient C_4 species need 20–30 t/GJ and many C_3 plants require over 100 t/GJ. Since the edible phytomass is usually no more than half of the synthesized total, the rates for plant food generally are 60–200 t/GJ. An average of 150 t/GJ for the annual global harvest of some 40 EJ of food and feed results in consumption of at least 6000 km^3 of water. In contrast, extraction of fossil fuels needs only 2–50 kg/GJ, and crude oil refining and coal-fired electricity generation require no more than 500 kg/GJ. Moreover, while most irrigation water is actually used in evapotranspiration, most of the water used by fossil fuel industries is neither incorporated nor evaporated.

The single largest need is for cooling in thermal electricity–generating stations and in refineries where evaporative losses can be kept to very low levels. Water losses in wet cooling towers are 400–600 kg/GJ of generated electricity; in dry cooling towers water consumption may be as low as 20 kg/GJ. Water requirement for once-through cooling are huge, about 29,000 kg/GJ of electricity with a 10°C temperature rise, 58,000 kg/GJ with a maximum 5°C increase; consumptive use is practically zero, but access to large volumes of water is necessary.

The global 1985 total water consumption for fossil fuel industries was 400–525 km^3. With 450 km^3 as the most likely value, this would be equal to about two-thirds of all global industrial water needs. However, consumptive water uses in fossil fuel industries would add up to no more than 40 km^3 of water, equal to about two-thirds of annual water requirements of domestic animals, and to less than half of global domestic and municipal consumptive water needs (L'vovich 1987). Clearly, fossil fuel energetics is not highly water-intensive in comparison with photosynthetic, livestock, and direct human needs.

In contrast, the necessity to secure large amounts of material inputs to construct and maintain modern energy industries means that mineral requirements of fossil fuel industries vastly surpass those of living systems. But the comparisons are not easy. While our knowledge of essential nutrients required by people and common domestic animals is fairly complete, there can be no easy generalizations about the material requirements of fossil-fueled energetics (Albers et al. 1980; National Academy of Sciences 1980c). A Saudi well may need just 1 g of steel for every gigajoule of extracted oil, whereas a giant Gulf of Mexico production platform alone may add 10 g/GJ of produced oil.

Similarly, an open-design oil-fired power plant may need no more than 100 g of concrete per gigajoule while a nuclear power station, with its massive foundations and reactor containment structure, may require up to 450 g/GJ and large hydrostations up to 20 kg/GJ for broad-based gravity dams. A very rough global estimate of the total emplaced mass of the two key materials furnishing the extractive, conversion, and transportation (including transmission) structure of fossil-fueled civilization would be at least 500 Mt of iron and steel and up to 800 Mt of concrete. These magnitudes would translate to about 50 g of iron and 30 g of calcium per watt of produced primary energy. For comparison, human metabolic requirements are, respectively, just around 0.1 μg of iron and about 10 mg of calcium per watt of digested food.

Heat is the only inevitable product of any metabolism but heat rejection in autotrophs is merely a matter of slightly delayed reradiation of the absorbed solar flux (as is the combustion of biomass). For many heterotrophs maintenance of desirable heat balance is a taxing matter, but heat released by their bodies is, again, only a very dilute and nearly instantaneous return of solar radiation. Heat generated by combustion of fossil fuels is in a different class of delay (up to 10^8 years) and occurs so often in such concentrated ways that local heat balances are profoundly disrupted and even regional effects may be discernible (Fortak 1979; Jäger 1983).

On the planetary level primary commercial energy consumption of the 1980s (including nonfossil electricity generation) prorated to 0.017 W/m^2 of the Earth's surface. Since this heat is overwhelmingly dissipated over the continents, a more meaningful terrestrial average is 0.058 W/m^2, equal to 0.03% of the global mean insolation absorbed by the continents (180 W/m^2), a negligible increment. The other extreme of anthropogenic heat releases is marked by microelectronic chips: the most concentrated VLSI circuits have to dissipate heat at rates approaching 10^7 W/m^2, a thermal loading higher than the space shuttle experiences when reentering the atmosphere and just an order of magnitude below the flux through the solar photosphere (Oktay et al. 1986). Between the extremes is a field of manmade heat rejections extending over 12 orders of magnitude in terms of total power (Fig. 12.3).

There is a clear general relationship displayed in Fig. 12.3 and highlighted in Table 12.3: as the power density of energy conversions increases the areal extent of the more intensive heat rejections declines at a considerably faster rate. This means that those heat-rejection phenomena constituting a significant portion of solar inputs (10^1 W/m^2) are limited to areas no larger than 10^8 m^2, whereas those continuously equaling (10^2 W/m^2) or even greatly surpassing (above 10^3 W/m^2) average insolation rates are restricted to areas smaller

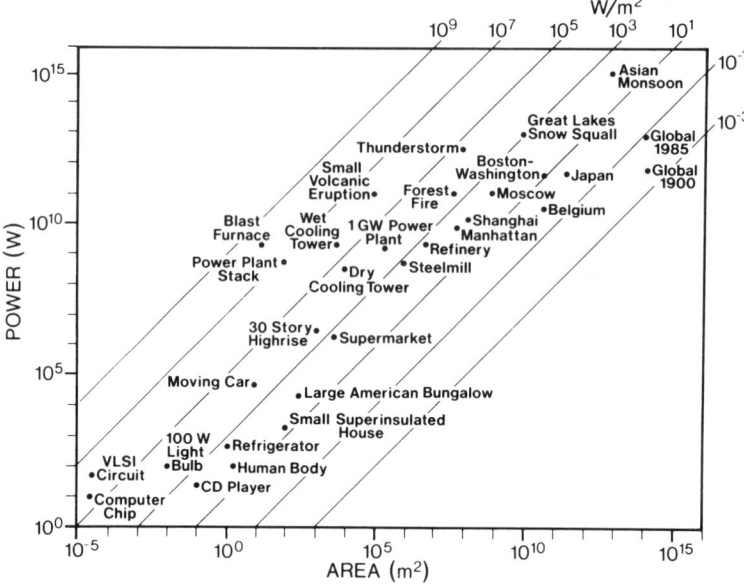

Figure 12.3 Power densities of heat rejection compared across a wide array of processes ranging from extraordinarily concentrated thermal fluxes of blast furnaces and microelectronic chips to the planetary mean arising from fossil fuel combustion.

Table 12.3 Anthropogenic Heat-Rejection Densities[a]

Heat Rejection	Power Density Magnitude (W/m^2)	Area Magnitude (m^2)
Planetary average	10^{-2}	10^{14}
Average for rich nations	10^{-1}	10^{13}
Average for densely populated rich nations	10^{0}	10^{10}
Large cities, industrial areas, transportation corridors	10^{1}	10^{8}
Downtowns of large cities	10^{2}	10^{6}
Iron and steel mills, refineries	10^{3}	10^{5}
Dry cooling towers	10^{4}	10^{3}
Wet cooling towers	10^{5}	10^{3}
Large power-plant stacks and blast furnaces	10^{6}	10^{2}

[a] From Smil (1987).

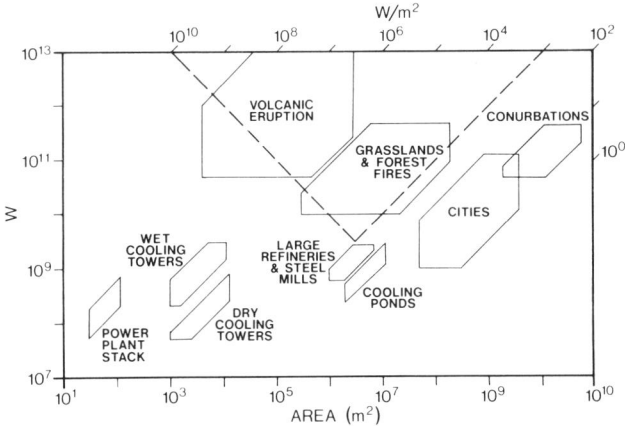

Figure 12.4 Another look at heat rejection. Power densities of human activities either are too small (conurbations, cities, cooling ponds, refineries) or, when large enough, are confined to limited areas (wet and dry cooling towers, plant stacks) so that all lie outside the zone (dash-outlined wedge) where Koenig (1979) found the heat rejections causing obvious cloudiness and precipitation anomalies.

than 10^6 m^2. Even within these relatively small areas the effects do not appear to be particularly disturbing.

Cooling towers and tall stacks will generate considerable clouds and some local fogging and icing but only infrequently will they create precipitation anomalies. Urban and industrial areas will create heat islands (with core temperatures up to 5–7°C higher and the annual means commonly 1°C above the surrounding countryside), but this modest warming does not bring persistent effects easily distinguishable from natural climatic variability.

Significantly greater cloudiness and higher rainfall in the temperate latitudes could follow only where heat rejection densities exceeded 10^3 W/m^2 over areas larger than 3×10^6 m^2 (a flux far above typical downtown rates) or when a product of smaller heat-rejection areas (in m^2) and their power production (in W) surpassed 10^{16} (Koenig 1979). Giant electricity nuclear power parks (10–50 GW in areas of 10^6–10^7 m^2) proposed during the 1970s would have been in this high-density category, somewhere between thunderstorms and squall lines. Prospects for such gargantuan projects remain remote and heat-rejection power densities of extensive conurbations, industrial facilities, cooling towers, and stacks are very much outside the undesirable region (Fig. 12.4).

Continuing growth of fossil fuel conversions or resumed expansion of nuclear power generation could substantially increase the existing heat-rejection densities before reaching an ultimate steady state. We have no way to determine these peaks but a look at a liberal allowance is revealing. Weinberg and Hammond (1970) assumed global population of 15 billion and used a purposely generous rate of 630 GJ per capita (twice the U.S. level in 1970) producing a total thermal load of 9.5×10^{21} J. At about 2 W/m^2 of continental area this is roughly 35 times the average 1985 flux but still only over 1% of mean insolation.

If nearly all of the waste heat from these conversions were dumped into streams, the runoff would be the limiting factor of global energetics. An acceptable rise of 5°C and annual global runoff of some 39,000 km^3 would accommodate heat rejection no greater than about 0.2 W/m^2, equivalent (with e_1 at 40%) to some 0.35 W/m^2 of total heat flux. But cooling towers have been easing this limitation for decades and coastal locations for heat release into the ocean would be preferable for power-generating concentrations.

With average capacities of 40 GW of electric power (that is 100 GW of thermal power), 3000 such sites would be needed and their heat rejection densities would have to be around 20 MW/m^2, a power flux akin to a relatively small, constantly burning forest fire (Fig. 12.4). These developments would bring significant modifications of local and regional climates and could also affect the position and movement of atmospheric long waves. But they would be unlikely to alter global climate beyond the range of observed natural fluctuations. All of those much more probable futures with lower populations and lower energy demand would have lesser thermal effects on the biosphere, especially with the growing success of direct solar conversions. A National Academy of Sciences (1977) study looking at energy and climatic change advocated neither panic nor complacency; this conclusion remains unchanged.

Particulates and gases emitted along with the heat are certainly more worrisome: since 1945 worldwide emissions of gases and solids have grown to such an extent that they are constituting significant fractions of natural fluxes and causing concerns about their long-term global effects. Emission factors for SO_2 are 500–2000 g/GJ in coal combustion, 200–1000 in oil burning; for nitrogen oxides 100–1000; for particulate matter up to 3000 for coal and rarely above 60 for oil. There is also a considerable mobilization of some trace elements: Pb, Zn, Cr, Mn, and Ni mostly fall between 100 and 200 mg/GJ; V from oil goes up to 2 g/GJ.

Coal ash is naturally the principal source of primary particulate matter. Uncontrolled emissions are generally between 2.5 and 3.0

g/MJ for pulverized coal combustion, but the now nearly universal diffusion of efficient electrostatic precipitators means that the annual global atmospheric input is only about 20 Mt of fly ash. Consequently, secondary particulates formed from the emissions of sulfur and nitrogen oxides are the dominant constituents of combustion-generated aerosols (Smil 1987).

Annual global releases amount to about 180 Mt of SO_2 and 40 Mt of NO_x; if two-thirds of these emissions are oxidized, annual fluxes of anthropogenic sulfates and nitrates would be, respectively, about 210 and 50 Mt. Conversions of hydrocarbons add about 20 Mt of particulates. Controls have made important inroads only in desulfurization of flue gases in new U.S. power plants (Rubin 1989). European efforts to halve SO_2 emissions by the mid-1990s will also require extensive desulfurization. Effective NO_x removal in stationary sources will be commercialized only in the 1990s.

Unburned hydrocarbons, together with nitrogen oxides, have been long recognized as the essential precursors of photochemical smog, now a frequent condition in most urban areas with large automobile concentrations. Highly effective controls (catalytic converters or modified combustion systems) have prevented further deterioration: since 1968 hydrocarbon releases from new cars sold in the United States were cut by 96% and those of NO_x by 76% (Motor Vehicle Manufacturers Association 1988).

Global production of primary and secondary particulates from combustion of fossil fuels amounted to roughly 300 Mt in the mid-1980s, less than half of the total anthropogenic generation of particulates; biomass combustion adds about 150 Mt, cropfield erosion, fertilization, and industrial processes over 200 Mt. A conservative account of natural particulate fluxes adds up to 2.6 Gt/y—and the share of combustion-generated aerosols would then be less than one-tenth of the total input (Smil 1987).

In contrast, emissions of trace heavy metals can rival and surpass natural airborne mobilization of these potentially hazardous elements. The best evidence suggests that mobilization factors (anthropogenic/natural flux) may be as high as 350 for Pb, over 20 for Zn, larger than 3 for V, and nearly 2 for Cr (Galloway et al. 1982). Fuel combustion accounts for most of the man-made releases of Pb (in leaded gasoline) and V (in fuel oils) as well as for the bulk of Hg.

Water pollution generated by fossil fuel industries has overwhelmingly localized impacts (acid drainage in coal mining, discharges of oil drilling liquids, storage tank spills) controllable by established techniques. Only very large oil spills following tanker accidents or

offshore well explosions can temporarily overwhelm natural seques-
tration and decomposition rate and cause severe damage to biota.

The worst tanker accidents thus far—*Atlantic Empress* offshore
Trinidad in July 1979 and *Castillo de Bellver* offshore South Africa in
August 1983—spilled over 250,000 t of crude, while the blow-up of
Mexican IXTOC 1 well in the Bay of Campeche spilled perhaps as
much as 1.4 Mt in 1979–1980 (*Exxon Valdez* offshore Alaska in 1989
released less than 50,000 t). Studies of such spills demonstrated
regular contamination of zooplankton and benthic invertebrates,
persistence of oil in anoxic sediments, and reduced abundance
and diversity of benthic communities (Teal and Howarth 1984).

Local or regional air and water pollution may be quite destructive
to affected ecosystems and risky to human health, but their long-term
consequences are minor compared to the effects of civilizational
metabolism on key biogeochemical cycles. Human activities always
influenced these cycles on local levels but only during this century
have these effects reached large regional or continental scales—and
in the case of the carbon cycle even global proportions.

12.4 INTERFERENCE IN GRAND BIOSPHERIC CYCLES

Most of the nutrients essential for photosynthesis and heterotrophic
metabolism do not really cycle (insoluble minerals stay put unless
moved by water or wind erosion, and soluble elements have just
one-way flows, from continents to oceans, with temporary interrup-
tions), and many are mobilized in minor quantities. The focus must
be on those elements that are introduced into the environment by
fossil-fueled civilization in large quantities and are doubly mobile—
that is, both water soluble and airborne. Only three elements are in
this class: carbon, nitrogen, and sulfur, All three must be cycled
together to sustain life and each element has a unique role in the
biota.

Carbon provides the basic matrix of life, accounting for nearly half
of the dry living mass; nitrogen's presence is essential in amino acids,
nucleic acids, enzymes, and chlorophyll; and sulfur is an indispensable building ingredient, the fortifier responsible for the three-
dimensional structure of proteins. The three elements are locked in
the lithosphere and hydrosphere as carbonates, nitrates, and sulfates,
and, to list just the principal members of airborne segments of their
respective cycles, as CO, CO_2, CH_4, N_2O, NO and NO_2, NH_3, NO_3,
SO_2, H_2S, and SO_4 in the atmosphere.

Combustion of fossil fuels has been reintroducing the long-dormant stores of carbon and sulfur into the atmosphere and generating nitrogen oxides (N in the fuel accounts for a minor part of NO_x emissions); nitrogen fertilization has affected the element's cycling on levels ranging from local to global. Anthropogenic fluxes of the three elements now form large shares of their total biospheric flows, and they are dominant in many densely industrialized or intensively farmed areas (Stumm 1977; Likens 1981; Bolin and Cook 1983; Smil 1985).

Of many possible environmental effects none is as far reaching and fascinating as the continuing increase of atmospheric CO_2 (Clark 1982; Bolin et al. 1986; Schneider 1989). This is because the atmospheric concentrations of the gas have played a major role in maintaining tropospheric temperatures within a relatively narrow range during more than 3 billion years of evolution. Kasting et al. (1988) offer a purely geochemical explanation of this remarkable homeostasis. Lovelock's (1979) Gaia hypothesis has the living organisms equilibrating an unstable composition of reactive atmospheric gases. Marine algae excreting dimethyl sulfide, the major source of condensation nuclei over remote oceans, may be critical in these feedbacks (Charlson et al. 1987).

Whatever the actual dynamics of homeostatic controls, post-1850 combustion of fossil fuels has released a large mass of CO_2 and these emissions have been the principal cause of steadily rising atmospheric concentrations. During the 1980s annual releases of carbon from fossil fuel combustion were about 5 Gt and the best estimate for the cumulative 1850–1990 total is roughly 180 Gt. Natural gas flaring, cement production, and combustion of wastes are smaller sources whose total is less than the uncertainty surrounding the net annual loss of carbon from the areal changes of ecosystems.

The best estimates put the net carbon emissions from tropical deforestation and conversions of forests, grasslands, and wetlands to fields at no less than 1 Gt and up to 2 Gt. Cumulative 1850–1990 carbon releases from ecosystemic changes caused by pioneer agriculture and deforestation were around 150 Gt. The grand total of all anthropogenic releases for the period 1850–1990 thus is about 330 Gt C, which would have increased atmospheric concentration by about 155 ppm. In reality, only about 40% of CO_2 input has been retained in the atmosphere. The rest has been sequestered in the ocean, whose mixed layer is involved in incessant equilibrating exchange with the atmosphere.

Air bubbles in ice cores show preindustrial levels of CO_2 at about

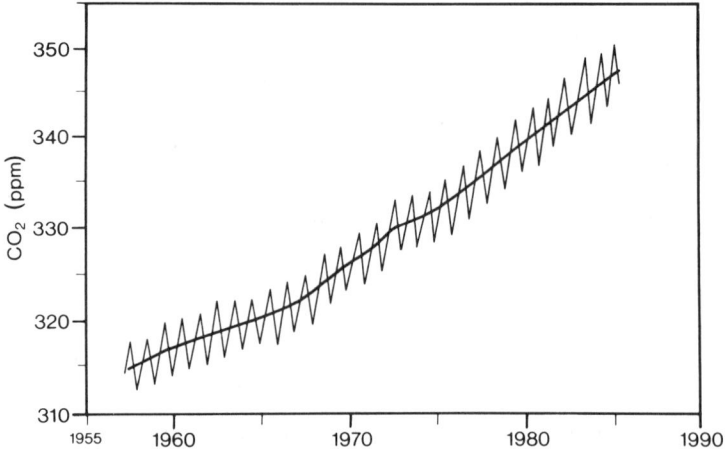

Figure 12.5 Rising levels of tropospheric carbon dioxide recorded since 1958 at the Mauna Loa monitoring station in Hawaii. Plotted from data in Keeling et al. (1982) and in the annual reports of the program for Global Monitoring for Climatic Change.

270 ppm, rising to around 300 ppm by 1920. Accurate monitoring started only in the late 1950s with the setting up of Mauna Loa and South Pole observatories; Barrow (Alaska) and Tutuila (American Samoa) were added later to complete the quartet of observatories of Geophysical Monitoring for Climatic Change. Mauna Loa records show an increase from an average of 315.66 ppm in 1959 to 346.73 ppm in 1986 (Fig. 12.5). There is the expected undulation reflecting the biospheric breath (late summer minima in the Northern Hemisphere) as well as the unexplained fluctuations (from just 0.5 to over 2 ppm) of annual increases.

Tyndall's first clear description of a CO_2-induced "greenhouse effect" in 1861 was followed by frequent but temporary revivals of concern about tropospheric warming; attention to this phenomenon was stimulated by the sudden post-1973 interest in energy affairs as well as by indications of a new global warming following the decades of post-1940 cooling. Thus far we have been unable to detect any incontrovertible sign of CO_2-induced planetary warming. Even in the midst of an unmistakable future warming trend, we may not be able to disentangle the effect ascribable to the tropospheric buildup of the gas from those of natural climatic fluctuations—or from the influence of other "greenhouse" gases, most notably CH_4, N_2O, and chlorofluorocarbons.

Worries about the rise of atmospheric CO_2 are based on the well-known physical properties of the gas (absorbing outgoing IR in several bands, most extensively at 12–15 μm) and on the results of increasingly complex but still relatively simplistic global climatic models. The consensus of these models is that the doubling of pre-industrial CO_2 levels to about 600 ppm will raise annual surface temperatures about 2–4°C; that this warming will be two to three times more pronounced in higher latitudes than in the tropics and greater in the Arctic than in the Antarctic; that it will be accompanied by stratospheric cooling; that it should bring general intensification of the global water cycle; and that it will cause some important but difficult to pinpoint shifts in frequency and amount of average precipitation, changes with considerable potential for economic disruption.

Shifted precipitation patterns would create regional winners and losers but CO_2 doubling would have a highly positive effect on the life of plants. Physiological expectations, a wealth of experimental data, and practical experience in CO_2-enriched greenhouses point to higher photosynthetic rates, higher dry weight per unit leaf area, better germination, earlier flowering, faster maturity, higher tolerance of some pollutants, higher rates of nitrogen fixation, and, most notably, lower transpiration losses and hence more efficient use of water (Lemon 1983).

For fertilized crops these desirable changes translate into yields 20–45% higher and as much as doubled efficiency of water use with CO_2 doubling. These beneficial effects are more marked for C_3 species, including most major food and feed crops, and less pronounced for C_4 plants, a group containing most of the world's principal weeds. The response of many natural ecosystems stressed by nutrient deficiencies and seasonal droughts will be much weaker than in well-fertilized and watered crops, but there is little doubt about the overall global increase of net photosynthesis.

Uncertainties concerning the future rates of fossil fuel combustion and atmospheric CO_2 retention open up a wide time span for eventual doubling of preindustrial CO_2 levels. The prevailing consensus of the 1970s—primary energy demand going up more than 4% a year and airborne fraction close to 0.6—would have brought the doubling before 2030; continuation of the 1980s growth rate of 2% and retention rate of 0.4 moves that date close to 2080. A process unfolding over some four generations should not precipitate any panicky reactions—even if the worst-case scenarios of today's imperfect modeling are correct.

There are no practical emission controls to prevent CO_2 releases.

Nor is it very likely that the United State, Soviet Union, and China, the world's three largest fossil fuel users, will rapidly and drastically limit their combustion to prevent unpredictable changes arriving at unknown dates sometime in the next century. Perhaps the best way to face the CO_2 challenge (calling it a "problem" brings an immediate bias: for most of the plants it will be a boon) is to see it as an ingredient of a complex process of climatic change. Our prehistorical ancestors were able to survive the changes which, considering the technical capabilities at their disposal, posed a much greater threat to their survival than the anticipated global warming poses to ours.

More recent historical evidence indicates on the one hand some striking synchronicity in the long-term fluctuations of temperature, crop yield, and population totals during the preindustrial era (Galloway 1986)—but on the other hand little demonstrable importance of climatic change as an agent of change in European economic advancement (Anderson 1981). Rate of change may be of greater importance than the eventual magnitude of climatic shift—but we are much better equipped than any previous civilization to cope with even drastic and rapid changes.

Even the ultimate long-term complication of a new wave of warming, the rise of the ocean level by 1–4 m, poses no catastrophic threats to global civilization determined to find efffective solutions. And there may be a most valuable benefit of this challenge: the unprecedented magnitude of these tasks calls for such a large-scale, long-term commitment of resources that these adjustments could be an important factor in moving away from militarization of modern societies.

Human interference with the nitrogen cycle also has a global component owing to the increase of atmospheric levels of N_2O—but that process is much slower than CO_2 buildup. Consequently, concerns about anthropogenic nitrogen will continue to have overwhelmingly local or regional dimensions. Nitrogen cycling is vigorous, relatively rapid, and extraordinarily complicated (Smil 1985). Human interference in the global nitrogen cycle took off only with the rapid expansion of synthetic nitrogen fertilizer production after 1950. Worldwide production of fertilizers (about 90 Tg N) may be currently equal to as much as 50% of natural terrestrial fixation.

More intensive fertilization will increase the rates of denitrification: the best available estimates for different agroecosystems show annual losses ranging from 1 to nearly 200 kg N/ha (Frissel and Kolenbrander 1977), representing less than five or as much as 35% of the applied nutrient. Higher denitrification should be increasing atmospheric levels of N_2O, leading to concerns about long-term effects on stratospheric O_3 as well as on planetary warming. Actual rates of increase have been only about 0.33% a year, or doubling roughly every two centuries.

Earlier concerns about nitrates in water—bringing the risks of infant methemoglobinemia and severe eutrophication of lakes and ponds—turned out to be exaggerated (Aldrich 1980; Organization for Economic Cooperation and Development 1982). Volatilization and leaching of nitrogen compounds can be greatly reduced by proper applications of ammoniacal fertilizers and manures and good agronomic practices; concentrated releases of nitrogen compounds in urban sewage and in feedlot wastes also have effective technical solutions.

Global emissions of nitrogen oxides from combustion of fossil fuels reached about 20 Tg N from stationary combustion and about 8 Tg N from transportation; chemical and other industries add at least 2 Tg N and vegetation burning in shifting cultivation releases about 5 Tg N. Anthropogenic aggregate of some 35 Tg N a year is thus a significant share of biogenic NO_x emissions estimated at anywhere from 10 to 90 Tg N. Anthropogenic NO_x emissions are concentrated in northern temperate latitudes where they have been one of the major causes of photochemical smog and one of the principal contributors to acidification.

Unlike the declining or stagnating emissions of SO_x, generation of NO_x is still increasing. Automotive emissions can be controlled by treatment of exhaust gases (U.S. choice of catalytical conversion) or by modification of the combustion process (Japanese way). The second approach is clearly superior, lowering the emission rates from precontrol levels of 2.5 g NO_2/km to as little as 0.13 g NO_2/km. Modified combustion is also the best choice for stationary NO_x controls, but achieving reductions of over 80% will require the still far from fully commercial flue gas treatment.

Extensive intervention in nitrogen cycling will remain unavoidable. Widespread and intensifying fertilization has no practical alternatives in the world of more than 5 billion people, a number increasing at nearly 2% a year. The only major reduction could come from lower meat consumption in the rich countries, a shift that would dispense with cultivation of feedgrains now dominating Western farming. Reducing stationary NO_x emissions will be a long-drawn process, as will be the effective controls of automotive emissions in the still growing world car market.

Interference in the sulfur cycle, largely in the form of acid deposition, is even more amenable to effective controls. This degradative syndrome was first noted in Scandinavia in the late 1960s and since the early 1970s it has been one of the most studied environmental alterations (Organization for Economic Cooperation and Development 1977; Swedish Ministry of Agriculture 1982; National Academy of Sciences 1983, 1986). Anthropogenic flux appears to be comparable to all biogenic releases and equivalent to between 33 and 40% of the element's total annual mobilization.

Combustion of fossil fuels accounts for over 80% of the anthropogenic flux, with coal contributing 75% of this total; nonferrous smelting accounts for most of the rest. With both combustion and smelting heavily concentrated in the temperate zone of the Northern Hemisphere, about 95% of all anthropogenic sulfur enters the atmosphere at 30–55°N; in areas of high power plant concentration within this zone SO_2 emissions overwhelm the local biogenic flux; for example, Czechoslovakia generates about 9 g SO_2/m^2 and Ohio over 10 g SO_2/m^2.

With post-1950 conversions to liquid and gaseous fuels SO_2 concentrations in urban areas of rich countries declined from levels in excess of 200 $\mu g/m^3$ to below 50 $\mu g/m^3$ but larger coal-fired power plants and taller stacks brought a new problem of long-range pollution as the photochemical and, above all, the wet-phase oxidation of the emitted SO_2 produce sulfates, the principal cause of increased precipitation acidity. Limited atmospheric residence time of sulfates (on the average no more than two to four days) and the abundant presence of neutralizing windswept alkaline matter over many continental areas preclude an emergence of acid deposition on a global scale.

But the residence time is long enough to affect areas up to about 1500 km downwind from large sources: eastern North America and western and central Europe are the regions experiencing the highest degree of acidification. Precipitation pH in these areas is commonly below 5—rain acidified by the natural presence of CO_2 has pH 5.6—and the lowest recorded values were less than pH 3. In some areas nitrates are a major, or even seasonally dominant, cause of acidity.

Initial concerns about acidification focused on the changes in sensitive lakes. Reproductive failure, higher morbidity, and eventual disappearance of some fish species can be ascribed to gradual decline of precipitation pH and to the high-acidity shocks accompanying spring meltdown, which are also highly injurious to amphibians. Phytoplankton makeup also changes considerably, but the phytomass total may be little affected. Liming of sensitive lakes has been an effective control technique. With acidification of groundwaters used for drinking, increased sulfate loading is less dangerous than elevated metal content arising from soil leaching (plenty of Al) and pipe corrosion (lots of Cu).

Effects on soil are beyond any simple generalizations as natural soil formation processes are often more important agents of acidification than is the deposition of sulfates (Krug and Frink 1983). Acidifying effects in farm soils are easily counteracted by liming. Crops are rather resistant to acid deposition—but forests are clearly at risk. Weakened growth and dieback, especially of coniferous species, have been noted both in eastern North America and in western and central Europe. But the injury process is clearly multifactorial (Hinrichsen 1986). The syndrome may include extreme weather, ozone, ammonia deposition, and heavy metals among its likely causes.

In contrast with uncontrollable CO_2 and difficult-to-control releases of various nitrogen compounds, SO_2 emissions can be minimized by fuel cleaning and can be effectively removed after or during combustion. Intensive mechanical coal cleaning can remove up to 80% of all inorganic sulfur but only the much more expensive chemical treatment could remove the organic sulfur. Wet alkali scrubbing with limestone or lime slurry has been the technique of choice for flue gas desulfurization; it can remove in excess of 90% of all SO_2 but it increases both the plant's capital cost and operating expenditures by 20–35%. The better way to go is fluidized bed combustion, a technique that is finally reaching the level of large-scale commercialization.

Scales, effects, and solutions of the three interventions are very different. Releases of CO_2 are the uncontrollable quintessence of fossil fuel combustion, the gas promptly becomes a global presence, and its tropospheric buildup carries a potential for significant global radiation changes and for planetary warming at an unprecedented rate: the future of fossil-fueled civilization may be influenced heavily by the need to moderate CO_2 emissions and to manage the effects of biospheric change. Only a minor part of nitrogen releases has a potential for global effects, most of them can be greatly reduced, and the effects can be confined to local or regional changes. Regional acidification caused by sulfate deposition is controllable, albeit at a substantial cost, by existing techniques—and it will be even more manageable in the future through fluidized bed combustion.

Controlling the interference in the three cycles has become an inescapable task for modern energy conversion. Available approaches range from mature and effective techniques to unprecedented attempts at international cooperation. Success of these efforts will be among the key determinants of maintaining habitable global environment—and hence also supporting viable economies and improving quality of life. These concerns are discussed in the last thematic chapter of this book.

13

ENERGETIC CORRELATES: COMPLEXITIES OF HIGH-ENERGY CIVILIZATION

Energy itself is not the problem, but rather appears as an instrument for social purpose misunderstood and misused.

—David J. Rose
Energy—More than Technology

Nowhere are these misuses clearer than in valuing energy as the measure of all things, in seeing entropic decay as the ultimate arbiter of all our actions. Taken to its limits, this position would result in complete inaction, for there is no way to complexify a civilization without increasing entropy. More practically, this attitude leads to preaching of austere ethics of simplicity and frugality in a new solar society. Besides looking at this extreme thermodynamic economics this chapter assesses more modest attempts at energy theory of value and reviews the evidence for closeness of links between energy use and economic performance and energy levels and quality of life before offering some generalizations concerning energy futures.

13.1 ENERGY AND VALUE

By far the most radical approach to valuing human affairs in energy terms is to put thermodynamics in command of economics. Georgescu-Roegen (1971) was the most passionate proponent of this radical shift. He called the second law of thermodynamics "the most economic of all physical laws" and argued that civilization's foremost goal should be to minimize entropic degradations. This led him to formulate the fourth law of thermodynamics—"In a closed system the material entropy must ultimately reach a maximum"—whose consequence is the elevation of accessible low entropy materials to "the most critical element from the bioeconomic viewpoint." The post–fossil fuel world will have the same share of solar energy—but not the same access to low-entropy materials.

Inevitably, this leads to a rejection of steady-state civilization (Daly 1973), a concept promoted in response to concerns about continuation of growth (Meadows et al. 1972). Such a civilization implies no clear limits on global population and standard of living, while the only thermodynamically acceptable population total is the one supportable by organic agriculture alone, a goal implying sizable reduction of the late-twentieth-century numbers. The most desirable prospect is thus not a steady society but a declining one.

Of course, Georgescu-Roegen is quite correct in principle. Unless our species eventually leaves this planet, the only possible strategy to maximize the duration of its terrestrial tenure is to minimize the entropic drift, a strategy that may require a gradually declining population in order to channel the finite amount of usable solar radiation into increasingly more energy-intensive procurement of materials. And yet the worship of entropy misses some key points.

On the most fundamental level Brooks and Wiley (1986) argue that life's evolutionary entropic behavior is not determined by the essentially unlimited energy flows.

These flows cannot explain the existence of organisms, their variability, or their structure. Changes in the organization of life "have their bases in mutation and there is no link between mutation and energy flow that is analogous to the role of energy flow in the organization of nonliving physical systems. It is an organism's intrinsic properties that determine how energy will flow, not the opposite." (Brooks and Wiley 1966) Organisms use energy in a relatively stochastic manner and their epigenetic information channels energy into maintenance, growth, differentiation, and reproduction; these irreversible, noncyclic, hierarchical transformations dissipate both matter and energy. Evolution is thus inevitably entropic but availability of energy is not its guiding force.

Layzer (1988) extends the caveat to material flows as well: "Of course, free energy and building blocks must be constantly supplied. But it would be misleading to regard the flow of free energy or of molecular building blocks as driving the evolutionary process. On the contrary, the ability of living organisms to mobilize free energy and organize matter is an evolutionary adaptation—a consequence of the reproductive instability of genetic material." While we can identify a large number of interactive processes which cause and direct evolution we still know virtually nothing about most of them (Endler and McLellan 1988).

Moreover, Georgescu-Roegen's entropic concerns extend far beyond any rational planning horizon, that is no further away than half a century. During that time we are much more likely to run into a variety of environmental constraints rather than to start running out of low-entropy energies and materials. Our inability to forecast even a decade ahead, indeed our incapacity to comprehend the existing wholes in their complexity, relegates any quarrels about the shape of very distant future to a category of (sometimes interesting) fairy tales.

Less radical attempts at energy valuation aim at substituting monetary appraisals which are either questionable or inadequate: exhaustion of nonrenewable resources, destruction of public goods, or valuations of biospheric services either have been beyond the realm of pricing or their valuation has neglected both the finite nature and low entropy of these resources. Yet all of all these processes involve measurable energy transformations and incur definable energy costs.

Not surprisingly, the energy standard of value had the appeal of rigor, universality, and rationality to those students of civilization who approached economics via natural sciences or engineering (Ostwald 1909; Soddy 1926; Technocracy 1937; Cottrell 1955; Odum 1971) and to some proponents of energy analysis during the 1970s. The link with ecological thinking has been a strong part of this tendency: by setting energy as the standard of value the society would realign itself with natural systems where energy's importance could be never in doubt.

Most advocates of the energy standard were content with assessing energy costs of goods and services. Howard T. Odum thought these attempts fundamentally mistaken because they excluded solar radiation as the initial universal input and did not convert different energy flows to equivalent energy costs expressed in the same form of energy. He proposed the use of energy quality (concentration) factors—solar "transformities" expressing "EMERGY" (embodied energy) in solar "emjoules" per joule—and calculated their values for a variety of biospheric and anthropogenic inputs (Odum 1975, 1984, 1988).

On the most general level it would be enough to cite Rose's (1986) dismissal of these efforts: "All these single-item theories of value—Marx's labor theory, or pure capital theories, for example—suffer from the disease of selective inattention to the interconnectedness of things and to the complexity of civilizations." Quite so: no single-variable valuation can serve satisfactorily, and energy is no exception. The first class of problems arises from the professed uniqueness of the variable. Costanza (1980) admits that energy valuation could have parallels in capital, labor, or government service theories of value but when looking to physical reality to determine which factors are inputs and which are internal transaction, "no one would seriously suggest that labor creates sunlight."

True, but sunlight did not create the Earth's crust and no one would seriously suggest that life could go on without relying on the crustal elements for both its fundamental functions and structures. Nor is solar energy the prime mover of constantly proceeding crustal rearrangements driven by the planet's internal heat—yet it is undeniable that the evolution of life has been critically affected by grand geotectonic processes. And surely time cannot be treated as a derivative of energy. Western civilization sees it as an obviously scarce entity and its value very often takes clear precedence over the levels and efficiencies of energy use (Spreng 1978). If time were of little or no value, then we could rely on low-power conversions, which would operate with superior efficiencies.

But even the gurus of the soft-energy path spread their visions at far-flung meetings and lectures to which they fly—rather than going by train or, best of all, by foot. They, too, value time and hence prefer high-speed transportation, which must necessarily forfeit the possibility of maximized efficiencies. Such tradeoffs are ubiquitous in modern society and only some of them are patently wasteful (i.e., very large power requirements needed solely for a few seconds of unnecessarily fast acceleration in many cars). Most of them have come to define the very fabric of modern society, although we may no longer think of them in that way: a refrigerator is primarily a time-saving device, as is a lightbulb. Conversely, this means that we could use time to save energy by slowing down the pace of life.

No valuation is just a matter of supply. In the real world there are always many relative scarcities and only demand can accomplish their effective valuation. By totally excluding demand considerations, energy valuations could be maximizing net energies but distorting markets and investment and resource allocation (Huettner 1976). Markets driven by demand and supply interplays also stimulate innovation, which can lead to major declines in embodied energy content of goods and services. Although the quest for thermodynamic thrift was not the reason for these changes, resulting energy savings may greatly surpass any gains achievable by efforts aimed solely at minimization of energy inputs. Many impressive examples fall into this category, from the development of electronic computers to the deployment of highly efficient turbofans in commercial aviation.

Actual execution of net energy assessments encounters the two most frustrating problems in its choice of boundaries and conversion to common denominators. Boundary dilemmas are very well illustrated by trying to decide between two common items, a cotton and a polyester shirt. Standard process energy analysis shows cotton lint costing three times as much primary energy as polyester fiber (van Winkle et al. 1978). But a cotton shirt's lower production energy cost is lost in the wearing: a lifetime comparison (including all washing, drying, and ironing) puts the cost of the polyester shirt about 35% lower. But cotton cultivation also yields cottonseed, the source of valuable oil, and if this oil's energy content is credited against the cost of the lint, the cotton shirt's energy cost will be only marginally higher than that of the polyester product.

The small remaining difference may be easily outweighed by generally superior wearing qualities of cotton and by the renewability of the natural fiber. Yet we may turn once again and point out that erosion in an average Texas cotton field removes annually about three times as much topsoil as is compatible with sustainable farming, that in Egypt cotton displaces food crops while the country imports bread flour, that the crop's irrigation in arid areas contributes heavily to soil salinization, and that residues from high doses of herbicides used to keep the crop weed-free pollute the local environment.

Decisions about where to stop the analysis have no acceptable universal solutions. Indeed, it may be argued that, like so many phenomena around us, energy flows have a fractal structure (Mandelbrot 1977): there may be no true net energy. Consequently, the crucial dilemma of the net energy concept "is not that we cannot tell when our estimates are close to the true value, but that there may be nothing to approximate in the first place" (Reaven 1984).

As long as the approximations are attempted, the most vexing boundary problem, closely tied with the conversion to a common denominator, is the matter of solar energy. Excluding the sunlight from net energy analyses is indefensible: sunlight will always be a vastly more important input than any combination of fuels and it should count. Yet to make it count in coherent ways is impossible. Odum's "EMERGY" of a flow or storage is the solar energy required to generate that flow or storage and solar transformity is the solar "EMERGY" per unit energy. But does the fact that in Odum's hierarchy items with longer turnover times have higher "transformities" imply that we should value oak trees up to 10^7 higher than bacteria? If not, what does the huge difference tell us? And what about the indisputable fact that bacteria rather than oak trees are indispensable in sustaining life-maintaining biospheric cycles?

What is the amount of solar energy needed to generate coal, oil, or gas? Is it just the solar radiation that energized the synthesis of original phytomass, or is it the radiation that energized the erosion and sedimentation processes which buried and transformed the phytomass? If the latter is the choice, we have to deal with fluxes acting intermittently over periods of 10^6–10^8 years. What portions of those flows—and over what areas—will be attributed to fuel formation? How will we include the prorated fractions of tectonic energies which contributed to the formation of fuel deposits? And what could be the solar equivalent of radioactive decay in the crust which powers these tectonic processes—and has nothing to do with the Sun?

Temporal boundaries challenge no less than spatial. How are we to assess real net energies of conversions leading, years to centuries in the future, to severe erosion, decline of water quality, climatic change, or radioactive contamination? In all such cases energy valuation offers no solutions superior to monetary estimates; multiple assumptions and arbitrary effect and time limits must be used in both analyses. And even should we somehow succeed in properly transforming energies to a common denominator, what is the utility of our findings? After all, techniques with lower net energy yields may be preferable if they provide more jobs, cause less environmental damage, are perceived as less risky, or find easier political or moral acceptance.

Original intents of energy valuations are laudable but their execution remains highly unsatisfactory. One does not have to extoll the standard economic analyses in order to concur with Mirowski's (1988) conclusion that the energetics movement brought more heat than light to our understandings of economic and social values. I agree with Reaven's (1984) view that all such energy valuations are subject to fundamental methodological problems typical of the social sciences where "generalizations that stand up are hard to come by, and jargon is not, and where theories tend to be trivial or obviously false." Even such a fundamental entity as energy cannot be an adequate surrogate for valuing space, raw materials, labor, ideas, social order, cultural riches, and morality.

Consequently, the next two sections on energy and the economy and energy and quality of life approach these inquiries from many different angles and with no preconceptions of grand underlying links, rules, and consequences.

13.2 ENERGY AND THE ECONOMY

The modern preoccupation with economic growth would seem to be the perfect foundation for extensive studies of the links between energy and economy, but until the 1970s such endeavors were surprisingly scarce. Generations of abundant energy supply and declining real fossil fuel and electricity prices are the best explanation of this widespread neglect. Post-1973 attention has been dominated by ephemeral microeconomic and pricing concerns and general treatments are still rather rare (Slesser 1978; Sonenblum 1978; Gordon 1981; Ramsey 1981).

There seems to be a strong relationship between aggregate economic performance and total primary energy inputs. Perhaps no image has been used so often to illustrate this closeness as a scattergram of per capita rates of energy use and gross economic product (GDP or GNP) on the global scale. Correlations for the complete global set are high (near 0.95), explaining about 90% of the variance and offering a seemingly near-perfect forecasting tool. But the link misleads as much as it enlightens. First are the important data limitations. On the energy side the most serious shortcomings are the exclusion of biomass, an omission substantially underrating actual fuel use in most poor countries, and the dilemma of converting primary electricity to a common energy denominator.

Inadequacies of national economic product figures are no less intractable. What is really growing in the measured economic growth? How can a critical observer of Western economies disagree with Rose's (1974) observation that "so far, increasingly large amounts of energy have been used to turn resources into junk, from which activity we derive ephemeral benefit and pleasure; the track record is not too good." Standard statistics omit the value of extensive subsistence production and barter, as well as the substantial contribution of black market transactions; are biased owing to fluctuating exchange rates, which reflect primarily the prices of internationally traded commodities; and have no acceptable solution for finding a common denominator for nonconvertible currencies.

Data quality aside, disaggregated analyses show that energy–GNP correlations are masking very large variabilities at all levels of the economic spectrum. Europe has certainly the most reliable data sets but even when using price-adjusted per capita GNP values (Summers and Heston 1984), absence of energy–GNP correlation is obvious (Fig. 13.1). As with the global energy–GNP sets for a single year, national long-term trends of economic growth and energy use indicate strong and stable relationship. This link was especially impressive during the years 1945–1973: the two growth rates marched everywhere in lock step, making energy/GDP elasticities (relative change in energy use in the numerator, change in the economic product in the denominator) a favorite forecasting tool.

Elasticities of rich economies were less than 1.0 and those of poor countries (where industrialization required large energy inputs not immediately equaled by corresponding increases of economic output) ranged well above 1.0. Historical comparisons, as well as post-1973 trends, show a more complex reality. By far the longest course of energy/GDP elasticity can be calculated for the United Kingdom; it shows a clear historic decline composed of waves of a rather large amplitude (Fig. 13.2).

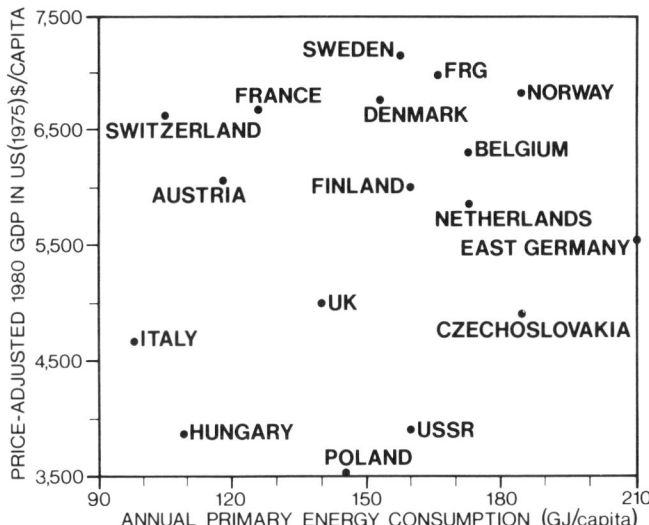

Figure 13.1 Energy–GDP correlations are quite weak for relatively homogeneous groups of countries. Even when European GDP values are expressed in price-adjusted terms (Summers and Heston 1984), some nations using essentially the same amount of primary energy have twofold difference in per capita GDPs (Poland and USSR vs. Denmark and Sweden)—while other countries with nearly identical levels of GDP have almost twofold difference in consumed energy (Switzerland vs. Norway, Italy vs. Czechoslovakia). Energy data from United Nations Organization (1980–).

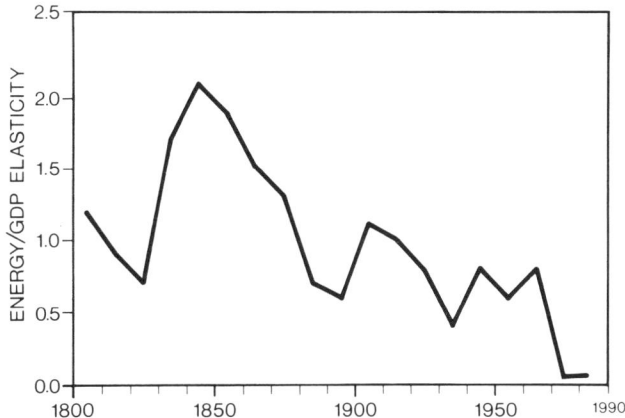

Figure 13.2 British data offer a unique opportunity to trace energy/GDP elasticity for nearly two centuries. The generally declining trend of long-amplitude waves is unmistakable. Based on data in Mitchell (1975), Humphrey and Stanislaw (1979), and United Nations Organization (1980–).

The large 1830–1850 rise is attributable to a rapid adoption of steam engines and to the diffusion of railways (Humphrey and Stanislaw 1979). Higher efficiencies in the iron industry and better performance of steam engines were the main causes of the subsequent decline. The two world wars and the intervening economic crisis make it difficult to interpret the shifts during the first half of the twentieth century but the rapid post-1970 fall is clearly a part of a worldwide trend.

American experience shows remarkable similarities with British trends (Fig. 13.3). The pronounced post-1940 rise in elasticity is best explained by the mass adoption of cars, extensive suburbanization, and increasing importance of such energy-intensive industries as aluminum smelting and petrochemical syntheses. Soviet elasticities appeared to lag about a decade behind the U.S. or British trends but instead of turning down in the early 1980s they rose substantially, an incredibly poor performance especially in comparison with China's rapid reductions (Fig. 13.3).

Another way to analyze this information is to look at national energy/GNP intensities expressed in constant monies (Fig. 13.4). Once again, American (and also Canadian) experience follows the British pattern with a lag of about 40 years. But while the post-1950 decline was continuous in Canada, the pre-1973 U.S. levels were stuck close to 65 MJ/(1972)$. In fact, for 17 years during that period the rate was within just 2% of its average value—while the real GNP grew 250%. As Socolow (1985) remarked, "such high correlations ... don't happen all that often in the social sciences" and "thus it did not seem all that surprising that intelligent people could believe with considerable conviction that this value ... had become an immutable fact of the U.S. economy."

But there were no immutable physical reasons for this temporary constancy and by 1985, with American GNP about one-fifth larger in constant monies than in 1973, the rate was down to 43 MJ/(1972)$, a third lower in a mere dozen years. Yet this impressive decline still left the country far behind Japan and the best European countries. Explanations of these differences range from climate to recreational habits, but most of the gap can be accounted for by the makeup of primary energy consumption and by the structure and efficiency of final conversions.

Economies heavily dependent on coal (U.S., U.K.) are handicapped in comparison with nations running on hydrocarbons and primary electricity (Japan, France)—as are the energy-exporting countries: energy self-sufficiency is not conducive to efficient conversions

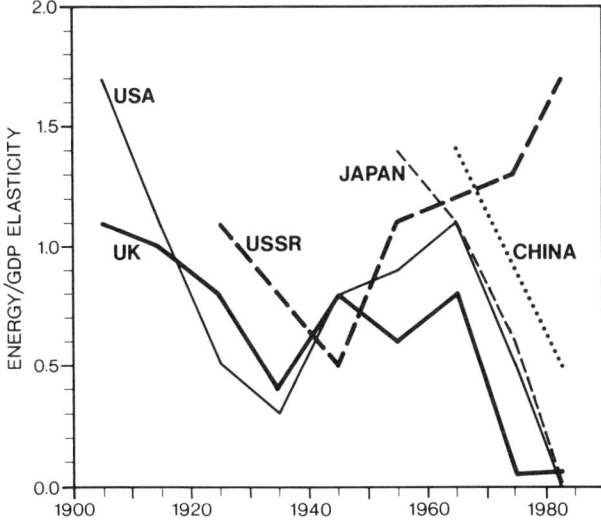

Figure 13.3 Long-term downward trends of energy/GDP elasticities for the world's largest economies resemble in the main the British experience. The Soviet Union has been the obvious exception. Data entries calculated from sources listed in Fig. 13.2 and from U.S. Bureau of the Census (1975) and the Central Statistical Office (1987).

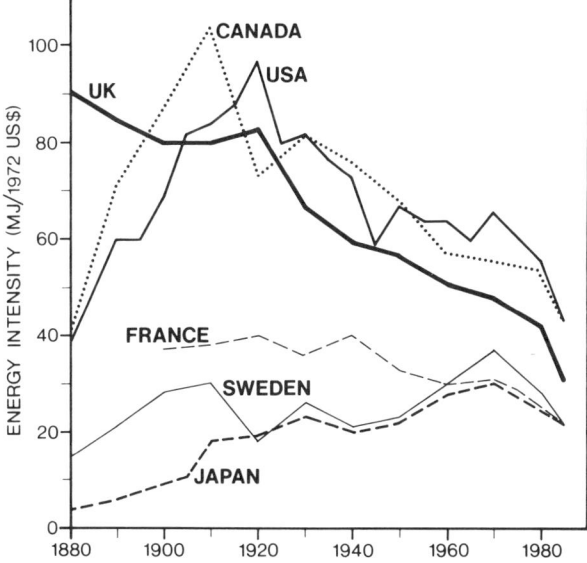

Figure 13.4 Long-term trends of energy/GDP intensity show an initial sharp rise followed by a long-term decline for early and large producers and users of coal and a gentler increase and relative stability for countries that were never as dependent on solid fuels. Same sources as for Fig. 13.3.

while high dependence on imports promotes savings. But no single factor is more responsible for variations in energy intensity than the combination of affluent energy uses—extensive commuting by car and leisure driving, ubiquitous air-conditioning, and excessive heating of oversize houses. The United States and Canada have the highest shares of liquid fuels used in private transportation (about 40% of the total) as well as the highest shares of residential uses of electricity (U.S. share of about 35% in 1985 was nearly equal to total industrial consumption). Energy burdens of integrating larger territories and differences in industrial structure explain the remaining disparities.

No fuel can confer the economic benefits of electricity (NRC 1986). Its high final conversion efficiencies, precise control, focused applications, and fractional uses offer an incomparable combination of advantages. Modern manufacturing—with its demands for flexibility, precision, expansion, and innovation—would be unthinkable without electricity, as would be the health services, and household comforts. Demand for electricity has been growing faster than that for all nonelectric energy combined: before 1973 it doubled nearly every decade. This persistent pattern was broken after 1973 but demand for electricity continued to grow while total primary energy use was declining or stagnating.

Electricity also revolutionized all kinds of information flows, and information links energy and the economy in a multitude of ways. Only a very small fraction of commercial energy is used directly to regulate the immense flows of fossil fuels and electricity—but a steadily increasing share of global energy consumption goes into generating, distributing and processing printed and electronic information which now suffuses all activities of modern civilization.

Shannon's formulation of the equivalence between energy and information opened the way for rigorous quantitative studies of these processes (Shannon and Weaver 1949). But applying this approach to economies at large (as well as to other biosystems) is neither easy nor necessarily useful. The quest for minimization of energy/information ratios (Joules/bit) in various devices and systems preparing, distributing and processing information keeps colliding with inherent inefficiencies of human decision making. Two broad trends have been aggravating these inefficiencies: on one hand the increasingly splintered expert knowledge and the need to accommodate a widening range of social and environmental concerns, on the other the ebbing of interest in science studies and the spreading functional illiteracy.

The last fundamental correlation that requires a closer look is the relationship between energy use and the quality of life, an inquiry made difficult by the complexity and fuzziness of the latter concept— but one which is obviously more meaningful to most individuals than the uncovering of links between aggregate economic performance, energy consumption and information flows.

13.3   ENERGY AND THE QUALITY OF LIFE

Economic growth and energy use are unreliable indicators of a process whose aim should be not only to satisfy basic physical human needs but also to develop human intellect—and to do so in ways that are least disruptive to the environment. "Quality of life" is a multi-dimensional concern embracing personal well-being, its wider environmental and social setting, and the vast intellectual aspect of human development. As such it cannot have a single meaningful indicator. Moreover, there is little connection between subjective appraisals of quality of life and personal satisfaction on the one hand and objective indicators on the other (Nader and Beckerman 1978; Andrews 1986).

But there is no doubt about links between per capita energy use and the physical quality of life characterized by health care, nutrition, housing, and education. This is made clear both by historical comparisons and by contemporary contrasts. However, nearly all unquantifiable or intangible ingredients of the "good life" do not appear to carry high energy costs. Common leisure activities, except for a few high-power American pastimes, entail no additional personal metabolic cost and only a modicum of embodied energy (in books, recordings, table games, radios, and television sets), only a temporarily elevated metabolic demand (in gardening or in scores of sporting activities), or a negligible use of electricity (be it in electronic gadgets or in furniture building).

Personal freedoms have been badly served by most low-energy societies in Africa, Asia, and Latin America—but cultural and historical forces have been much more important: European communist dictatorships used significantly more energy than European democracies. Perhaps most importantly, all of the fundamental personal freedoms and democratic institutions were introduced and codified by our ancestors at times when their energy use was a mere fraction of ours. And so the search for minimum per capita energy use needed to support good life can rest on health, nutritional, and

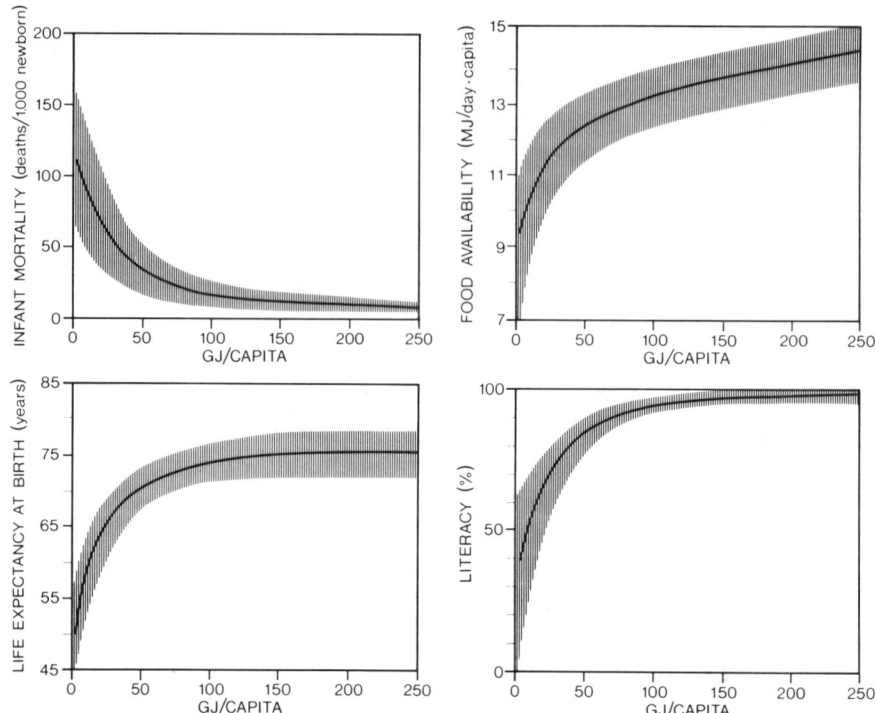

Figure 13.5 Four physical quality of life indicators plotted against the consumption of primary energy. Gains in longevity, declines in infant mortality, improvements in food availability and in literacy rapidly diminish with per capita primary energy consumption averages higher than 40–50 GJ/yr. Data from United Nations Organization (1980–) and World Bank (1988).

environmental variables; they require much higher energy flows than the preservation of personal freedoms or exercises of creativity.

Life expectancy at birth and infant mortality are perhaps the two most revealing data sets (Fig. 13.5). The first figure subsumes decades of nutritional, health-care, and environmental effects and the second finesses these factors for the most vulnerable age group. Average life expectancy of 70 years corresponds to annual per capita use of 40–50 GJ of primary energy—and so does the infant mortality rate below 40. Increases in energy use beyond this range bring first rapidly diminishing returns and soon a leveling off with no further gains.

National means of food energy availability should not be used for "the higher the better" comparisons but rather as indicators of relative abundance and variety of food, two considerations associated with the notion of good life in virtually every culture. Minimum per capita availabilities satisfying the conditions of adequate supply and good variety are over 12 MJ/day, the rates corresponding, once again, to the average inflection range of 40–50 GJ of primary energy per year (Fig. 13.5). And although literacy statistics (generally measured as shares of pupils completing grade school) convey no qualitative distinctions between the functionally illiterate and those who can continue their studies, as indicators of general availability of basic education they have their midway inflection point corresponding, again, to primary energy use of 40–50 GJ/y (Fig. 13.5).

A purposeful society could thus guarantee decent physical well-being and longevity, reasonably varied nutrition, and fair education opportunities with annual use of 40–50 GJ of primary energy. This range refers to energy conversion efficiencies prevailing during the 1980s, the rates leaving much room for additional improvements. Higher minima may be desirable. Life expectancy of 75 years requires at least 70 GJ/y, as does infant mortality below 10; there is no benefit in pushing food energy above 13 MJ/day, but very good educational opportunities (literacies above 90%) also require at least about 70 GJ.

This rate is substantiated by looking at the share of young adults receiving tertiary education. Available statistics do not convey important qualitative differences, but providing more than 20% of young adults with postsecondary education is generally connected with annual per capita consumption of over 70 GJ. Conversion savings on the order of 25%, quite realistic to achieve over a period of 15–20 years, would lower the total to around 50 GJ/y and the rate of 50–70 GJ would then appear to be the desirable minimum for any society where satisfaction of essential physical needs is combined with adequate opportunities for intellectual advancement.

By coincidence, 50 GJ/y equals almost exactly the worldwide mean of the late 1980s. Egalitarian redistribution of global primary energy production, coupled with a realistic conversion efficiency improvement, would thus provide every inhabitant of this planet with as much useful energy as the Swedes had at their disposal in the early 1950s, the French around 1960, and the Japanese in the late 1960s. This comparison demonstrates that an impressively high worldwide standard of living could be achieved with virtually unchanged global energy consumption—but, given the enormous size of the existing consumption inequities, such a possibility is predicated on substantial cuts of average consumption in all rich nations.

Labor and leisure implications arising from higher use of energy have been profound, proceeding in two grand waves. First, increasing energy subsidies in farming cut labor requirements and released peasants for urban factories. These enterprises, thanks to cheap supplies of concentrated high-quality energies, grew in size to levels unprecedented in human history, exerting a pull on the rural labor force rendered obsolete by machines and fertilizers—clearly a matter of deviation-amplifying development. In the early decades of this shift manufacturing required a large share of heavy exertion and drudgery and, much like traditional farming systems, it retained the extensive participation of child labor.

It is only the second transformation wave, moving the labor force from manufacturing to services, that has nearly eliminated heavy labor. All economies, at different rates, have been following this general path of economic transformation which has two other important quality-of-life implications. Because services have considerably lower energy intensity per unit of value added, growth of economy can proceed with lower environmental impact. Moreover, the rising affluence associated with this shift means that households tend to get saturated with high-energy-intensity products and consumption moves on to more structured items of lower energy intensity (Roberts 1981), a change of preferences often reinforced by higher education.

Again, electricity has played the critical role in this transformation: regardless of the availability of other energy forms, often it was only the introduction of electricity that did away with exhausting and dangerous labor. Hundreds of millions of peasants throughout the poor world are still waiting for this phenomenal liberation. When it comes, women are its greatest beneficiaries, but the best available time allocation studies show that even in high-energy societies their total labor time has not decreased significantly in comparison with their traditional counterparts (Minge–Klewana 1980). Because of the forgone labor of children (and of adolescents undergoing longer education) women work more in modern households than they did in traditional ones; tasks are easier but there have been no great gains of leisure time.

When looking at energy-related timesaving and leisure activities, the automobile must be singled out for special attention—both because of its increasingly contradictory contribution to these goals and because of its inordinately high social valuation. Even the most efficient cars are useless in frequent, or nearly permanent, tie-ups afflicting virtually all large urban areas whose average peak traffic hour speeds are commonly lower than the speeds of horse-drawn omnibuses and electric streetcars before the end of the nineteenth century. And the stunning energy losses in automotive transport—rolling efficiencies of 5% of the initial crude oil input—are a major contributor to local and regional environmental degradation.

Continuing expansion of the increasingly irrational private car ownership has explanations beyond the rapidly disappearing time advantage. The car, in Boulding's (1974) perfect description, "turns its driver into a knight with the mobility of the aristocrat.... The pedestrian and the person who rides public transportation are, by comparison, peasants looking up with almost inevitable envy at the knights riding by in their mechanical steeds. Once having tasted the delights of a society in which almost everyone can be a knight, it is hard to go back to being peasants."

This dependence, causing extensive spatial and social restructuring of modern society, presents a formidable obstacle to rationalization of energy use and introduction of alternative supplies. Driving has also become a leading cause of death and injuries in most rich societies. The realm of risks associated with extraction, transportation, and use of energies is immense, ranging from frequent but rarely fatal exposures (presence of natural toxins in food or of *Legionella* bacteria in air-conditioning ducts) to rare but potentially very dangerous mishaps (fires in refineries, failures of large dams, releases of radiation from nuclear power plants). In addition, there are the catastrophes whose first instance is yet to happen, such as explosion of a tanker carrying liquefied natural gas.

This possibility exemplifies the challenge of assessing risky events purely on the basis of complex assumptions. A five-minute burn of a single 25,000 m^3 LNG tank (ships usually carry five such tanks) would release energy equivalent to about 10 Hiroshima bombs. Yet all we can do to estimate the risks of such a catastrophic event is to extrapolate from the known collision risks in harbors, then assume probabilities of subsequent tank ruptures and LNG spills, and then make more assumptions about the actual modes of LNG burning (Fay 1980).

Perhaps the most prominent risk whose assessment relies largely on compounded assumptions is an accidental release of ionizing radiation from nuclear power plants: the Three Mile Island mishap in 1979 and the Chernobyl disaster in 1986 caused some profound reappraisals of probabilities with which such events can occur (Hohenemser 1988). Differences between low and high estimates of their frequency range over several orders of magnitude and the same degree of disagreement applies to long-term health effects of such disasters.

This is the critical point for risk appraisals of both nuclear and fossil-fueled systems: most of the diseases and fatalities will be caused by long-term, cumulative exposures, and so the relevant risk analyses must go beyond the immediate effects of accidental events. Assessments of nuclear risks have been particularly affected by these complex considerations, above all owing to fundamental uncertainties concerning the dose–effect relationship at low exposures. Assumptions of threshold sigmoid response, of no-threshold linear effect, or of linear–quadratic relationship will result in very different numbers of potential lifetime fatalities. Selective reading of publications on the risks of ionizing radiation may lead to an adoption of a combative no-threshold posture—or to the consideration of beneficial effect (hormesis) of very low radiation doses (Sagan 1987).

Estimates of the chances of the most severe accident, core meltdown, in an American nuclear power plant during a period of 20 years differ by a factor of 200 (Hively 1988). Yet this is not a unique uncertainty, as our understanding of long-term effects of air pollution is similarly uncertain. Attempts to quantify the number of premature deaths caused by emissions from a 1 GW coal-fired power plant have produced totals between 0.07 and 400,000 (Ricci and Molton 1986). Comparative risk assessments will then almost invariably rest on dubious foundations. Awareness of these complications is more important than the results of many available calculations of mortality and morbidity risks for various energy techniques or complete supply systems (Goodman and Rowe 1979; Travis and Etnier 1983; International Atomic Energy Agency 1984; Morone and Woodhouse 1986; Fremlin 1987).

Risk comparisons are complicated by differences in perception and tolerance of voluntary and involuntary exposures (Starr 1976): people are willing to voluntarily assume risks about three orders of magnitude higher than those arising from involuntary exposures. Although the risks of involuntary pollution or radiation exposures appear to be no higher than those of common natural hazards, they often are seen as totally unacceptable. Nevertheless, a critical evaluation of the evidence shows that the risks associated with modern energy supply and conversion have been acceptably low. Of course, future developments may change this judgment.

But rather than offering dubious scenarios of various energy futures I will just outline a few constants that have been limiting our understanding of coming changes and highlight several key realities that will mold our quest for energy supply in the twenty-first century.

13.4 ENERGY FUTURES

Before looking ahead it is necessary to look back at long-range energy assessments and forecasts. By exposing the recurrent pattern of large errors such exercises remind us of the very limited utility of current and future forecasting. On the most general level one must note the recurrent influence of a Western intellectual tradition favoring a catastrophic view of the future. A particularly interesting subset of this vision is the fear of running out of energy resources.

A great nineteenth-century classic of this genre (Jevons 1865) anticipated exhaustion of Britain's coal resources followed by decline of the empire. British fortunes declined in the twentieth century—but exhaustion of coal certainly was not the reason. In the early 1920s American geologists predicted an early end of oil era—but in 1930 came the discovery of the supergiant East Texas oilfield. Putnam (1954) forecast global depletion of recoverable fossil fuel reserves for the years 2000–2025, and the Central Intelligence Agency (1979) concluded that "the world does not have years in which to make a smooth transition to alternative energy resources." Yet in 1990 the global oil reserve/production ratio stood at a record level. Earth's crust holds finite amounts of recoverable minerals but timing their exhaustion has been much like beholding a mirage: on closer approach the specter disappears.

The other most powerful factor molding our perception is the vision of the future as a replica of the recent past. This influence of prevailing moods was particularly strong from 1960 to 1980, when it resulted in forecasts overestimating actual energy use commonly by 20–30% in less than a decade (Basile 1976; Perry and Streiter 1977). Reasons for the large numbers of wrong forecasts can be found in the herdlike behavior of forecasters (Leydon 1987). Another shared trait is the infatuation with new energy sources and conversions. Basalla (1982) illustrated this propensity by pointing out the recurrent myth of new sources as ultimate solutions. Coal was to solve the shortage of wood; within a century its mystique was transferred to hydro-electricity, whose large-scale development appealed to Lenin as much as it did to Ford; nuclear power, the next carrier of the utopian dream, was soon displaced by solar conversions.

Nuclear fission has been the most obvious case of unfulfilled expectations, but there are many other items on the list of conversion techniques with elusive breakthroughs: hydrogen economy and fusion electricity generation represent the two most elusive advancing targets. The Office of Technology Assessment (1987) concluded that after 40 years and some \$20 billion it will take at least 20 more years and another \$20 billion just to reach the point where engineering and economic feasibility assessment can begin—with no guarantee of eventual viability.

The virtues of a hydrogen-powered society have been extolled for decades (Hoffmann 1981; Veziroglu and Getoff 1986), but costs and complications of producing this secondary energy carrier and setting up the requisite distribution infrastructures have turned all proposals for hydrogen-fueled civilization into fictional accounts whose dates of realization keep moving further into the future. And there is no shortage of utterly unrealistic ideas, ranging from Goela's (1979) wind harnessing with huge kites flying 1.5 km aloft to the tapping of meltwaters from the eastern Greenland glaciers and transmission of the electricity across the North Atlantic to western Europe (Partl 1977).

In searching for energy futures we have to accept the limits of our understanding. Thus far, fossil-fueled civilization has enjoyed relatively easy access to coals and hydrocarbons recoverable with high net energy returns. Indeed, the ease of discovery, cheapness of extraction, and size of reserves of many supergiant oil and gas fields provided extraordinary energetic boons. But falling net energy returns may not be extraordinarily restrictive: what we will do with the available energy will matter more than its net costs.

And since energy concerns span a vast realm of human and natural affairs, sensible forecasts cannot be confined to energy. The search is not for output or demand figures or for energy return ratios. The challenge is to find out what the whole society will be like—and given our powers this really means what we want it to look like and how much we are willing to sacrifice to bring about the requisite changes. Within the confines of resource availability and thermodynamic laws lies a host of possible futures: civilization's course is not preordained but remains open to our choices.

Long-term perspectives on the order of 10^2–10^4 years are impossible. Hubbert (1962) popularized the notion of fossil-fueled civilization as a brief interlude between the solar past and the solar future. Continuation of historic rates of extraction would exhaust more than 90% of fossil fuel resources in less than 500 years. That is a short span even in historic terms but what could be accomplished during that period is best appreciated by trying to look at the end of the twentieth century from the vantage point of the late fifteenth century. All we can do with some semblance of rationality is to map our tasks for the next few generations.

Energy decisions involve often irreconcilable concerns about thermodynamic efficiency, personal comforts, resource depletion, economic well-being, environmental degradation, national security, social stability, and democratic values. This reality is incompatible with any strategic optima; it merely admits a choice of practical alternatives. Imperatives of infrastructural inertia and contrasts of conversion and consumption power densities will limit our choices. In the long run we may deurbanize as lower, stabilized populations with access to advanced communication techniques may make the largest cities obsolete. But during the twenty-first century we will have to cater to scores of huge cities and conurbations, still rapidly growing everywhere in the poor world.

Further extension of these energy-intensive structures and their high power consumption densities—rates in the poor world's cities are already very similar to those of many of their affluent counterparts—mandate that large-scale high power density conversions producing electricity and liquid fuels will remain essential around the world. Any possibilities for rapid substitutions by low power density renewable flows are illusory without the sudden collapse of urban societies. The presence of costly and extensive infrastructures of fossil fuel energetics will also tend to lengthen our reliance on coals, oils, and gases. Escaping the imperatives of scale built into the rich world's energy systems will not be easy.

Already stabilizing populations of the rich countries combined with the still enormous conservation opportunities and continuing shifts toward less energy-intensive economies make it easy to imagine not only stationary but even significantly declining needs for primary energy among the billion richest people. A gradual shift of extraordinarily meaty Western diets toward more balanced nutrition would greatly reduce the needed energy subsidies while helping to preserve many agroecosystems. Growing concerns about planetary effects of fossil fuel combustion and apprehensions about safety of nuclear energetics will further strengthen the trend toward limited and rational uses of energy.

In contrast, the rising numbers of still very poor rural populations in Asia, Latin America, and Africa will require considerably larger energy supplies. Their traditional reliance on biomass energies has been causing deforestation and erosion. Perhaps only their cooking needs can be rationally accommodated by improved cultivation of suitable fuelwood species and by diffusion of better stoves. Use of crop residues for fuel should be discouraged everywhere, since their value is infinitely higher when recycled to replenish nutrients and organic matter and to prevent erosion. And production of biomass-derived alcohols is sensible only in limited local settings in order to avoid cutting into the tightening availability of farmland for food crops.

Some parts of the poor world have considerable untapped hydro-electricity potential but solar radiation has been seen as the best choice for starting the transition to efficient renewable energetics. Such a transformation will be a slow one. For a long time direct solar conversions will do little to supply the critical energy inputs needed for farming intensification—for synthesis of fertilizers and for making and powering the farm machinery. These energies will have to be delivered mostly as high power density flows coming overwhelmingly from fossil fuels and thermal electricity. Even small- and medium-scale uses will be complicated by the source's seasonality, intermittency, and unequal spatial distribution and by the ownership of land (a key consideration in using a flow of low power density).

These realities belie any simplistic notions about the efficacy of two different technical fixes often contrasted under the misnomers of hard- and soft-energy paths. Advocacy of the decentralized, small-scale, solar-powered soft path (Lovins 1976) ignored the continuing needs for high power density flows in urban areas of rich and poor countries, as well as for their provision in the production of key industrial and farming inputs, again in both rich and poor nations.

In contrast, visions of extremely concentrated, large-scale installations increasingly dominated by huge coal-conversion plants and breeder reactors (Energy Systems Program Group 1981) hardly took note of enormous conservation opportunities and proceeded to fill the exaggerated forecast needs with little regard for the environment. These two extreme views of energy futures led to two unrealistic global expectations. By the year 2030 the soft path was to reduce the world's 1980 energy use by half—while the hard option was to lead to as much as a fourfold rise in primary energy conversion. In spite of this difference, both visions shared the key bias of a quick technical fix, a misplaced and mistaken faith in a particular set of techniques as *the* solution for a complex energy challenge (Smil 1987).

But we have to shun exclusive grand designs of perfect futures. We need any workable, reliable, economical, and environmentally acceptable approaches. We cannot run away from our primary energizers without profoundly reshaping our whole way of life. But we also need to begin the inevitable transition to the post–fossil fuel world—and so we have to diversify with appropriate renewable or nuclear conversions. Inevitably, there are apprehensions—and there are hopes. On the debit side are the concerns about the availability of fossil fuels extractible with high net energy returns, environmental consequences of energy conversions, and the bridging of the enormous gap between the rich and poor economies.

On the credit side is our increasingly better understanding of biospheric functions, our technical ingenuity, and our social adaptability. Above all, we should not encourage questionable demands, equating them with needs—and fill the gap by providing more energy. And we should not believe that higher demand for many energy services must be equated with higher supply of primary energy. Most of our ingenuity should be devoted to the reduction of final uses rather than the expansion of supply.

But we do not know a priori what will work best. As Sophocles wrote, "One must learn by doing the thing, for though you think you know it, you have no certainty until you try." The necessity of tolerating uncertainty, that essential openness of human future, demands that we avoid foreclosing any useful options. John von Neumann (1955) summarized the task perfectly:

The one solid fact is that the difficulties are due to an evolution that, while useful and constructive, is also dangerous. Can we produce the required adjustments with the necessary speed? The most hopeful answer is that the human species has been subjected to similar tests before and seems to have a congenital ability to come through, after varying amounts of trouble. To ask in advance for a complete recipe would be unreasonable. We can specify only the human qualities required: patience, flexibility, intelligence.

14

GRAND PATTERNS: ENERGETIC—AND OTHER —ESSENTIALS

Constantly consider how all things such as they now are, in time past also were, and consider that they will be the same again. And place before thy eyes entire dramas and stages of the same form, whatever thou hast learned from thy experience or from older history. . . .

—Marcus Aurelius
Meditations

Everything in the observable universe can be seen in energy terms. The environment of the Earth is an intricate, interactive assembly of energy storages and flows; life maintains itself only through incessant imports and conversions of external energies; civilizational development has been a quest for higher energy throughputs transformed into larger anthropomass and greater complexity. And yet this essential attribute of being is an abstraction: energy is an intellectual construct, a concept evolved by a small group of nineteenth-century scientists in order to analyze and explain a variety of natural phenomena ranging from the hue of arterial blood to the efficiency of mechanical engines. They elucidated energy's permanence as well as its changing quality and diminishing availability inherent in all of its conversions.

Since then science has delved into myriads of energy phenomena, reaching all the way into the ephemeral world of atomic particles and into the stunning information storage of human genome. As a counterpoint, broadly based syntheses attempted to present generalized images of energetics, or at least of its principal parts—but this interdisciplinary genre has been surprisingly neglected. This book has attempted such a synthesis and the following sections review most of its key facts, conclusions, and interpretations.

14.1   ENERGY IN THE BIOSPHERE

The Sun floods the Earth with a surfeit of energy. Of 174 PW intercepted by the planet about 122 PW (240 W/m^2) is absorbed by the biosphere where water's unique energetic attributes—high specific heat, high heat capacity, and high heat of vaporization—are decisive in storing and redistributing the absorbed radiation. Slightly more than a third of the absorbed flux drives the planetary water cycle; only a small percentage of it (3.5 PW) is needed to keep the air in motion (Fig. 14.1). Power densities of solar flows range from 10^2 W/m^2 for direct radiation to 10^0 W/m^2 in ocean wave, river runoff, and wind transforms.

Kinetic energies of air and water are frequently destructive as vertical power densities of cyclones, floods, and avalanches commonly surpass 10 kW/m^2. These violent events have had immense influence on the evolution of ecosystems. Yet in aggregate terms all rain-carrying cyclones release much more energy as latent heat, although relatively small kinetic energies of precipitation and

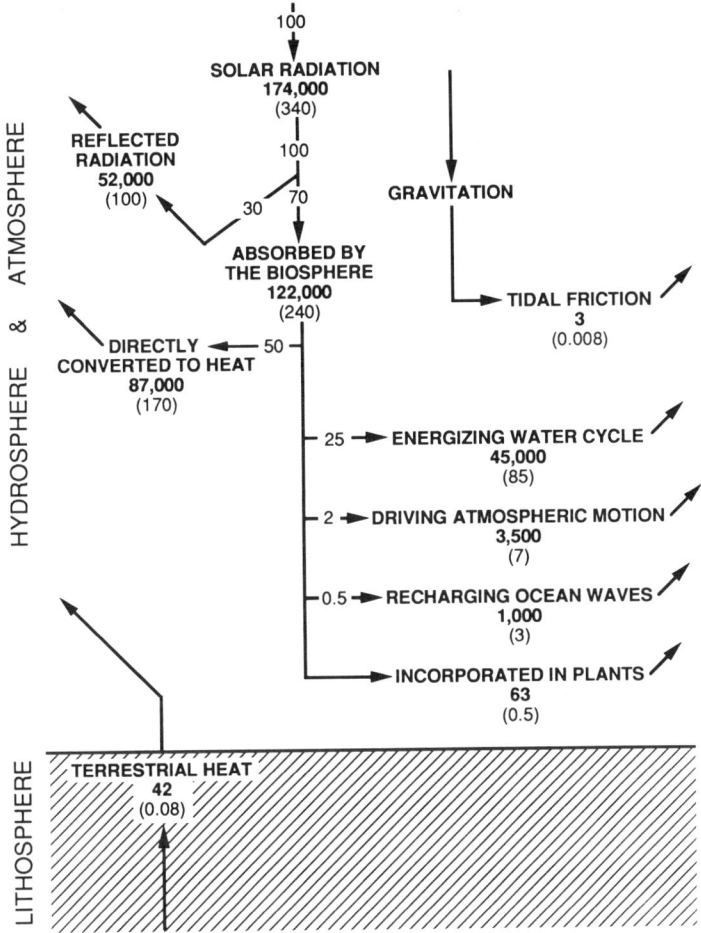

Figure 14.1 Biospheric solar and terrestrial energy flows. The upper number is the annual flux in terawatts, the lower one average power density in watts per square meter.

potential energies of water have been key agents of geomorphic and ecosystemic change.

The Earth's solid crust is made up of huge rigid plates undergoing long-term cycles of creation and destruction energized by basal cooling of the planet and by radioactive decay of several isotopes. Lunar and solar gravity are responsible for tidal friction. In comparison with solar flows, terrestrial heat aggregates and average power densities are minuscule (some 42 TW or 80 mW/m²). But the durability of these low-power flows eventually reshapes ocean floors and breaks apart and reassembles the continents. Immediately more dramatic effects of concentrated energy releases in earthquakes and volcanic eruptions have also had great importance in biospheric evolution.

In general, there is a surfeit of free energy in the biosphere. This is equally true about the maintenance of life on Earth: although less than half of incoming radiation is in wavelengths activating leaf pigments, photosynthesis is rarely limited by the availability of solar energy. Relatively low concentration of atmospheric carbon dioxide is its most fundamental global restriction while water and temperature stresses and shortages of nutrients reduce the efficiency of photosynthesis and shape vegetation patterns on regional and local scales. Inevitable reaction rate losses and tissue respiration limit the photosynthetic conversion efficiency to no more than 5%.

Species with suppressed light respiration using C_4 photosynthetic pathway have an obvious efficiency advantage; C_4 plants also have substantially higher water utilization efficiencies. Unfortunately, most of the major agricultural crops are C_3 species. A CAM pathway is a an excellent adaptation to extreme temperature and water stresses but its productivity is necessarily very low. Assessments of phytomass stores—whose energy densities range, in dry terms, from 17 to 20 kJ/g—remain approximate. Global aggregate is most likely around 1 Tt, or 20,000 EJ, with more than 80% of the total in woody tissues (Fig. 14.2). Small sizes and rapid turnovers mean that the oceanic phytomass equals less than 0.1% of continental stores.

Primary productivities are usually expressed in net terms (i.e., after respiration) and their large-scale estimates are even more uncertain than those of the standing phytomass. A global sum of over 100 Gt, or some 63 TW, is divided more equally among major ecosystems than is the phytomass: forests contribute about 40%, grasslands roughly 25% (Fig. 14.2). Marine production is especially uncertain but it equals at least half of the continental total. Large-scale efficiencies of net primary productivity are highest in wetlands, where over 2% of incoming radiation is fixed in new phytomass, and rarely surpass 1.5% in forests. Terrestrial mean is no higher than about 0.3%, oceanic average (reflecting the nutritional impoverishment of euphotic zone) is an order of magnitude lower.

Since biospheric evolution tends to channel gradually larger shares of available energy into maintenance, net ecosystem productivities (rates after heterotrophic respiration, yields in farming and forestry) are highest in young and simple formations and lowest in forests. Energetic imperatives of forest growth favor evergreens in xeric and boreal environments, limit the number of surviving mature trees

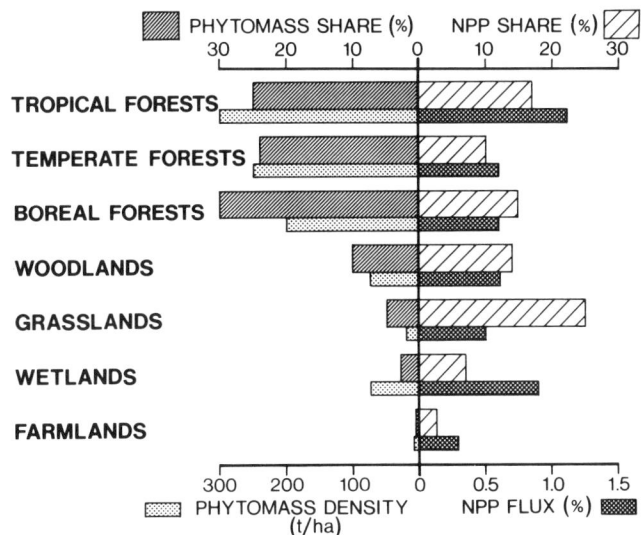

Figure 14.2 Approximate distribution of planetary phytomass and net primary production. Dominance of forest ecosystems—80% of all phytomass and 40% of all NPP—is obvious. Extensive areas and fairly high productivities of grasslands translate into a large contribution to the NPP total, but their phytomass is only a fraction of forest stores.

per unit area, and determine the efficiency of nutrient cycling and utilization.

Heterotrophs found it profitable to eschew fixation in favor of feeding on the abundant polymers synthesized by plants. Of the two principal metabolic choices aerobic respiration has a decisive energetic advantage over anaerobic fermentation: all higher organisms use it. Complex plant carbohydrates are the primary source of heterotrophic energy, followed by lipids; proteins, critical building blocks of animal tissues, are used as energy sources only when carbohydrates and fats are limited.

As with autotrophs, ATP is the energy carrier of intricate heterotrophic metabolism whose overall efficiency can go up to 60%. Basal metabolic rates of vertebrates scale with the 0.75 power of their body mass. The best explanation of this regularity is in mechanical requirements of animal bodies. There are some notable departures for various groups of vertebrates—but the slope remains constant or at least very close to 0.75, even for most invertebrates.

Energy requirements of reproduction, growth, and locomotion raise the total need substantially above the resting basal metabolic rate, but thermoregulation puts the highest constant energy burdens

Table 14.1 Comparison of Major Energetic Characteristics of Ectothermic and Endothermic Vertebrates

Energetic Characteristic	Ectotherms	Endotherms
Body mass (g)	10^0-10^5	10^1-10^7
Basal metabolic rates (mW/g)	0.25–0.30	1–10
Metabolic scopes	2–10	6–30
Growth efficiencies (%)	20–40	1–4
Maximum predator prey ratios (%)	10–80	1–3
Daily movement distances (km)	0.1–1	0.5–12
Social behavior	Weak	Developed

on heterotrophs. Ectothermic thermoregulation—the search for optimal microenvironments—is overwhelmingly behavioral. Inevitably, heterotherms cannot be active in thermally extreme environments, nor can they be as competitive in optimal temperatures as endotherms (Table 14.1).

Carrying a uniform thermal environment confers enormous survival and competitive advantages on endotherms, which have radiated to virtually every terrestrial niche. But there is high energetic cost to this successful strategy: in order to remain above the ambient temperature most of the time, body temperatures of endotherms are relatively high and their maintenance, which requires much higher feeding rates than that of ectotherms, limits the share of energy that can be diverted to reproduction and growth. Endothermy also limits the size of the smallest animals as the rising specific metabolic rates with falling mass require higher frequency of feeding in smaller creatures.

But endotherms are unsurpassed masters of both rapid and long-distance locomotion. Their metabolic scopes are mostly around 10 for mammals and up to 15 for birds. Maximum aerobic power of endotherms is commonly an order of magnitude higher than in ectotherms. Larger animals have lower specific locomotion costs but both massive mammals and tiny birds are capable of astonishingly long seasonal migrations powered by lipid stores. In terms of minimal transportation cost, swimming is the most efficient way of locomotion, running the most demanding. Most of the time spent in motion is in search of feed but foraging is not simply a matter of minimized energy expenditures. Selection of specific nutrients rather than maximization of net energy is often the objective.

Inevitably large losses in energy transfers between successive trophic levels limit the biomasses of heterotrophs to small fractions of standing phytomass. Inverse relationship between body size and density (exponent of -0.75 when scaling body mass, -2.25 when scaling body length) means that there may be a few megajoules per square meter of decomposer biomass, about one-tenth as much of soil invertebrates, and usually less than 1 kJ/m^2 of birds and insectivorous and carnivorous mammals. Global zoomass may be nearly 10 Gt (200 EJ), or roughly 1% of all phytomass.

But practical understanding of biospheric processes benefits more from the appreciation of numerous factors limiting energy flows rather than from further inquiries into the conversions themselves. Low conversion efficiencies in photosynthesis—commonly an order of magnitude below the genetic potential—arise overwhelmingly from the presence of such inhibiting triggers, above all numerous nutritional and widespread water availability deficiencies. Not surprisingly, rates of photosynthetically active radiation are a vey poor predictor of the spatial distribution of low-productivity ecosystems.

Animal ecology provides extensive experimental and field evidence of limits to breeding populations of vertebrates even when there are no shortages of available feed energy. Behavior is frequently a major limiting factor, especially as increasing densities promoting territorial strife and aggression reduce fecundity of many fish, bird, and mammalian species. Physical features of the environment—above all availability of cover and nesting or denning sites—also limit the spacing, density, and dispersion of vertebrate species even in the presence of abundant feed.

Diversity of heterotrophs also demonstrates clearly that there is no best energetic strategy assuring evolutionary competitiveness. Although endotherms have an undisputed adaptive edge, ectotherms remain both very abundant and highly diversified. And although larger bodies entail both thermoregulatory and locomotive advantages, energy harvested daily per unit area is independent of the unit mass of feeding heterotrophs and hence no herbivorous species can become more successful only because of its bigger size. Energetic considerations alone are also insufficient to explain spatial behavior of heterotrophs—and the degree of their explanatory power clearly decreases with the advancing complexity of behavior.

Evolution of heterotrophs provides clear evidence of deviation-amplifying changes: there was a span of some 2.5 Gy between the emergence of the first prokaryotic cells and more complex eukaryota but metazoa were present only 300 My later. Similarly, in human

Table 14.2 Comparison of Major Energetic Characteristics of Large Terrestrial Mammals and *Homo sapiens*

Energetic Characteristic	Large Terrestrial Mammals	Homo sapiens
Adult body weights (kg)	10–2700	40–100
Gestation periods (days)	60–450	266–294
Basal metabolic rates (mW/g)	0.8–4.6	1–2.7
Metabolic scopes	10–32	15–25
Maximum perspiration rates (g/m²·h)	100–250	300–500
Encephalization quotients	0.3–2.4	7.3–7.7
Longevity (years)	1–50	50–80

evolution there is a huge disparity between the duration of stone-tool cultures (about 2 My) and the time between first agricultures and industrial civilization (less than 10 ky).

Rapid ascent of *Homo sapiens*, one of whose consequences has been the decline of wild vertebrate zoomass, is an impressive but worrisome testimony to the success of the most versatile and the most adaptive of all heterotrophs (Table 14.2). Indeed, among terrestrial mammals humans have no equal as great generalists. High metabolic scopes and perspiration rates made them a outstanding runners and facilitated radiation into extreme environments—and unrivaled encephalization enabled humans to construct and use a myriad of exosomatic aids which have elevated human existence far above the plane of mammalian heterotrophy. Yet human energetics is still far from perfectly understood. The array of essential nutrients—water, carbohydrates, lipids, proteins, vitamins, and minerals—required for human growth, tissue maintenance, and activity is well known but uncertainties surround many intake recommendations.

Carbohydrates dominate nearly all human diets, proteins supply 11 essential amino acids required for growth, and lipids contain essential fatty acids. Of these three macronutrients lipids have by far the highest energy density (38 kJ/g); the digestion of carbohydrates and proteins provides 17 kJ/g. Overall food requirements are a complex function of age, sex, body size, activity, climate, and individual metabolism.

Even basal metabolic rates, ranging between 60 and 90 W for most adults, show considerable variation, especially after adolescence. The early peak of specific metabolic rates (2.7 W/kg reached during the first six months of life) is followed by a steady decline to levels around 1 W/kg by the age of 70 years. Minimum survival needs are about 1.25 times the basal rate and different levels of normal activity raise the multiple to 1.6–2.1.

The energy cost of growth ranges from 15 to 35 kJ/g of lean tissue; pregnancy, which also requires higher basal rate and cardiorespiratory effort, costs about 27 kJ/g. Lactation, drawing on fat reserves deposited during pregnancy, claims an additional 35 W in Western women but its cost among many poor populations of Asia and Africa was found to be minimal. The same discrepancy applies to pregnancy: superior metabolic efficiency of pregnant women in nutritionally limited circumstances is the only plausible explanation.

Expenditure measurements show optimum speed for walking (5–6 km/h) but the energetic cost of running remains nearly identical for a wide range of speeds. This feat, enabling man to outrun many ungulates, is the result of bipedalism and efficient heat dissipation. Core temperature is maintained by dilation of peripheral vessels and shifting of blood to feet and hands—but above all by copious sweating. Human sweating rates surpass those of other efficient perspirers—camels and horses—and can sustain exertions of up to 600 W. The best acclimatized individuals can remove over 1300 W of heat by sweating.

Limits of human physical performance are set by hydrolysis of high-energy compounds. Anaerobic processes support very high but brief exertions (up to 8–12 kW for trained individuals) but, as long-term efforts are energized overwhelmingly by aerobic recharge, limits of human power are largely a function of maximum oxygen intakes. Most people can support rates of 600–900 W; best performers can go up to 2 kW, or about 25 times the basal metabolic rate. Among mammals this metabolic scope is surpassed only by canids. Efficiency of converting digested food to work is 10–13% in anaerobic processes and usually 15–20% in aerobic exertions. This means that during an eight-hour spell the maximum sustained output of 300–350 W translates into 1.5–2 MJ of useful work at rates of 50–70 W.

Men living as simple heterotrophs in foraging societies favored gathering seeds and nuts (15–25 kJ/g) and often prolific, though less nutritious, roots and tubers. This was an energetically highly rewarding strategy—net returns of 5–15, and up to 40—and not highly time-consuming. But in most ecosystems seasonal fluctuations of phytomass availability and a variety of natural hazards made foraging a perilous experience. Great diversity of habitats and subsistence patterns precludes sweeping generalizations but a few energetic imperatives are obvious. All foraging societies were omnivorous but, except for maritime groups, plant foods were dominant.

Much of the plant gathering could fit the optimal foraging pattern but the process was complicated by the necessities of securing water and sources of vitamins and minerals as well as the presence of large carnivores and competing foragers. Gathering in forests required more frequent residential moves, whereas the seasonal surfeit of grains, nuts, and roots in grassland ecosystems allowed for fewer, but longer, camp relocations. Hunting was largely for herbivores and its energy returns were often very low, especially in tropical forests with their small folivorous fauna. Larger grassland herbivores were preferred because of their higher fat content (lean wild meat has only 6 kJ/g). Group hunting improved the chances of success and hence the rates of energy return.

Population densities of foragers ranged over two orders of magnitude, from 1–2/100 km^2 in forests to around 100/km^2 in maritime groups, which were the first societies to adopt sedentary living. Evolution of sedentism—with its food storage, increase of private property, and more complex social structures—was multifocal and protracted as incipient agricultures coexisted for millennia with gathering and hunting.

14.2 ENERGY IN CIVILIZATION

With agriculture humans ceased to be simple heterotrophs and became increasingly sophisticated manipulators of solar flows and builders of complex societies. The transition from foraging to farming, a multifocal, evolutionary process, was driven largely by increasing population densities. Energy returns in early farming were typically no better than the gains in foraging, but larger populations could be supported from smaller areas of land (Fig. 14.3). This was a deviation-amplifying process as higher densities promoted further intensification of farming. Resulting growth of permanent settlements, diversification of manufactures, accumulation of knowledge, aggradation of private ownership, and evolvement of hierarchical structures were the key ingredients of civilizational development.

The most extensive forms of managed food production—nomadic pastoralism and shifting agriculture—have persisted alongside the intensifying farming for millennia, a clear testimony to the appeal of minimized energy inputs. With stocking rates ranging typically from 0.05 to 0.15 animal units/ha the grasslands can carry 0.8–2.7 people/

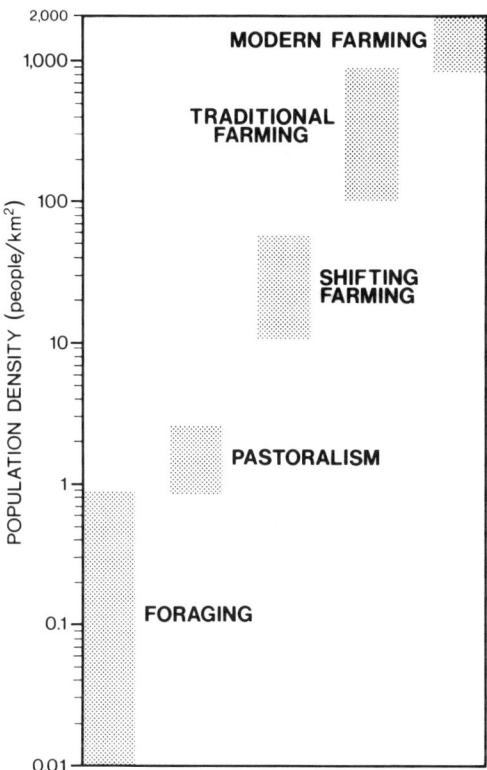

Figure 14.3 Intensification of agriculture has been driven by the necessity to support higher population densities. Each intensification step has been able to accommodate roughly 10 times as many people as its precursor.

km^2. Shifting agriculture follows a sequence of clearing the natural or secondary growth, burning the undergrowth, planting a variety of food, feed, and medicinal species, weeding the plots or gardens, and staggered harvesting. Energy inputs range between 1 and 2 GJ/ha, net energy returns vary mostly between 15 and 30, and they can support 10–30 people/km^2.

Continuing population increases shortened the fallow periods of shifting agriculture and led to intensive field cropping with plowing, requiring animal draft, as its energetic hallmark. Working animals range from donkeys to large horses, delivering sustained power of 100–800 W. Horses are the most powerful choice—but not unless properly harnessed and adequately fed. A good horse can easily do the work of six to eight men, but its average feeding requirements of 110–120 MJ/day mean that relatively large shares of farmland had to be devoted to pastures and feed crops. In agricultures relying on

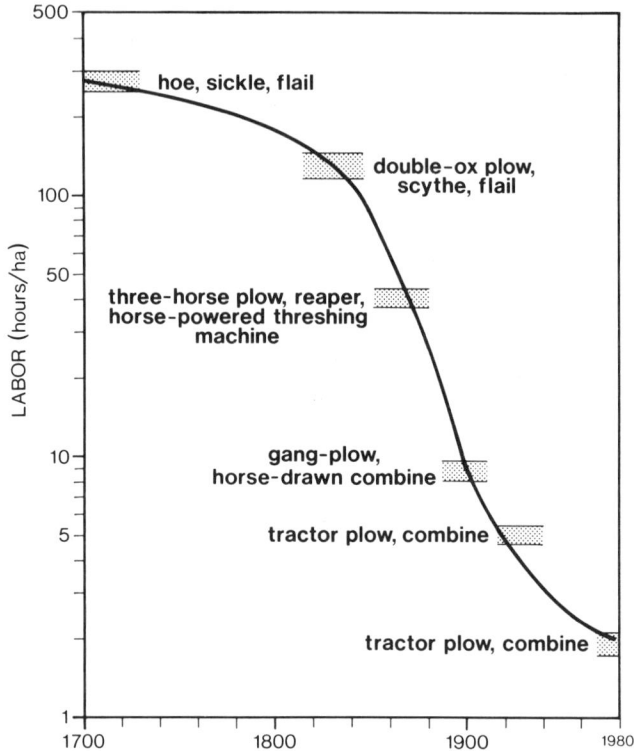

Figure 14.4 Global decline of labor needs in staple cropping is impressively illustrated by the shrinking of typical requirements for U.S. wheat harvests. The most primitive sequence of pioneer farming involving hoeing, harvesting with sickles, and flailing required up to 200 times more labor per hectare than does modern cultivation with tractor-drawn implements and combining.

oxen and water buffalo the shares were considerably lower. Better implements revolutionized traditional farming and sharply reduced inputs of human labor (Fig. 14.4).

Irrigation's antique origins are attested by a variety of simple water-lifting machines tediously powered by men or animals. Lifts were mostly low (0.5–2.5 m), efficiencies no better than 20–30%, hourly capacities 3–15 m^3, and energy costs ranged between 100–250 kJ/m^3 for men-powered lifts and 4.5–6.5 MJ/m^3 for animal water raising. Energy returns in terms of additional crop yields increased by a factor of 10–20. Fertilization had to rely solely on the recycling of organic matter. Nitrogen was nearly always the limiting nutrient but in all common organic fertilizers it is just 0.5–1.5% of fresh weight.

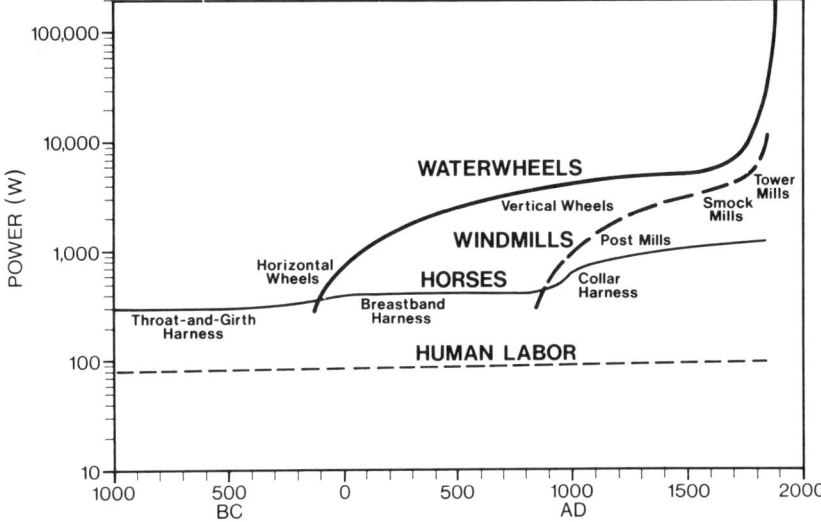

Figure 14.5 Before the introduction of simple horizontal waterwheels typical unit rates of preindustrial power inputs were limited to a few hundred watts in poorly harnessed horses. Better harnessing, better feed, and heavier breeds eventually raised the top horsepower to just over 1 kW—but by that time the early nineteenth-century windmills could deliver nearly 10 times and waterwheels 10–100 times more useful power.

Consequently, effective fertilization required laborious applications of huge amounts of organic matter. But energy gained in additional yields was 10–20 times higher than the cost of fertilization.

Crop rotation was the third key ingredient of farming intensification. Leguminous species were rotated with staple grains in every traditional agriculture and they were also an indispensable source of food and feed protein complementing incomplete cereal proteins. Typical subsistence agricultures stagnated for centuries between 2 and 3 people/ha, while the most intensive agricultures of China and western Europe had annual yields of up to nearly 50 GJ/ha and overall energy returns of 15–20; they could support more than 7 people/ha and, in the best environments, even over 12 people on modest diets.

For millennia, complexification of traditional urban societies rested on animate power and combustion of biomass fuels. Considerable ingenuity went into the design of mechanical devices but the maximum power inputs were raised substantially only with the widespread diffusion of waterwheels and windmills (Fig. 14.5). Human labor can sustain only 50–100 W/person and a pair of properly harnessed horses around 1 kW. Most of the land transportation was thus dominated by unit inputs of 100–1000 W.

Unit ratings of waterwheels remained limited for centuries—even in the eighteenth century the typical size did not exceed 4 kW—but multiple installations could deliver 10–20 kW even in antiquity and close to 100 kW by the eighteenth century. Major technical improvements during the nineteenth century led to the largest unit sizes of up to 400 kW. Simple medieval post windmills could deliver just a few kilowatts of useful power; tower and smock mills of the eighteenth and nineteenth centuries, 6–14 kW.

Thermal needs of preindustrial civilization came overwhelmingly from combustion of phytomass. Absolutely dry wood has 17.5–21 MJ/kg and crop residues 17–18 MJ/kg. But when burned (air-dry) these fuels have around 15 MJ/kg and most of their heat content was wasted in open fires and simple fireplaces, which deliver no more than 5–10% useful energy. In warm climates traditional societies annually consumed less than 10 GJ of phytomass per capita; in the middle latitudes with increasing manufacturing uses, up to 50 GJ; and in the most advanced wooden-age society—the mid-1800s United States—close to 100 GJ per capita. Since the late Middle Ages production of high-energy-density charcoal for iron smelting accounted for a rising share of consumed phytomass.

Traditional charcoaling was very wasteful, as was primitive smelting, which needed up to 600 MJ/kg of pig iron. These demands led to extensive deforestation throughout Europe and Asia. By the end of the eighteenth century the smelting rate was below 250 MJ/kg. Substitution of coke for charcoal was one of the two major reasons for rapid expansion of coal mining; diffusion of the steam engine was the other.

This was a profound change. Earlier societies had been energized by virtually instantaneous solar flows. Coal mining laid the foundation for a civilization living off the accumulated stores of fossil fuels. Escalating withdrawals of this fuel capital have energized the exponential increase of global population through unprecedented improvements of agricultural productivity and have led to enormous increases of material affluence among the richest 20% of mankind as well as to globalization of human affairs and concerns.

Around 1750 biomass combustion surpassed the burning of coal roughly tenfold; by 1900 gross energy content of biomass energies was equal to those used as fossil fuels; by 1985 fossil fuels contributed an order of magnitude more than biomass (Fig. 14.6). Considering

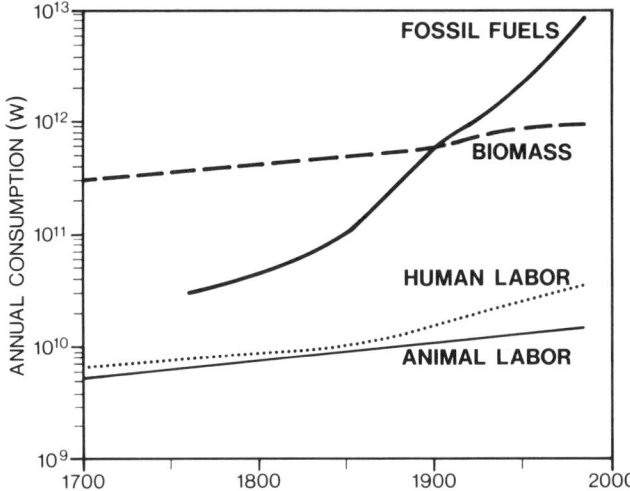

Figure 14.6 Around 1900 global consumption of some 600 GW of fossil fuels sur-passed the total combustion of biomass energies; four generations later the expon-ential ascent of coal and hydrocarbon extraction brought fossil fuel consumption to nearly 9 TW, a total roughly 250 times larger than useful human labor and 600 times greater than animal power.

the superior conversion efficiencies of fossil fuel combustion—on the average at least four times higher than burning of phytomass—coals, oils, and gases contribute about 40 times more useful energy than plant matter. This dominance has had far-reaching effects on the structure and functioning of human societies. Our dependence on coals and hydrocarbons cannot be shed without profoundly reshaping society.

Resources which molded our civilization are mineraloids of organic origin (though perhaps some are abiogenic), containing variable shares of moisture, trace elements (most notably sulfur and nitro-gen), and incombustible ash. Combustible carbon in coals releases as little as 8 MJ/kg in the poorest lignites and over 30 MJ/kg in the best anthracites. Standard bituminous coal has 29 MJ/kg, most steam coals around 20 MJ/kg. Complex hydrocarbons give more homo-geneity to crude oils; their energy density varies just between 42 and 44 MJ/kg. Natural gases have 30–45 MJ/m^3.

Total resources of these fuels are uncertain. Global coal reserves recoverable with the late twentieth-century techniques amount to some 250 years of the 1985 production. Uncertainties regarding ulti-mately recoverable hydrocarbon reserves result in estimated reserve/production ratios of 50–120 years for crude oils and 200–380 years for natural gases. The fossil fuel era will be relatively short even when measured on civilizational scale.

Spatial distribution of fossil fuels is highly unequal, with three nations (the United States, Soviet Union, and China) claiming more than two-thirds of all coal, and the stunning resource singularity of the Persian Gulf–Zagros basin storing more than half of all recoverable oils and gases. Proved recoverable stores of all conventional fossil fuels add up to some 3×10^{22} J, and ultimate recovery estimates are more than twice as large.

Coal's ascent was based on an enormous amount of hard and dangerous human labor. Mechanization of underground mining spread widely only after 1945 and it remains very low in many poor coalmining nations. Daily productivities of 0.5–2 t per miner were substantially increased by a gradual shift to surface mining. Its inherent safety, high coal recoverability (over 90% compared to 50% with the traditional room-and-pillar method), and high productivity (commonly over 20 t per shift) make it superior even to longwall extraction, the best modern underground technique. Power densities of coal extraction range from 1–2 kW/m^2 in underground mines up to 20 kW/m^2 in surface operations. The United States, Soviet Union, and China produce nearly 75% of global output.

Although very little coal is exported (about 10% of global extraction), crude oil is the world's most valuable traded commodity (40% of its output is exported). Worldwide growth of its use has been driven by the diffusion of internal combustion engines, conversion of coal-fired industrial boilers, and expansion of petrochemical industries. Improvements in exploratory and production drilling, transportation (larger and longer pipelines, increased tanker sizes peaking at just over 500,000 t), and processing (larger and more efficient refineries with catalytic cracking) made the fuel the world's most important source of primary energy. Its production densities are generally 10–20 kW/m^2, about 10% of the total output goes for nonenergy uses (lubricants, petrochemical feedstocks), and although the Soviet Union is the leading producer, the Middle Eastern fields continue to dominate the world market.

Outside of the United States natural gas became an important fuel only after World War II, with the advent of long-distance, large-diameter pipelines. Exports as LNG remain limited. As with the crude oil, the fuel is also an important feedstock and its cleanliness makes it the preferred choice for household and urban uses. The two superpowers are by far the largest producers.

Since the 1890s increasingly larger amounts of fossil fuels (more than 25% by the late 1980s) have been converted to electricity. Edison's brilliant creation of a whole new energy generation, distribution, and conversion system started a still continuing expansion of the most convenient, cleanest (at point of use), and productively most rewarding source of energy. Growth rates of electricity production have consistently outperformed increases in fossil fuel extraction.

Most of the installed global capacity remains in fossil-fueled, especially coal-fired, power plants but almost 25% of total output comes from hydrogeneration and nearly 20% from nuclear fission. Efficiencies of thermal generation rose from an initial 5% to just over 40% in the best plants, long-distance transmission links (often DC) move gigawatt blocks of electricity over distances exceeding 1000 km (there is a lively international trade), and electric lighting and motors are as indispensable in industries as in households and agriculture.

Exponential growth of fossil energy use has been accompanied—and promoted—by several systematic technical improvements of new prime movers: increase of unit sizes, decline of power intensities, and improvement of conversion efficiencies. Without these developments there could not have been such a pronounced concentration of production, processing, and conversion capacities; emergence of extensive personal mobility and growth of global trade; and the revolution in agricultural productivity. In turn, these changes were the keys opening the gates to rapid urbanization, increasing affluence, and continuing integration of the global economy.

For millennia animate power limited unit work inputs to 10^2 W. Waterwheels and windmills raised the power—locally and sporadically —to 10^3 W. During the 1720s an inefficient steam engine moved for the first time ahead of the two mechanical solar prime movers. The steam engine era lasted until the beginning of the twentieth century: by 1905 there could be no doubt about the supremacy of steam turbines. Nearly a century later they still remain the most powerful continuous-load prime movers (only rocket engines surpass them during their brief firings): the largest ones, delivering up to 1.5 GW, are about 200,000 times more powerful than the largest prime mover of the pre-steam era (Fig. 14.7).

Evolution of power intensities (weight/power ratios) presents an inverse image (Fig. 14.8). Men and draft animals needed at least 500 g/W of useful power and the earliest steam engines were no better. Eventually, locomotive steam engines rated well below 100 g/W. Soon this performance was vastly surpassed by internal combustion engines, whose power intensities dropped by more than two orders of magnitude in less than a century, to just 1 g/W for the

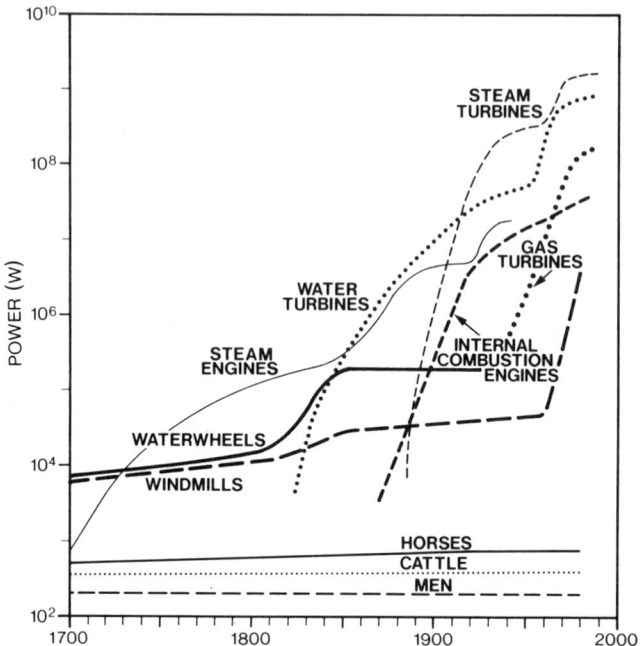

Figure 14.7 Three centuries of rising unit powers of prime movers. Steam engines passed watermills by 1730 and, in turn, they were made rapidly obsolete by steam turbines after 1905. The largest prime movers of the late twentieth century are about 200,000 times more powerful than their counterparts in 1700.

best automotive engines and nearly half the value for the best aeroengines.

These innovations ushered the second revolution in land transport —the automotive revolution—and made it possible to realize the ancient dream of powered flight. As the reciprocating aeroengines were nearing the limit of their performance in the 1940s the rapid development of turbojets and turbofans took over and pushed the global trend of power intensities below 0.1 g/W, opening the way to mass air travel and freight. Gas turbines powering trains and ships also transformed important segments of land and water transport.

Increased unit ratings, lower power intensities, and higher efficiencies of prime movers led to impressive growth of machines and plants and to enormous concentration of extractive and processing facilities. Coal mines grew from small eighteenth-century pits producing a few hundred kilowatts to surface giants shipping over 1 GW; numerous oil fields extract more than 100 MW, some Middle Eastern giants over 50 GW; the largest thermal electricity–generating plants surpass 5 GW, the largest hydrostations 10 GW.

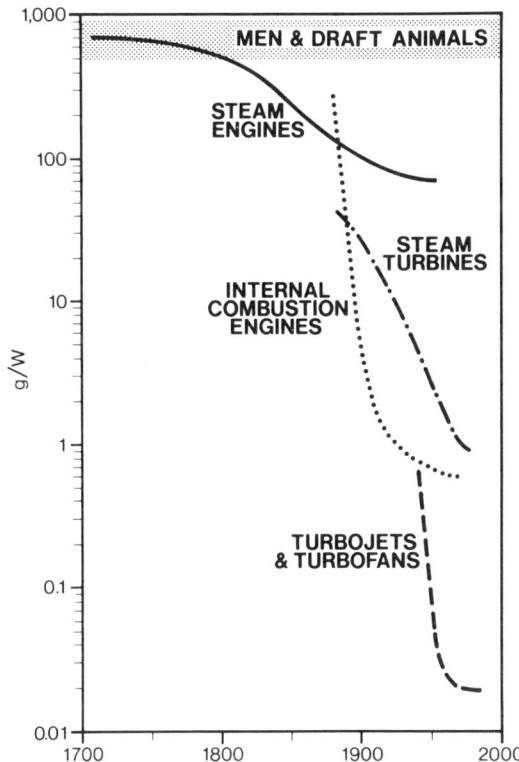

Figure 14.8 Declining power intensity of common prime movers has spanned four orders of magnitude in two centuries. Power intensities of the first commercial steam engines—660–650 g/W—were no better than those of human and animal muscles. A quarter millennium later the best steam turbines and automotive internal combustion engines rate just 1 g/W, the best jet engines less than 0.05 g/W.

The importance of personal transportation is perhaps best appreciated by noting the exponential growth of car ownership and its diffusion beyond the once heavily dominant United States. This comes at a major energy cost: building a car takes 80–150 MJ/kg; running most cars requires 2.5–3.5 MJ/km, considerably down from 6 MJ/km a generation ago, but still far from the best achievable rates of just over 1 MJ/km.

Declining costs and better performance of heavy-duty prime movers—diesel engines and gas turbines—were critical for the sustained expansion of global economic integration (foreign trade accounted for about 20% of the gross world product in the late 1980s, compared to less than 1% in 1945) and for the explosion of air travel. Considering their large structural mass, machines have been generally fairly energy-efficient providers of transportation (Fig. 14.9).

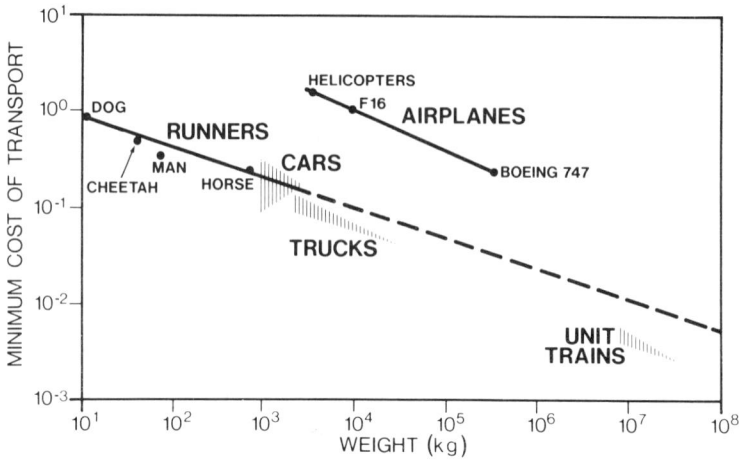

Fugure 14.9 Flying is certainly no energy bargain—minimum costs of airplane transport are about an order of magnitude higher than those of land locomotion—but automobile performances fit right along the line for large mammalian runner, costs of truck transport fall clearly below it, and heavy unit trains are even more efficient. Minimum costs of transport are calculated by dividing the product of heterotrophic or machine weight (in kg), and gravitational constant and optimum speed (in m/s) into the power input (in W).

Availability of increasingly more powerful, lightweight, and efficient prime movers also revolutionized field farming. Where irrigated crops are grown the cost of irrigation is the largest energy input (up to 20 GJ/ha). In nonirrigated fields fertilizers are the largest energy subsidy. Mining and relatively simple processing of potash (energy costs 4–10 MJ/kg of potassium) and phosphates (20–30 MJ/kg phosphorus) add up to a fraction of energy costs of synthesizing ammonia from atmospheric nitrogen and hydrogen derived mostly from methane.

The Haber–Bosch process was launched in 1913 and its initially very high energy costs (in excess of 100 MJ/kg nitrogen) have been brought down by a series of innovations to a low of 35 MJ/kg nitrogen in the 1980s. Nitrogen fertilizers subsequently synthesized from ammonia cost 50–100 MJ/kg, and the maximum fertilizer applications represent embodied energy costs of up to 40 GJ/ha.

Agricultural machinery has been the third indispensable ingredient of farming energy subsidies. A century after the trial of the first gasoline tractor (in 1892) there were nearly 30 million machines worldwide, averaging 30–50 kW (maxima around 300 kW), costing 70–120 MJ/kg to make, and consuming 1.5–3.5 GJ/ha in a variety of field operations. Pesticides are the latest, post–World War II, major subsidy, but since they are applied in much smaller amounts than fertilizers, their high unit energy costs (150–250 MJ/kg) do not add up to a high total.

Rates of carbon fixation are fixed but the new high-yielding cultivars—especially the short-stalked varieties of wheat and rice—have been bred to store more of their photosynthate in the grain and less in the stem and leaves (harvest indices rose from 25–30% in traditional varieties to 45–55% in the improved cultivars). They also respond vigorously to water, nutrient, and pesticide subsidies, which free them from channeling much of the photosynthate into growth and maintenance of more root biomass and into the repair and replacement of tissues. But the peak national harvests of staple cereals (around 100 GJ/ha in terms of edible grain) are still only a fraction of recorded maxima.

Worldwide, synthetic fertilizers account for slightly more than half of all farming energy subsidies, which prorate from just a few gigajoules per hectare in extensive cropping to about 10 GJ/ha in the United States, nearly 30 GJ/ha in China, and 80 GJ/ha in Israel. Their global total (some 12 EJ in 1985) has expanded more than a hundredfold since 1900 and it has been responsible (as cultivated area increased only by about a third) for feeding an additional 2.5 billion people (Fig. 14.10).

Doing without these subsidies would necessitate, even should the rich nations revert to a largely vegetarian preindustrial diet, a drastic reduction of the global population. With them the world as the whole has a surplus of agricultural production (daily average of some 20 MJ per capita, compared to the existential needs of some 9 MJ) and the continuing shortages and even famines are products of unequal distribution and national mismanagement.

Rising energy subsidies in farming released the rural labor force for the growing manufacturing sector and started a rapid transition to urban society, which is now essentially completed in the rich countries. In turn, cities and conurbations with populations in the multimillions could not have arisen without high energy densities of fossil fuels, portability of refined oil products, great convenience of natural gas, and superior flexibility of electricity.

The highest levels of this dependence—prorating annually to close to 300 GJ per capita in the United States and Canada, and ranging mostly between 100 and 180 GJ per capita in Europe—are so high that the richest tenth of global population accounts for 40% of all primary energy use while the poorest half of the mankind consumes less than 15% of the annual fuel and electricity flow. The post–World

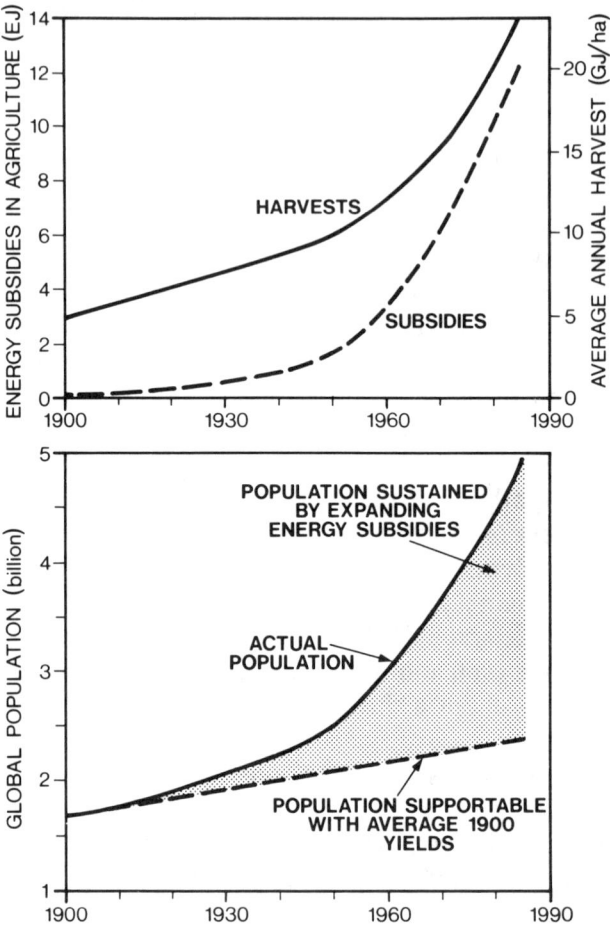

Figure 14.10 More than a hundredfold increase in agricultural energy subsidies between 1900 and 1985 led to a nearly fivefold rise in average yields. These expanded harvests could not only sustain a nearly three times larger global population—they could also provide much higher nutritional variety and quality to a growing number of people. Without higher energy subsidies the perpetuation of average 1900 yields on a slowly rising cultivated area could support less than 2.5 billion people by 1985; return to the overwhelmingly solar farming of the late nineteenth century would result in halving the world's population.

War II decades have seen only a slight reduction of this enormous inequality and although there is no need to raise average consumption rates to over 100 GJ/y in order to enjoy a high quality of life, there is an obvious need to increase substantially the 1980s poor world average of just around 20 GJ per capita.

While rich nations can contemplate generations of stabilized and even significantly declining energy needs, the poor world's demand, even when reducing many conversion inefficiencies, must increase if the still growing populations are to be provided with basic physical necessities and with at least a modicum of intellectual advancement.

Higher pollution and more widespread environmental degradation will be an inevitable consequence of this needed expansion. A large variety of environmental problems is caused by extraction and conversion of fossil fuels, most of them having merely local (surface mining, hot-water discharges) or regional (visibility reduction, acid deposition) impacts and, once the investment is available, ready technical solutions. Similarly, soil erosion, the most widespread environmental price exacted by energy-intensive farming, is manageable by a combination of managerial and technical measures.

Nor is the increased generation of heat a major worry. Although it is nearly a quarter of the planetary heat flow, the total anthropogenic power flux of some 10 TW is less than 0.01% of the radiation absorbed by the biosphere, still far from being able to influence global climate. Intractable long-term concerns arise from the growing interference in grand biospheric cycles of carbon and nitrogen.

Potentially the most worrisome aspect of these changes is the rapidity and the extent of future climatic changes influenced by accumulation of tropospheric CO_2—at over 350 ppm in 1990 and rising about 0.05%/y—and by slowly increasing concentrations of other "greenhouse" gases (prominent among them being N_2O from higher fertilizer-induced denitrification). There is no simple fix: complete combustion must oxidize C to CO_2 and there is, fortunately, no way to prevent denitrification (otherwise the biosphere would run out of N). Only modifications of future fuel conversion and fertilizer application rates can break the long-term trend of steady increases, but such changes could be expected only in a world of much greater economic equality and effective international cooperation. Energy challenges thus become environmental challenges, which, in turn, are largely the tests of social arrangements.

Gradual transition from fossil-fueled energetics to a civilization running, again, on instantaneous solar flows (but now converted with efficiencies superior to those of preindustrial societies) is the most obvious solution to energy-induced global environmental threats—as well as a necessity dictated by the limited crustal stores of fossil fuels. But such a transition will be neither fast nor easy.

Every society could be thought of as being composed of the energy it consumes: a different set of primary energizers must necessarily remold it in many profound ways. Fossil fuels shaped twentieth-

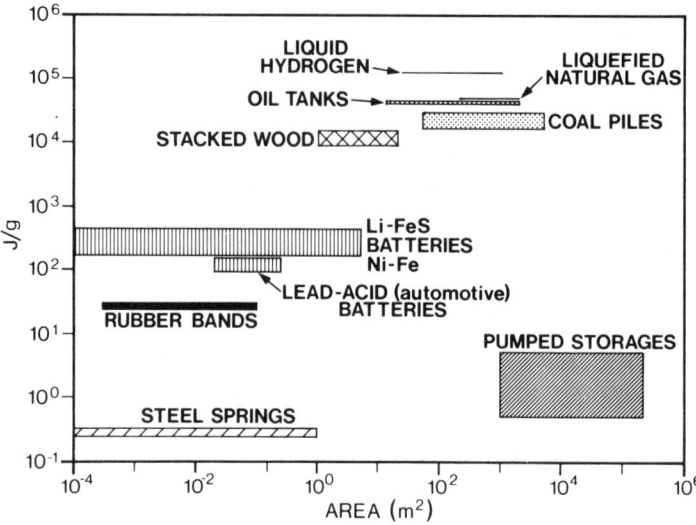

Figure 14.11 Energy in fossil fuels can be stored with densities two orders of magnitude higher than in various batteries and about four orders of magnitude higher than in water reservoirs. Only liquid hydrogen's density is higher—but its storage is obviously much more difficult than simple coal piles or oil tanks.

century civilization. No other energies (save nuclear fission with its questionable future) can be made available with such high power densities—and no other energies can be so conveniently stored with such high energy densities as fossil fuels (Fig. 14.11). Renewable energies have an enormous resource base, but this enormity is reduced to much lower magnitudes when considering their low power densities and their stochasticity. Decentralization will be imperative, and no small challenge for a still centralizing (i.e., still rapidly urbanizing) civilization. Can we, social creatures *par excellence*, opt for major decentralization?

Agricultural adjustment will be equally profound. Intensive field cropping is a space-reduction technique and land use can be further intensified through higher energy subsidies. But since power densities for producing these inputs are fixed, there is a clear limit to these gains. This limitation would be specially acute in a purely solar society using such low power density techniques as hydro or solar-generated electricity to decompose water in order to get hydrogen for ammonia synthesis.

The available resource base guarantees that the first half of the new century can still be comfortably energized by fossil fuels—although environmental consequences may greatly modify future recovery rates. But unless global civilization finds and acceptable energy base in advanced nuclear techniques, the second half of the century will be marked by a determined transition to solar energetics.

There is little difficulty in envisaging limitations and adjustments of the coming transition—as well as many of its advantages. But there are no decision-making shortcuts to aid us in selecting sensible strategies. Careful process energy analyses are valuable for better management of energy use but thermodynamic efficiency should not become the overriding arbiter in social decisions. Our monetary valuations are undoubtedly very unsatisfactory on many scores but substituting energy valuations would merely install another misleading denominator.

Calculations of net energy returns are desirable—although not necessarily decisive—in assessing the costs of energy provision, but they reveal merely the obvious and the irrelevant when applied to food eaten for its protein, vitamins, and minerals. Moreover, in every society eating is much more than merely a sustenance of heterotrophic metabolism.

Perhaps the key to future success is to break with the long past of increasing energy availabilities. Of course, Ethiopians or Chinese need more energy services and hence an efficiently expanded supply. But most of the world's low-entropy flux is used by nations that could derive great benefits from seriously examining their longstanding pursuit of higher energy inputs.

At the beginning of the twentieth century Ostwald was the first scientist to tie the availability of energy, substitution of mechanical prime movers for human labor, and increased efficiency of energy conversions to cultural progress—and the extension of Lotka's maximum power–optimum efficiency principle to human affairs would mean that the most competitive societies would strive for the highest possible energy fluxes converted at relatively low, but somehow optimal, efficiencies.

Yet if competitiveness is defined broadly as a combination of social stability, domestic economic strength, and vigorous export performance, then the indisputable primacy near the end of the twentieth century belongs to Japan—a nation of frugal consuming habits, minimized energy flows in industries and households, and maximized conversion efficiencies and material recycling.

Historical perspective casts more doubt on the validity of the maximum power stratagem. Expansions of empires may be seen as perfect examples of striving for maximized power flows. But societies commanding prodigious energy flows—be it late imperial Rome or late twentieth-century superpowers—can be highly vulnerable to internal malaise and display the loss of direction incompatible with resources at their command. And even at the peak of their physical powers these high-energy societies may be unable to deal with weaker but determined enemies—be they Germanic tribes, Vietnamese peasants, or Afghan tribesmen.

Linkages between energy use and accomplishment are suspect on many grounds. Higher energy use by itself does not guarantee anything—except for inevitably higher environmental burdens. Higher energy use does not make a country more secure. The Soviet case—with nearly doubled postwar per capita energy use—is perhaps the most striking example. Achievement of de facto nuclear parity with the United States during the 1970s was followed by open admissions of economic prostration in the 1980s, by a fundamental reappraisal of the Soviet strategic posture, and by an initiation of radical economic changes.

Higher energy use does not guarantee economic prosperity. Per capita energy consumption of Japan and West Germany—the two leading global economies—are, respectively, less than 40% and 60% of the U.S. rate. Higher energy use does not bring greater cultural flowering. If this self-evident fact needs illustrating it is enough to juxtapose Greek urban civilization of 450 B.C. with today's Athens, or Florence of the late fifteenth century with Los Angeles of the late twentieth. In both comparisons there is about an order of magnitude difference in per capita use of primary energy—coupled with immeasurable *inverse* disparity in terms of respective cultural legacies.

Greater energy use does not create superior quality of life. Above the desirable annual minima there is very little, if any, correlation between energy use and any sensible indicator of good living. Higher energy flows erode quality of life, first for populations immediately affected by extraction or conversion, eventually for everybody through worrisome global environmental changes. Higher energy use does not promote social stability. Rather, a case could be made for greater social disintegration and malaise accompanying the evolution of high-energy societies. None of the social ills—abuse of children and women, abortion, marriage breakdown, violence, alcohol and drug abuse—has ebbed in affluent societies; in most they have only grown worse.

Higher energy use in farming does not guarantee prosperous agriculture, for a large share of fossil fuel and electricity subsidies may be used with poor efficiency (in irrigation or fertilization), may go to support unhealthy diets, or may be responsible for environmental degradation incompatible with permanent farming (soil erosion, salinization, pesticide residues). Higher energy use in industry does not lead to needed modernization in poor nations. Here China's post-1949 example is most instructive. With huge energy flows powering decades of heavy, inefficient Stalinist industrialization, China's relatively high per capita energy use has done much less for the country's advancement than a balanced strategy of industrial development combined with promotion of the service sector.

Higher energy use does not necessarily bring higher system efficiencies. Evolution tends to increase the ecological efficiency of ecosystems and there have been impressive efficiency increases of individual prime movers and combustors during the recent centuries of civilizational advance. But as a large part of the total primary energy consumption goes into frivolous generation of short-lived junk and into dubious pleasures promoted by advertising, the overall ecological efficiency of modern high-energy societies is hardly an improvement on the earlier state of human development.

Along the same lines, higher energy use does not meaningfully increase civilization's diversity. Unlike the case of natural ecosystems, with their clear link between useful energy throughputs and species diversity, it would be an audacious generalization to interpret the greater availability of consumer goods and services in high-energy societies as signs of diversity. Rather, with rampant materialism, with increasing numbers of functionally illiterate people, and with powerful appeal of the mass media striving for the diffusion of the lowest common denominator, human intellectual diversity may be nearing a historically low point.

For all desirable gains other inputs and conditions are no less essential than the requisite energy flows. National security is unattainable without social cohesion, purposeful striving for a more fulfilling future, and a sound economy. Economic security comes when nations do not live beyond their means. Quality of life arises from awareness of history, from strong cultural values, and from preservation of nature's irreplaceable services as opposed to mere extraction of its goods. Social stability rests on the cohesiveness of family, on the sense of belonging, on shared moral values. Satisfactory performance in agriculture comes from farming without excess. Wise investment of energy in a nation's modernization requires diversification and flexibility. And true human diversity is impossible without elevating human efforts above the worship of material goods.

We must hope that during the next century people will work out a new balance between the necessity of providing adequate energy to sustain decent quality of life for increasing proportions of humanity and the equal imperative of not affecting the essential biospheric functions in ways inimical to human survival. Hope is necessary, for achieving this grand compromise is not inevitable. Possibilities of other outcomes were on Samuel Brody's mind when, in 1945, he was concluding his monumental *Bioenergetics and Growth*:

> The current confusions and maladjustments may be transitory, temporary disharmonies associated with rapid growth, in the nature of "growing pains," which may be resolved into a harmonious unity. But this is not certain. The future trends of human behavior are not predictable because human behavior is modifiable by intelligence, by education, by laws of human making.... This relative indeterminacy of human behavior and social phenomena is the despair of the social scientist and perhaps the hope of humanity; for this gives man an opportunity to mold his destiny that is not given to other species that are subject to more determinate, orderly, laws.

Our hope must be that we will find the determination and intellect confirming the Linnean designation of our species—*sapiens*.

APPENDIX

newton - metre

Power is simply the rate of flow of energy: 1 joule per second is 1 watt. In contrast, no energy definition can be this crisp. I prefer Rose's (1986) evasive solution: *energy* "is an abstract concept invented by physical scientists in the 19th century to describe quantitatively a wide variety of natural phenomena." Textbooks and encyclopedias define energy as "the capacity for doing work"—but this definition fits only when *work* is understood as the "act of producing a change of configuration in a system in opposition to a force which resists that change" (Maxwell 1872). The standard physical derivation of the basic energy unit via Newton's second law of motion—1 joule is the force of 1 newton (i.e., mass of 1 kg accelerated by 1 m/s^2) acting over a distance of 1m—is impeccable but relates only to kinetic energy, and it hardly fosters an intuitive grasp of the elusive concept.

Other energies were originally perceived as independent entities and defined accordingly. The classical derivation of the unit of thermal energy—1 calorie as the amount of heat needed to raise the temperature of 1 g of water from 14.5 to 15.5°C—describes a process that may be easier to imagine than acceleration of a small object. Ever since the acceptance of the first law, heat also has been quantifiable in dynamical units (1 cal = 4.184 J). Potential energy is, in the strict physical sense, only a property of a mass raised through a distance. Potential energy of a lump of coal lying on the ground is thus nil—but its heat content is 25 MJ/kg. Electric energy is frequently, and incorrectly, called power. Perhaps the best approach to expand the universal definition of energy as the capability of doing work is to set down the matrix of energy conversions (Figure A1).

Power ratings are more widely understood than energy storages and heat contents. Tables A1 and A2 help in setting some basic power levels, while Table A3 lists energy contents of common foodstuffs,

FROM / TO	ELECTRO-MAGNETIC	CHEMICAL	NUCLEAR	THERMAL	KINETIC	ELECTRICAL	GRAVITATIONAL
ELECTRO-MAGNETIC		CHEMILUMINES-CENCE	GAMMA REACTIONS / NUCLEAR BOMBS	THERMAL RADIATION	ACCELERATING CHARGE / PHOSPHOR	ELECTRO-MAGNETIC RADIATION / ELECTRO-LUMINESCENCE	
CHEMICAL	PHOTO-SYNTHESIS / PHOTO-CHEMISTRY	CHEMICAL PROCESSING	RADIATION CATALYSIS / IONIZATION	BOILING / DISSOCIATION	DISSOCIATION BY RADIOLYSIS	ELECTROLYSIS	
NUCLEAR	GAMMA-NEUTRON REACTIONS						
THERMAL	SOLAR ABSORPTION	COMBUSTION	FISSION / FUSION	HEAT EXCHANGE	FRICTION	RESISTANCE HEATING	
KINETIC	RADIOMETERS	METABOLISM / MUSCLES	RADIOACTIVITY / NUCLEAR BOMBS	THERMAL EXPANSION / INTERNAL COMBUSTION	GEARS	MOTORS / ELECTRO-STRICTION	FALLING OBJECTS
ELECTRICAL	SOLAR CELLS / PHOTO-ELECTRICITY	FUEL CELLS / BATTERIES	NUCLEAR BATTERIES	THERMO-ELECTRICITY / THERMIONICS	CONVENTIONAL GENERATOR		
GRAVITATIONAL					RISING OBJECTS		

Figure A1 Conversion matrix of transformations among the energies. Where more possibilities exist only the two most important conversions are listed; fields of unknown transforms are left empty.

their constituents, and their metabolic products, Table A4 ranks energy storages and fluxes over more than 30 orders of magnitude, and Table A5 lists gross and net heat contents of common fuels.

Different energies can be easily converted to a common denominator but the results ignore critical qualitative differences. On the most general level, common denominators ignore the fundamental distinction between low-entropy stores or flows and high-entropy states. Yet the practical difference between equivalent totals of available energy in fuels, steam, or electricity and dissipated heat is all too obvious. Nor do the common denominators differentiate between renewable and fossil fuels energies. Yet these distinctions have fundamental implications for the nature and sustainability of a given energy system. In societies where animate labor is still an important source of mechanical energy there is no easy way to compare these inputs with those contributed by fuels and electricity.

Even when the comparisons are limited to fossil fuels there is a considerable loss of information in conversion to a common denominator. Differences of energy form, thermodynamic availability,

Table A1 Power Levels of Continuous Phenomena

Energy Flows	Power Ratings[a]	
	Watts	Exponent
Global intercept of solar radiation[b]	1.7	17
Wind-generated waves on the ocean[b]	9	16
Solar radiation received by China[b]	2	15
Global gross primary productivity[b]	1	14
Global Earth heat flow[b]	4.2	13
Worldwide fossil fuel combustion[b]	9.5	12
Global earthquake activity	3	11
Florida Current between Miami and Bimini[b]	2	10
Rotating turbogenerator	1	9
Midsize nuclear reactor	5	8
Gas pipeline compressors	2	7
Electricity for a 30-story high-rise[b]	1.5	6
Energy needs of a typical supermarket[b]	2	5
Large waterwheel	1	4
Japanese per capita energy use[b]	3	3
Emergency exit light[b]	2	2
Basal metabolism of a 70 kg man[b]	8	1
Net productivity per square meter of tropical forest[b]	1	0
Metabolic rate of neonate heart[b]	4	−1
Mean global rate of erosion per square meter[b]	5	−2

[a] All ratings are either quoted or calculated in Chapters 2–10.

[b] Marks continuous conversions.

energy density, extractability, transportability, ease of conversion, pollution potential, and convenience and safety of usage can never be contained by a single denominator. Equally intractable are comparisons of different electricity generation modes. In the case of fossil-fueled generation the primary energy equivalent of electricity is clearly the heat content of the fuel burned. But what is the primary energy content of nuclear electricity when the fission reactors use only a small portion of uranium's thermal potential?

And when converting hydroelectricity should one start with the potential energy of precipitation in the watershed, with all the water behind the dam, with only that part of the storage which is above the penstock intakes, or with only that small share of stored water which is actually used in generation? Or is it simply the straight thermal equivalent of electricity (i.e., 1 kWh = 3.6 MJ)?

Table A2 Power Levels of Ephemeral Phenomena

Energy Flows	Duration (s)	Power Ratings (W)[a] Actual Multiple	Order of Magnitude
Richter magnitude 8 earthquake	30	1.6	15
Large volcanic eruption	10^4	1	14
Giant lightning	10^{-5}	2	13
Rainstorm's latent heat	1200	1	12
Thunderstorm's kinetic energy	1200	1	11
Large World War II bombing raid	3600	2	10
Average U.S. tornado	160	1.7	9
Mount St. Helens seismic waves	10^4	5	8
Small avalanche with 500 m drop	20	1.1	7
Large coal unit train shuttle	10^3	5	6
Intercity truck trip	10^4	3	5
Gasoline for a 20 km drive	1200	4	4
Running 100 m dash	10	1.3	3
Machine-washing laundry	1500	5	2
CD player spinning Mozarts' last symphony	2238	2.5	1
Small candle burning to the end	1800	3	0
Hummingbird flight	300	7	−1

[a] Calculated largely from data in Chapters 2–10.

Total primary energy inputs are the most commonly used indicators of energy utilization—but they bear no simple relationship to the final inputs of useful energy generating comfort and wealth. The difference between the two measures is the result not only of national disparities in engineering and managerial prowess but also of a host of social and environmental factors.

Another simple measure hiding considerable complexities is conversion efficiency, the standard yardstick to assess the performance of energy transformations. The most commonly used ratio is not necessarily the most revealing one, the quest for the highest rate is not always the most desirable goal, and inevitable preconversion energy losses may be far greater than any conceivable conversion improvements. The ratio of energy output or transfer of the desired kind achieved by a converter to the initial energy input to the device, organism, or system is by far the most commonly used value (Table

Table A3 Energy Equivalents of Common Foodstuffs, Their Constituents, and Their Metabolic Products[a]

	Energy Content (kJ/g)	
	Raw	Cooked
Foodstuff		
Cereal grains	15.2–15.4	5.0–10.0
Legume grains	14.2	4.4–4.9
Potatoes	3.2–4.8	3.2–4.9
Sugar	16.1	16.1
Plant oils	37.0	37.0
Vegetables	0.6–1.8	0.6–1.5
Fruits	1.5–4.0	1.5–4.1
Red meats	5.6–23.1	7.0–18.0
Poultry	4.9–13.6	6.9–13.3
Milks	1.5–2.9	1.5–2.9
Butter	30.0	30.0
Eggs	6.8–8.0	6.8–8.0
Fish	2.9–9.3	2.9–7.6
Alcoholic beverages	1.7–12.3	
	Total	Digestible
Nutrients		
Carbohydrates	17.0	17.0
Protein	23.0	17.0
Lipids	39.0	38.0
Ethanol	29.6	—[b]
	Fresh	Dry
Metabolic products		
Muscle	5.9–9.5	23.8
Fat	31.0–37.0	36.0–38.0
Urine	0.1–0.2	—[c]
Feces	1.8–3.0	7.2–12.0

[a] Food values recalculated from Watt and Merrill (1975); values for cooked foodstuffs are without any additions of fats or sugars.

[b] On digestion of ethanol see Chapter 5.

[c] Urea, the most important component of dry matter in urine, has energy content of 10.6 kJ/g.

Table A4 Energy Flows and Storages

Energy Flow or Storage[a]	Actual Multiple (J)	Order of Magnitude
Solar radiation intercepted by the Earth[b]	5.5	24
Global coal resources	2	23
Global plant mass	2	22
Global net photosynthesis[b]	2	21
Global fossil fuel production[b]	3	20
Typical Caribbean hurricane	3.8	19
Global lightnings[b]	3.2	18
Largest H-bomb tested in 1961	2.4	17
Global zodiacal light[b]	6.3	16
Latent heat of a thunderstorm	5	15
Kinetic energy of a thunderstorm	1	14
Hiroshima bomb (1945)	8.4	13
Coal load in a 100 t hopper car	2.5	12
Good grain corn harvest (8 t/ha)	1.2	11
Gasoline for a compact car[b]	4	10
Barrel of crude oil	6.5	9
Basal metabolism of a large horse	1	8
Daily adult food intake	1	7
Bottle of white table wine	2.6	6
Large hen egg	4	5
Vole's daily basal metabolism	5	4
Small chickpea	5	3
Baseball (140 g) pitched at 40 m/s	1.1	2
Tennis ball (50 g) served at 25 m/s	1.5	1
Full teacup (300 g) held in hand	2.6	0
Falling 2 cm hailstone	2	−1
Striking a typewriter key	2	−2
Fly on a kitchen table	9	−3
Small bird's 5 second song	5	−4
A 2 mm raindrop falling at 6 m/s	7.5	−5
The same drop on a blade of grass	4	−6
Flea hop	1	−7

[a] Calculated from data in Chapters 2–10 and from measurements of common activities and objects.
[b] Annual total.

Table A5 Comparisons of Gross and Net Heat Values for Common Fuels[a]

Fuel	Heat Value (MJ/kg)	
	Gross	Net
Anthracite	32	32
Standard bituminous coal	28–29	27–28
Typical steam coals	22–24	19–21
Good lignites	18–20	15–17
Coke	29	28
Gasoline	47	44
Diesel oil	46	43
Fuel oils	43–44	40–41
Natural gases	36–47	39–52
Hydrogen	114	93
Air-dried wood	14–15	11–12
Crop residues	12–15	8–13

[a] Based on United Nations Organization (1988), Long (1978), and Smil (1983). All values are rounded to the nearest whole number.

A6). This first-law efficiency (e_1) has at least three important drawbacks: its maximum values may be less than, equal to, or greater than unity (in the last case it is usually called a coefficient of performance); it does not adequately express the efficiency limitations imposed by the second law; and it is not readily applicable to systems whose desired output is a combination of work and heat.

For these reasons the American Institute of Physics favors the use of the second-law efficiency (e_2), which is the ratio of the least available work that could perform the task to the available work actually used in performing the job with a given device or system (Ford et al. 1975.) Obviously, this efficiency cannot surpass unity and it offers direct insight into the quality of performance relative to the ideal (the goal of energy management) and it focuses attention on the desired task, not just on a device or a system currently used for that purpose.

Where the available work is nearly equal to the heat of combustion—as is the case with large electricity generating stations— e_1 and e_2 will be very close. But whereas e_1 of home gas furnaces is typically around 60%, e_2 will be just short of 10% because the equivalent heating could be done much more efficiently by a heat pump. All conversions where high-temperature combustion is used to

Table A6 Efficiencies of Common Anthropogenic and Natural Energy Conversions[a]

Conversion	Energies[b]	Efficiencies
Large electricity generators	$m \to e$	98–99
Large power plant boilers	$c \to t$	90–98
Large electric motors	$e \to m$	90–97
Best home natural gas furnaces	$c \to t$	90–96
Dry-cell batteries	$c \to e$	85–95
Human lactation	$c \to c$	75–85
Overshot waterwheels	$m \to m$	60–85
Small electric motors	$e \to m$	60–75
Best bacterial growth	$c \to c$	50–65
Glycolysis maxima	$c \to c$	50–60
Large steam turbines	$t \to m$	40–45
Improved wood stoves	$c \to t$	25–45
Large gas turbines	$c \to m$	35–40
Diesel engines	$c \to m$	30–35
Mammalian postnatal growth	$c \to c$	30–35
Best photovoltaic cells	$r \to e$	20–30
Best large steam engines	$c \to m$	20–25
Internal combustion engines	$c \to m$	15–25
High-pressure sodium lamps	$e \to r$	15–20
Mammalian muscles	$c \to m$	15–20
Milk production	$c \to c$	15–20
Pregnancy	$c \to c$	10–20
Broiler production	$c \to c$	10–15
Traditional stoves	$c \to t$	10–15
Fluorescent lights	$e \to r$	10–12
Beef production	$c \to c$	5–10
Steam locomotives	$c \to m$	3–6
Peak field photosynthesis	$r \to c$	4–5
Incandescent light bulbs	$e \to r$	2–5
Paraffin candles	$c \to r$	1–2
Most productive ecosystems	$r \to c$	1–2
Global photosynthetic mean	$r \to c$	0.3

[a] All ranges are the first-law efficiencies in percent.
[b] Energy labels: c = chemical, e = electrical, m = mechanical (kinetic), r = radiant (electromagnetic, solar), t = thermal.

provide low-temperature heat will be exposed as similarly wasteful by using e_2 as the true measure of their performance.

Calculations of e_2, always more difficult than appraisals of e_1, often meet social and behavioral concerns. How can we factor in the unoccupied seats in commuter cars or what is indeed the meaning of e_2 in the case of a weekend driving? And using a measure equivalent to e_2 in comparing efficiencies of photosynthetic or heterotrophic conversions is largely of theoretical interest. Comparison of food grains reveals that wheat converts solar radiation less efficiently than sorghum—but the two cereals have different photosynthetic pathways and sorghum will never yield excellent bread-making flour. The worship of high abstract energy efficiencies is a dubious faith.

Where e_2 values make sense—for evaluating the performance of countless converters sustaining modern civilization—their meaningful calculation on regional and national levels runs into data shortages. Rough estimates show that national averages of e_2 are below 10% for all common household energy conversions, just over 10% for driving, and generally 25–40% for industrial applications, leaving considerable margins for conservation improvements. Infinitesimally slow conversions would produce maximum efficiencies but transformations done at rewarding rates inevitably carry high heat waste penalties.

REFERENCES

Adair, L. S., and E. Pollitt. 1982. Energy balance during pregnancy and lactation. *Lancet* **2**:219.

Adler, U., et al., eds. 1986. *Automotive Handbook*. Robert Bosch GmbH, Stuttgart.

Ajtay, G. L., et al. 1979. Terrestrial primary production and phytomass. In B. Bolin et al., eds., *The Global Carbon Cycle*. Wiley, New York, pp. 129–181.

Albers, J., et al. 1980. *Demand and Supply of Nonfuel Minerals and Materials for the United States Energy Industry, 1975–90*. USGPO, Washington, D.C.

Aldrich, S. R. 1980. *Nitrogen in Relation to Food, Environment, and Energy*. University of Illinois Press, Champaign–Urbana.

Alexander, R. M. 1984. Elastic energy stores in running vertebrates. *American Zoologist* **24**:85–94.

Allan, W. 1965. *The African Husbandman*. Oliver & Boyd, Edinburgh.

Anderson, D. L. T. 1983. The oceanic general circulation and its interaction with the atmosphere. In B. Hoskins and R. Pearce, eds., *Large-Scale Dynamical Processes in the Atmosphere*. Academic Press, New York, pp. 305–336.

Anderson, J. L. 1981. Climatic change in European economic history. *Research in Economic History* **6**:1–34.

Andrews, F. M., ed. 1986. *Research on the Quality of Life*. Institute for Social Research, University of Michigan, Ann Arbor.

Anthes, R. A. 1982. *Tropical Cyclones*. American Meteorological Society, Boston.

Ardrey, R. L. 1894. *American Agricultural Implements*. Published by the author, Chicago.

Ayensu, E. S., ed. 1980. *Jungles*. Crown, New York.

Bailey, S. M. 1982. Absolute and relative sex differences in body composition. In R. L. Hall, ed., *Sexual Dimorphism in Homo sapiens*. Praeger, New York, pp. 363–390.

Baird, G., et al. 1984. *Energy Performance of Buildings*. CRC Press, Boca Raton, Fl.

Baker, R., ed. 1981. *The Mystery of Migration*. Viking Press, New York.

Baker, V. R., ed. 1981. *Catastrophic Flooding*. Dowden, Hutchison & Ross, Stroudsburg, Pa.

Bakker, R. T. 1975. Experimental and fossil evidence for the evolution of tetrapod bioenergetics. In D. M. Gates and R. B. Schmerl, eds., *Perspectives of Biophysical Ecology*, Springer–Verlag, New York, pp. 365–399.

Bannister, R. L., and G. J. Silvestri. 1989. The evolution of central station steam turbines. *Mechanical Engineering* **11**(2):72–78.

Barber, F. M. 1900. *The Mechanical Triumphs of Ancient Egyptians*. Tribner, London.

Barlowe, R. 1979. In M. T. Beatty et al., eds., *Planning the Uses and Management of Land*. American Society of Agronomy, Madison, Wisc., pp. 3–25.

Barnes, R. D. 1989. Diversity of organisms: how much do we know? *American Zoologist* **29**:1075–1084.

Bartels, H. 1982. Metabolic rate of mammals equals the 0.75 power of their body weight. *Experimental Biology and Medicine* **7**:1–11.

Basalla, G. 1982. Some persistent energy myths. In G. H. Daniels and M. H. Rose, eds., *Energy and Transport*. Sage, Beverly Hills, Calif., pp. 27–38.

Bascom, W. 1959. Ocean waves. *Scientific American* **201**(2):74–84.

Basile, P. S., ed. 1976. *Energy Demand Studies: Major Consuming Countries*. MIT Press, Cambridge, Mass.

Bassham, J. A., and M. Calvin. 1957. *The Path of Carbon in Photosynthesis*. Prentice-Hall, Engelwood Cliffs, N.J.

Batty, J. C., and J. Keller. 1980. Energy requirements for irrigation. In D. Pimentel, ed., *Handbook of Energy Utilization in Agriculture*. CRC Press, Boca Raton, Fla., pp. 35–44.

Bazilevich, N. I., et al. 1971. Geographical aspects of biological productivity. *Soviet Geography* **12**:293–317.

Beech, G. A. 1980. Energy use in bread baking. *Journal of Science of Food and Agriculture* **31**:289–298.

Bell, I. L. 1884. *Principles of the Manufacture of Iron and Steel*. George Routledge & Sons, London.

Benedict, F. G. 1938. *Vital Energetics: A Study in Comparative Basal Metabolism*. Carnegie Institution, Washington, D.C.

Benedict, F. G., and E. P. Cathcart. 1913. *Muscular Work*. Carnegie Institution, Washington, D.C.

Benefice, E., et al. 1984. Nutritional situation and seasonal variations for pastoralist populations of the Sahel (Senegalese Ferlo). *Ecology of Food and Nutrition* **14**:229–247.

Berner, E. K., and R. A. Berner. 1987. *The Global Water Cycle*. Prentice-Hall, Englewood Cliffs, N.J.

Bernstein, I. S., and E. O. Smith. 1979. *Primate Ecology and Human Origins*. Garland STPM Press, New York.

Bertalanffy, L. von. 1932–1942. Theoretische Biologie (II vols.). Borntraeger, Bonn.

Bertalanffy, L. von. 1968. *General System Theory*. George Braziller, New York.

Bethe, H. A. 1939. Energy production in stars. *Physical Review* **55**:434–456.

Bhattacharya, A. K. 1986. Protein-energy malnutrition (kwashiorkor-marasmus syndrome): Terminology, classification and evolution. *World Review of Nutrition and Dietetics* **47**:80–133.

Black, J. N. 1971. Energy relations in crop production—a preliminary survey. *Annals of Applied Biology* **67**:272–278.

Blem, C. R. 1980. The energetics of migration. In S. A. Gauthreaux, ed., *Animal Migration, Orientation and Navigation*, Academic Press, New York, pp. 175–224.

Block, G. 1982. A review of validations of dietary assessment methods. *American Journal of Epidemiology* **115**:492–505.

Blouin, G. M., and C. H. Davis. 1975. *Energy Requirements for the Production and Distribution of Chemical Fertilizers in the United States*. TVA, Muscle Shoals, Ala.

Bolin, B., and R. B. Cook, eds. 1983. *The Major Biogeochemical Cycles and Their Interactions*. Wiley, New York.

Bolin, B., et al., eds. 1986. *The Greenhouse Effect, Climatic Change, and Ecosystems*. Wiley, Chichester.

Borrini, G., and S. Margen. 1985. *Human Energetics*. IDRC, Ottawa.

Boserup, E. 1965. *The Conditions of Agricultural Growth: The Economics of Agrarian Change under Population Pressure*. Aldine, Chicago.

Boserup, E. 1976. Environment, population, and technology in primitive societies. *Population and Development Review* **2**(1):21–36.

Bouchard, C., et al. 1981. Advances in selected areas of human work physiology. *Yearbook of Physical Anthropology* **24**:1–36.

Boulding, K. E. 1974. The social system and the energy crisis. *Science* **184**:255–257.

Boustead, I., and G. F. Hancock. 1979. *Handbook of Industrial Energy Analysis*. Ellis Horwood, Chichester.

Brafield, A. E., and M. J. Lllewellyn. 1982. *Animal Energetics*. Blackie, Glasgow.

Brantly, J. E. 1971. *History of Oil Well Drilling*. Gulf Publishing, Houston.

Bray, F. 1984. *Science and Civilisation in China. Volume 6, Part II: Agricul-*

ture. Cambridge University Press, Cambridge.

Bray, W. 1977. From foraging to farming in early Mexico. In J. V. S. Megaw, ed., *Hunters, Gatherers and First Farmers Beyond Europe.* Leicester University Press, Leicester, pp. 225–250.

Breuel, A., ed. 1981. *Oil Spill Cleanup and Protection Techniques for Shorelines and Marshlands.* Noyes Data, Boca Raton, Fla.

Broda, E. 1978. *The Evolution of Bioenergetic Processes.* Pergamon Press, Oxford.

Brody, S. 1945. *Bioenergetics and Growth.* Reinhold, New York.

Brooks, A. N., et al. 1986. Human-powered watercraft. *Scientific American* **255**(6):120–130.

Brooks, D. R., and E. O. Wiley. 1986. *Evolution as Entropy.* University of Chicago Press, Chicago.

Buck, J. L. 1930. *Chinese Farm Economy.* University of Nanking, Nanking.

Buck, J. L. 1937. *Land Utilization in China.* University of Nanking, Nanking.

Budyko, M. I. 1982. *The Earth's Climate: Past and Future.* Academic Press, New York.

Budyko, M. I., et al., eds. 1963. *Atlas of the Heat Balance of the Earth.* MGK, Moscow.

Burchell, R. W., and D. Listokin. 1982. *Energy and Land Use.* Center for Urban Policy Research, Piscataway, N.J.

Burris, R. H., and C. C. Black, eds. 1976. *CO_2 Metabolism and Productivity of Plants.* University Park Press, Baltimore.

Butzer, K. W. 1976. *Early Hydraulic Civilization in Egypt.* University of Chicago Press, Chicago.

Caine, N. 1976. A uniform measure of subaerial erosion. *Geological Society of America Bulletin* **87**:137–140.

Calder, W. A. 1983. Ecological scaling: Mammals and birds. *Annual Review of Ecology and Systematics* **14**:213–230.

Caldwell, J. C. 1976. Toward a restatement of demographic transition theory. *Population and Development Review* **2**(3–4):321–366.

Calow, P. 1977. Conversion efficiencies in heterotrophic organisms. *Biological Reviews* **52**:385–409.

Campbell, H. R. 1907. *The Manufacture and Properties of Iron and Steel.* Hill Publishing, New York.

Cardwell, D. S. L. 1971. *From Watt to Clausius.* Cornell University Press, Ithaca, N.Y.

Carlson, P. S., ed. 1980. *The Biology of Crop Productivity.* Academic Press, New York.

Carnot, S. 1824. *Reflexions sur la puissance motrice du feu et sur les machines propres a developper cette puissance.* Paris.

Carrier, D. R. 1984. The energetic paradox of human running and hominid evolution. *Current Anthropology* **25**:483–495.

Carter, G. F. 1977. A hypothesis suggesting a single origin of agriculture. In C. Reed, ed., *Origins of Agriculture*. Mouton, The Hague, pp. 123–138.

Cataldi, R. 1986. Geothermal energy. In World Energy Conference, *Survey of Energy Resources*. World Energy Conference, London, pp. 183–186.

Central Intelligence Agency. 1979. *The World Oil Market in the Years Ahead*. CIA, Washington, D.C.

Central Statistical Office. 1987. *Narodnoye khozyaystvo SSSR*. Statistika, Moscow.

Chabot, B. F., and D. J. Hicks. 1982. The ecology of leaf life spans. *Annual Review of Ecology and Systematics* **13**:229–259.

Chapman, P. F. 1974. Energy costs: A review of methods. *Energy Policy* **2**:91–103.

Chapman, P. F., et al. 1974. The energy cost of fuels. *Energy Policy* **2**:231–243.

Chapman, P. F., and D. F. Hemming. 1976. Energy requirement of some energy sources. In A., Verbraeck, ed. *The Energy Accounting of Materials, Products, Processes and Services*. TNO (Netherlands Institute for Applied Scientific Research), Rotterdam, pp. 119–129.

Charlson, R. J., et al. 1987. Oceanic phytoplankton, atmospheric sulphur, cloud albedo and climate. *Nature* **326**:655–661.

Chatterson, E. K. 1977. *Sailing Ships: The Story of Their Development from the Earliest Times to the Present*. Gordon Press, New York.

Childe, V. G. 1951. The Neolithic revolution. In V. G. Childe, ed., *Man Makes Himself*. C. A. Watts, London, pp. 67–72.

Christ, K. 1984. *The Romans*. University of California Press, Berkeley.

Christensen, E. H. 1953. Physiological valuation of work in the Nykroppa iron works. In W. F. Floyd and A. T. Welford, eds., *Ergonomics Society Symposium*. Lewis, London, pp. 93–108.

Clark, C., and M. Haswell. 1970. *The Economics of Subsistence Agriculture*. Macmillan, London.

Clark, D. A., and D. B. Clark. 1984. Spacing dynamics of a tropical rain forest tree: Evaluation of the Janzen-Connell model. *American Naturalist* **124**:769–788.

Clark, E. L. 1986. Cogeneration—Efficient energy source. *Annual Review of Energy* **11**:275–294.

Clark, J. D., and S. A. Brandt, eds. 1984. *From Hunters to Farmers*. University of California Press, Berkeley.

Clark, W. C., ed. 1982. *Carbon Dioxide Review: 1982*. Oxford University Press, London.

Clausius, R. 1867. *Abhandlungen über die mechanische Wärmetheorie*. F. Vieweg, Braunschweig.

Cockrill, W. R., ed. 1974. *The Husbandry and Health of the Domestic Buffalo*. FAO, Rome.

Cohen, M. N. 1977. *The Food Crisis in Prehistory: Overpopulation and the Origins of Agriculture*. Yale University Press, New Haven.

Coltman, J. W. 1988. The transformer. *Scientific American* **258**(1):86–95.

Committee for the Compilation of Materials on Damage Caused by the Atomic Bombs in Hiroshima and Nagasaki. 1981. *Hiroshima and Nagasaki*. Basic Books, New York. (cited as Committee for Compilation)

Constant, E. W. 1981. *The Origins of the Turbojet Revolution*. Johns Hopkins University Press, Baltimore.

Costanza, R. 1980. Embodied energy and economic valuation. *Science* **210**:1219–1224.

Cottrell, F. 1955. *Energy and Society*. McGraw-Hill, New York.

Coughenour, M. B., et al. 1985. Energy extraction and use in nomadic pastoral ecosystem. *Science* **230**:619–624.

Coupland, R. T., ed. 1979. *Ecosystems of the World: Analysis of Grasslands and Their Uses*. Cambridge University Press, Cambridge.

Crawley, M. J. 1983. *Herbivory: The Dynamics of Animal-Plant Interactions*. University of California Press, Berkeley.

Croft, T. A. 1976. Nighttime images of the Earth from space. *Scientific American* **239**(1):86–98.

Crowley, T. J. 1983. The geologic record of climatic change. *Review of Geophysics and Space Physics* **21**:828–877.

Czaya, E. 1981. *Rivers of the World*. Van Nostrand Reinhold, New York.

Daly, H. E., ed. 1973. *Toward a Steady-State Economy*. W. H. Freeman, San Francisco.

Damuth, J. 1981. Population density and body size in mammals. *Nature* **290**:699–700.

Danks, S. M., et al. 1983. *Photosynthetic Systems*. Wiley, Chichester.

Darwin, C. 1888. *The Formation of Vegetable Mould through the Action of Worms*. John Murray, London.

Daumas, M., ed. 1969. *A History of Technology & Invention*. Crown Publishers, New York.

Davies, G. F. 1980. Review of oceanic and global heat flow estimates. *Reviews of Geophysics and Space Physics* **18**:718–722.

Davis, C. G., et al. 1982. Direct-reduction technology and economics. *Ironmaking and Steelmaking* **9**:93–129.

de Duve, C. 1984. *A Guided Tour of the Living Cell*. Scientific American Library, New York.

De Niro, M. J. 1987. Stable isotopy and archaeology. *American Scientist*

75:182–191.

Dent, A. 1974. *The Horse*. Holt, Rinehart and Winston, New York.

des Noëttes, R. J. E. C. L. 1931. *L'Attelage et le Cheval de Selle a travers les Âges*. Picard, Paris.

Devine, W. D. 1983. From shafts to wires: Historical perspective on electrification. *Journal of Economic History* **63**:347–372.

DeZeeuw, J. W. 1978. Peat and the Dutch Golden Age. *AAG Bijdragen* **21**:3–31.

Dickinson, H. W. 1939. *A Short History of the Steam Engine*. Cambridge University Press, Cambridge.

Dieffenbach, E. M., and R. B. Gray. 1960. The development of the tractor. *Agricultural Yearbook* **1960**:24–45.

Dinneen, G. U., and G. L. Cook. 1974. Oil shale and the energy crisis. *Technology Review* **76**(3):26–33.

Doering, O., et al. 1977. *Current Energy Use in the Food and Fiber System*. USDA, Washington, D.C.

Donald, C. M., and J. Hamblin. 1976. The biological yield and harvest index of cereals as agronomic and plant breeding criteria. *Advances in Agronomy* **28**:361–405.

Donaldson, E. C., et al., eds. 1985. *Enhanced Oil Recovery*. Elsevier, Amsterdam.

Doorenbos, J., et al. 1979. *Yield Response to Water*. FAO, Rome.

Dovring, F. 1984. *Energy Use for Midwest Agriculture*. Agricultural Experiment Station, Urbana-Champaign, Illinois.

Dovring, F. 1985. Energy use in United States agriculture: a critique of recent research. *Energy in Agriculture* **4**:79–86.

Dregne, H. 1983. *Desertification in Arid Lands*. Harwood Academic Publishers, New York.

Drela, M., and J. S. Langford. 1985. Human-powered flight. *Scientific American* **253**(5):144–151.

Duby, G. 1968. *Rural Economy and Country Life in the Medieval West*. Edward Arnold, London.

Dudal, R. 1987. Land resources for plant production. In D. J. McLaren and B. J. Skinner, eds., *Resources and World Development*. Wiley, Chichester, pp. 659–670.

Dumas, M. J., and M. J. B. Boussingault. 1844. *The Chemical and Physiological Balance of Organic Nature*. Saxton & Miles, New York.

Düring, I. 1966. *Aristoteles Darstellung und Interpretation seines Denkens*. Carl Winter, Heidelberg.

Durnin, J. V. G. A., and R. Passmore. 1967. *Energy, Work and Leisure*. Heinemann Educational Books, London.

Earl, D. 1973. *Charcoal and Forest Management*. Oxford University Press,

Oxford.

Eastman, J. T., and A. L. De Vries. 1985. Antarctic fishes. *Scientific American* **255**(5):106–114.

Edmonds, R. L., ed. 1982. *Analysis of Coniferous Forest Ecosystems in the Western United States.* Van Nostrand Reinhold, New York.

Edwards, G., and D. Walker. 1983. *C₃, C₄: Mechanisms and Cellular and Environmental Regulation of Photosynthesis.* University of California Press, Berkeley.

Einstein, A. 1905. Zur Elektrodynamik bewegter Körper. *Annalen der Physik* **17**: 891–921.

Einstein, A. 1907. Relativitätsprinzip und die aus demselben gezogenen Folgerungen. *Jahrbuch der Radioaktivität* **4**:411–462.

Eisenberg, J. F. 1981. *The Mammalian Radiations.* University of Chicago Press, Chicago.

Elder, J. 1976. *The Bowels of the Earth.* Oxford University Press, Oxford.

Ellett, D., ed. 1982. *Time Series of Ocean Measurements.* WMO, Geneva.

Elton, C. 1927. *Animal Ecology.* Macmillan, New York.

Emsley, J., and D. Hall. 1976. *The Chemistry of Phosphorus.* Harper & Row, London.

Endler, J. A., and T. McLellan. 1988. The processes of evolution: toward a newer synthesis. *Annual Review of Ecology and Systematics* **19**:395–421.

Energy Systems Program Group. 1981. *Energy in a Finite World.* Ballinger Publishing Company, Cambridge, Mass.

Engelsted, O. P., ed. 1986. *Fertilizer Technology and Use.* Soil Science Society of America, Madison, Wis.

Esch, G. W., and R. W. McFarlane, eds. 1976. *Thermal Ecology.* Technical Information Center, Springfield, V.

Evangelou, P. 1984. *Livestock Development in Kenya's Maasailand.* Westview Press, Boulder, Colo.

Evans, L. T. 1980. The natural history of crop yield. *American Scientist* **68**:388–397.

Evans, O. 1795. *The Young Mill-wright & Miller's Guide.* O. Evans, Philadelphia.

Faegri, K., and L. van der Pijl. 1979. *The Principles of Pollination Ecology.* Pergamon Press, Oxford.

Falk, J. H. 1980. The primary productivity of lawns in a temperate environment. *Journal of Applied Ecology* **17**:689–695.

Farey, J. 1827. *A Treatise on the Steam Engine.* Longman, Rees, Orme, Brown and Green, London.

Fay, J. A. 1980. Risks of LNG and LPG. *Annual Review of Energy* **5**:89–105.

Ferguson, E. S. 1971. The measurement of the "man-day." *Scientific Amer-*

ican **225**(4):96–103.

Fettweis, G. B. 1979. *World Coal Resources*. Elsevier, Amsterdam.

Flink, J. J. 1985. Innovation in automobile industry. *American Scientist* **73**:151–161.

Fluck, R. C. 1979. Energy productivity: a measure of energy utilisation in agricultural systems. *Agricultural Systems* **4**:29–37.

Fluck, R. C., and C. D. Baird. 1980. *Agricultural Energetics*. Avi Publishing, Westport, Conn.

Folk, G. E. 1976. *Textbook of Environmental Physiology*. Lea & Febiger, Philadelphia.

Food and Agriculture Organization. 1977. *The State of Food and Agriculture*. FAO, Rome.

Food and Agriculture Organization. 1980. *A Global Reconnaisance Survey of the Fuelwood Supply/Requirement Situation*. FAO, Rome.

Food and Agriculture Organization. 1988. *Production Yearbook*. FAO, Rome.

Forbes, R. J. 1958. Power to 1850. In C. Singer, et al., eds., *A History of Technology*, vol. 4. Clarendon Press, Oxford, pp. 148–167.

Forbes, R. J. 1964–1972. *Studies in Ancient Technology (IX vols.)*. E. J. Brill, Leiden.

Forbes, R. J. 1965. *Studies in Ancient Technology. Volume 2. Irrigation and Drainage*. E. J. Brill, Leiden.

Ford, K. W., et al., eds. 1975. *Efficient Use of Energy*. American Institute of Physics, New York.

Fortak, H. G. 1979. Entropy and climate. In W. Bach et al., eds., *Man's Impact on Climate*. Elsevier, Amsterdam, pp. 1–14.

Fowler, J. M. 1975. *Energy and the Environment*. McGraw-Hill, New York.

Fox, W., et al. 1976. *The Mill*. McClelland and Stewart, Toronto.

Foyer, C. H. 1984. *Photosynthesis*. Wiley, New York.

Francis, P. 1983. Giant volcanic calderas. *Scientific American* **246**:60–70.

Francis, W., and M. C. Peters, eds. 1980. *Fuels and Fuel Technology*. Pergamon, New York.

Freedman, J. L. 1980. Human reactions to population density. In M. N. Cohen et al., eds., *Biosocial Mechanisms of Population Regulation*. Yale University Press, New Haven, pp. 189–207.

Freeman, J. D. 1955. *Iban Agriculture*. Colonial Office, London.

Fremlin, J. H. 1987. *Power Production: What Are the Risks?* Oxford University Press, Oxford.

French, A. R. 1988. The patterns of mammalian hibernation. *American Scientist* **76**:569–575.

Fretwell, S. D. 1987. Food chain dynamics: The central theory of ecology? *Oikos* **50**:291–301.

Friedman, H. 1986. *Sun and Earth*. Scientific American Library, New York.

Frissel, M. J., and G. J. Kolenbrander. 1977. The nutrient balances. *Agro-Ecosystems* **4**:277–292.

Fröhlich, C. 1987. Variability of the solar "constant" on time scales of minutes to years. *Journal of Geophysical Research* **92**:796–800.

Fussell, G. E. 1972. *The Classical Tradition in West European Farming*. Fairleigh Dickinson University Press, Rutherford, N.J.

Galloway, J. N., et al. 1982. Trace metals in atmospheric deposition: A review and assessment. *Atmospheric Environment* **16**:1677–1700.

Galloway, P. R. 1986. Long-term fluctuations in climate and population in the preindustrial era. *Population and Development Review* **12**:1–24.

Garland, T. 1983. Scaling the ecological cost of transport to body mass in terrestrial mammals. *American Naturalist* **121**:571–587.

Gary, J. H., and G. E. Handwerk. 1984. *Petroleum Refining*. Marcel Dekker, New York.

Gates, D. M. 1980. *Biophysical Ecology*. Springer-Verlag, New York.

Gates, D. M. 1985. *Energy and Ecology*. Sinauer Associates, Sunderland, Mass.

Gautier, C., and M. Fieux, eds. 1984. *Large-Scale Oceanographic Experiments and Satellites*. D. Reidel, Dordrecht.

Gebhardt, M. R., et al., 1985. Conservation tillage. *Science* **230**:625–630.

Georgescu-Roegen, N. 1971. *The Entropy Law and the Economic Process*. Harvard University Press, Cambridge, Mass.

Gerking, S. D., ed. 1978. *Ecology of Freshwater Fish Production*. Wiley, New York.

Gersmehl, P. J. 1978. No-till farming: the regional applicability of a revolutionary agricultural technology. *The Geographical Review* **68**:66–79.

Gibbons, J. H., and W. U. Chandler. 1981. *Energy: The Conservation Revolution*. Plenum, New York.

Gibbons, J., and P. Blair. 1989. *Energy Efficiency: Its Potential and Limits to the Year 2000*. National Academy of Engineering, Washington, D.C.

Goela, J. S. 1979. Wind power through kites. *Mechanical Engineering* **101**(6):42–43.

Gold, T. 1987. *Power from the Earth: Deep Earth Gas—Energy for the Future*. J. M. Dent and Sons, London.

Golding, D., and R. E. Kasperson. 1988. Emergency planning and nuclear power. *Land Use Policy* **5**:19–36.

Golley, F. B. 1972. Energy flux in ecosystems. In J. A. Wiens, ed., *Ecosystem Structure and Function*. Oregon State University Press, Corvallis, pp. 69–88.

Golley, F. B., and E. Medina, eds. 1975. *Tropical Ecological Systems*. Springer-Verlag, New York.

Good, N., and D. H. Bell. 1980. Photosynthesis, plant productivity and crop yield. In P. S. Carlson, ed., *The Biology of Crop Productivity*. Academic Press, New York, pp. 3–51.

Goodman, G., and W. Rowe, eds. 1979. *Energy Risk Management*. Academic Press, London.

Gordon, R. L. 1981. *An Economic Analysis of World Energy Problems*. MIT Press, Cambridge, Mass.

Goudie, A. 1984. *The Nature of the Environment*. Basil Blackwell, Oxford.

Govindjee, ed. 1982. *Photosynthesis*. Academic Press, New York.

Greenewalt, C. H. 1975. The flight of birds. *Transactions of American Philosophical Society* **65**:1–67.

Greenwood, W. H. 1907. *Iron*. Cassell & Company, London.

Grigg, D. B. 1974. *The Agricultural Systems of the World*. Cambridge University Press, Cambridge.

Grossling, B. 1977. A critical survey of world petroleum opportunities. In *Project Interdependence: U.S. and World Energy Outlook Through 1990*. USGPO, Washington, D.C., pp. 645–658.

Gunston, B. 1986. *World Encyclopaedia of Aero Engines*. Patrick Stephens, Wellingborough.

Gustavson, M. R. 1979. Limits to wind power utilization. *Science* **204**:13–17.

Gutenberg, B., and C. F. Richter. 1942. Earthquake magnitude, intensity, energy and acceleration. *Bulletin Seismological Society of America* **32**: 163–191.

Häfele, W., and W. Sassin. 1977. The global energy system. *Annual Review of Energy* **2**:1–30.

Hairston, N., et al. 1960. Community structure, population control and competition. *American Naturalist* **94**:421–425.

Hall, D. O., et al. 1982. *Biomass for Energy in the Developing Countries*. Pergamon Press, Oxford.

Hall, P. 1984. *The World Cities*. St. Martin's Press, New York.

Hamakawa, Y. 1987. Photovoltaic power. *Scientific American* **256**(4):87–92.

Hamdan, G. 1970. *Shakhsiyyat Masr: A Study on the Genius of a Place*. Anglo-Egyptian Bookshop, Cairo.

Hanna, J. M., and D. E. Brown. 1983. Human heat tolerance: An anthropological perspective. *Annual Review of Anthropology* **12**:259–284.

Hanson, H., et al. 1982. *Wheat in the Third World*. Westview Press, Boulder, Colo.

Harako, R. 1981. The cultural ecology of hunting behavior among Mbuti Pygmies of the Ituri Forest, Zaire. In R. S. O. Harding, and G. Teleki, eds., *Omniverous Primates*. Columbia University Press, New York, pp. 499–555.

Harding, R. S. O., and G. Teleki, eds. 1981. *Omnivorous Primates*. Columbia University Press, New York.

Hardy, J. D., et al. 1971. Man. In G. C. Whittow, ed., *Comparative Physiology of Thermoregulation*. Academic Press, New York, pp. 327–380.

Harold, F. M. 1986. *The Vital Force: A Study of Bioenergetics*. W. H. Freeman, New York.

Harris, J. R. 1974. The rise of coal technology. *Scientific American* **233**(2):92–97.

Hassan, F. A. 1984. Environment and subsistence in Predynastic Egypt. In J. D. Clark and S. A. Brandt, eds., *From Hunters to Farmers*. University of California Press, Berkeley, pp. 57–64.

Haudricourt, A. G., and M. J. B. Delamarre. 1955. *L'Homme et la Charrue à travers le Monde*. Gallimard, Paris.

Hayden, B. 1981. Subsistence and ecological adaptations of modern hunter/gatherers. In R. S. O. Harding and G. Teleki, eds., *Omnivorous Primates*. Columbia University Press, New York, pp. 344–421.

Haywood, R. W. 1980. *Analysis of Engineering Cycles*. Pergamon Press, New York.

Head, J. W., and S. C. Solomon. 1981. Tectonic evolution of terrestrial planets. *Science* **213**:62–76.

Heal, D. W. 1975. Modern perspectives on the history of fuel economy in the iron and steel industry. *Ironmaking and Steelmaking* **2**:222–227.

Heichel, G. H. 1973. *Comparative Efficiency of Energy Use in Crop Production*. Connecticut Agricultural Experiment Station, New Haven, Conn.

Helland, J. 1980. *Five Essays on the Study of Pastoralists and the Development of Pastoralism*. Universitet i Bergen, Bergen.

Helmholtz, H. 1847. *Über die Erhaltung der Kraft*. G. Reimer, Berlin.

Helsel, Z., ed. 1987. *Energy in Plant Nutrition and Pest Control*. Elsevier, Amsterdam.

Hemmingsen, A. M. 1960. Energy metabolism as related to body size and respiratory surfaces, and its evolution. *Reports of Steno Memorial Hospital* **9**:1–110.

Hill, R. D. 1979. A survey of lightning energy estimates. *Review of Geophysics and Space Physics* **17**:155–164.

Hill, K., and A. M. Hurtado. 1989. Hunter-gatherers of the New World, *American Scientist* **77**:436–443.

Hinrichsen, D. 1986. Multiple pollutants and forest decline. *Ambio* **15**:258–265.

Hitchcock, R. K., and J. I. Ebert. 1984. Foraging and food production among Kalahari hunter/gatherers. In: J. D. Clark and S. A. Brandt, eds., *From Hunters to Farmers*. University of California Press, Berkeley, pp. 328–348.

Hively, W. 1988. Nuclear power at risk. *American Scientist* **76**:341–343.

Ho, Ping-ti. 1975. *The Cradle of the East*. Chinese University of Hong Kong Press, Hong Kong.

Hoffmann, P. 1981. *The Forever Fuel*. Westview Press, Boulder, Colo.

Hohenemser, C. 1988. The accident at Chernobyl: Health and environmental consequences and the implications for risk management. *Annual Review of Energy* **13**:383–428.

Holland, H. D. 1985. The oxygen content of our atmosphere. *Earth and Mineral Sciences* **55**(2):14–17.

Hopfen, H. J. 1969. *Farm Implements for Arid and Tropical Regions*. FAO, Rome.

Houghton, J. T. 1984. *The Global Climate*. Cambridge University Press, Cambridge.

Howard, J. R., ed. 1983. *Fluidized Bed Combustion and Applications*. Elsevier, Amsterdam.

Hu, S. D. 1983. *Handbook of Industrial Energy Conservation*. Van Nostrand Reinhold, New York.

Hubbard, H. M. 1989. Photovoltaics today and tomorrow. *Science* **244**:297–304.

Hubbert, M. K. 1962. *Energy Resources: A Report to the Committee on Natural Resources*. NAS, Washington, D.C.

Huettner, D. A. 1976. Net energy analysis: An economic assessment. *Science* **192**:101–104.

Huey, R. B., et al., eds. 1983. *Lizard Ecology*. Harvard University Press, Cambridge, Mass.

Hughes, T. P. 1983. *Networks of Power*. Johns Hopkins University Press, Baltimore.

Humphrey, W. S., and J. Stanislaw. 1979. Economic growth and energy consumption in the UK, 1700–1975. *Energy Policy* **7**:29–42.

Hyde, C. K. 1977. *Technological Change and the British Iron Industry 1700–1870*. Princeton University Press, Princeton, N.J.

Hytten, F. E. 1980. Nutrition. In F. E. Hytten and G. Chamberlain, eds., *Clinical Physiology in Obstetrics*. Basil Blackwell, Oxford, pp. 163–192.

Ikoku, C. U. 1984. *Natural Gas Production Engineering*. Wiley, New York.

International Atomic Energy Agency. 1984. *Risk and Benefit of Energy Systems*. IAEA, Vienna.

International Federation of Institutes for Advanced Study. 1974. *Energy Analysis Workshop on Methodology and Conventions*. IFIAS, Stockholm.

Isaacs, J. D., and W. R. Schmitt. 1980. Ocean energy: Forms and prospects. *Science* **207**:265–273.

Jäger, J. 1983. *Climate and Energy Systems*. Wiley, Chichester.

Janzen, D. H. 1970. Herbivores and the number of tree species in tropical forests. *American Naturalist* **104**:501–528.

Jevons, W. S. 1865. *The Coal Question*. Macmillan, London.

Johannsen, O. 1953. *Geschichte des Eisens*. Stahleisen, Düsseldorf.

Joint FAO/WHO/UNU Expert Consultation. 1985. *Energy and Protein Requirements*. WHO, Geneva.

Jordan, C. F., and R. Herrera. 1981. Tropical rain forests: Are nutrients really critical? *American Naturalist* **117**:167–180.

Joule, J. P. 1850. *On Mechanical Equivalent of Heat*. R. and J. E. Taylor, London.

Judson, S. 1968. Erosion of the land. *American Scientist* **56**:356–374.

Kanamori, H., and E. Boschi, eds. 1983. *Earthquakes: Observation, Theory and Interpretation*. North-Holland Publishing, Amsterdam.

Kapitza, P. 1976. Physics and the energy problem. *New Scientist* **72**(1021): 10–12.

Karr, J. R. 1975. Production, energy pathways, and community diversity in forest birds. In F. B. Golley and E. Medina, eds., *Tropical Ecological Systems*. Springer-Verlag, New York, pp. 161–177.

Kasting, J. F., et al. 1988. How climate evolved on the terrestrial planets. *Scientific American* **259**(2):90–97.

Keeling, C. D., et al. 1982. Measurements of the concentration of carbon dioxide at Mauna Loa Observatory, Hawaii. In W. C. Clark, ed., *Carbon Dioxide Review: 1982*. Oxford University Press, London, pp. 377–385.

Kelly, R. L. 1983. Hunter-gatherer mobility strategies. *Journal of Anthropological Research* **39**:277–306.

Kendeigh, S. C., et al. 1977. Avian energetics. In J. Pinowski and S. C. Kendeigh, eds., *Granivorous Birds in Ecosystems*, Cambridge University Press, Cambridge, pp. 127–204.

Kessler, A. 1985. *Heat Balance Climatology*. Elsevier, Amsterdam.

Khazanov, A. M. 1984. *Nomads and the Outside World*. Cambridge University Press, Cambridge.

King, C. D. 1948. *Seventy-five Years of Progress in Iron and Steel*. American Institute of Mining and Metallurgical Engineers, New York.

Kinne, O., ed. 1983, *Marine Ecology*. Wiley, Chichester.

Kirk, R. L. 1981. *Aboriginal Man Adapting*. Clarendon Press, Oxford.

Kittleman, L. R. 1979. Tephra. *Scientific American* **241**(6):160–177.

Kleiber, M. 1932. Body size and metabolism. *Hilgardia* **6**:315–353.

Kleiber, M. 1961. *The Fire of Life*. Wiley, New York.

Klemm, F. 1964. *A History of Western Technology*. MIT Press, Cambridge, Mass.

Klinge, H., et al. 1975. Biomass and structure in a Central Amazonian rain forest. In F. B. Golley and E. Medina, eds., *Tropical Ecological Systems*. Springer-Verlag, New York, pp. 115–122.

Koenig, L. R. 1979. *Anomalous Cloudiness and Precipitation Created by Industrial Heat Rejection*. Rand Corporation, Santa Monica, Calif.

Kreith, F., and R. T. Meyer. 1983. Large-scale use of solar energy with central receivers. *American Scientist* **71**:598–605.

Krug, E. C., and C. R. Frink. 1983. Acid rain on acid soil: A new perspective. *Science* **221**:520–525.

Landels, J. G. 1980. *Engineering in the Ancient World*. Chatto & Windus, London.

Landsberg, J. J. 1986. *Physiological Ecology of Forest Production*. Academic Press, New York.

Lanly, J-P. 1982. *Tropical Forest Resources*. FAO, Rome.

Layzer, D. 1988. Growth of order in the universe. In B. H. Weber et al., eds., *Entropy, Information, and Evolution*, The MIT Press, Cambridge, Mass., pp. 23–39.

Leach, G. 1975. *Energy and Food Production*. IIED, London.

Lee, R. B. 1979. *The !Kung San: Men, Women and Work in a Foraging Society*. Cambridge University Press, Cambridge.

Lee, R. B., and I. De Vore, eds. 1968. *Man the Hunter*. Aldine, Chicago.

Lemon, E. R., ed. 1983. *CO_2 and Plants*. Westview Press, Boulder, Colo.

Leser, P. 1931. *Entstehung und Verbreitung des Pfluges*. Aschendorff, Münster.

Leydon, K. 1987. So forecasting is easy! *Energy in Europe* **9**:17–25.

Liebig, J. 1843. *Die Chemie in ihrer Anwendung auf Agricultur und Physiologie*. F. Vieweg, Braunschweig.

Lieth, H. 1975. Modeling the primary productivity of the world. In H. Lieth and R. H. Whittaker, eds., *Primary Productivity of the Biosphere*. Springer-Verlag, New York, pp. 237–263.

Lieth, H., and R. H. Whittaker, eds. 1975. *Primary Productivity of the Biosphere*. Springer-Verlag, New York.

Likens. G. E., ed, 1981. *Some Perspectives of the Major Biogeochemical Cycles*. Wiley, Chichester.

Lindeman, R. 1942. The trophic-dynamic aspect of ecology. *Ecology* **23**:399–418.

Lindsay, J. 1974. *Blast-Power & Ballistics*. Harper & Row, New York.

Lizot, J. 1977. Population, resources and warfare among the Yanomami. *Man* **12**:497–517.

Lobban, C. S., and M. J. Wynne, eds. 1981. *The Biology of Seaweeds*. University of California Press, Berkeley.

Long, J. N. 1982. Productivity of Western coniferous forests. In R. L. Edmonds, ed., *Analysis of Coniferous Forest Ecosystems in the Western United States*. Van Nostrand Reinhold, New York, pp. 89–125.

Long, T. V., ed. 1978. *Energy Analysis Methodology*. U.S. Department of

Energy, Washington, D.C.

Lopreato, J. 1984. *Human Nature and Biocultural Evolution.* Unwin Hyman, London.

Lorenz, E. N. 1976. *The Nature and Theory of the General Circulation of the Atmosphere.* WMO, Geneva.

Lotka, A. 1925. *Elements of Physical Biology.* Williams and Wilkins, Baltimore.

Lovelock, J. E. 1979. *Gaia.* Oxford University Press, Oxford.

Lovins, A. B. 1976. Energy strategy: The road not taken. *Foreign Affairs* **55**(1):65–96.

Lovins, A. B. 1977. *Soft Energy Paths: Toward a Durable Peace.* Friends of the Earth and Ballinger, Cambridge, Mass.

Lowrance, R., et al., eds. 1984. *Agricultural Ecosystems.* Wiley, New York.

L'vovich, M. I. 1987. Ecological foundations of global water resources conservation. In D. J. McLaren and B. J. Skinner, eds., *Resources and World Development.* Wiley, Chichester, pp. 831–839.

MacArthur, R., and E. O. Wilson. 1967. *The Theory of Island Biogeography.* Princeton University Press, Princeton, N.J.

Macdonald, K. C., and B. P. Luyendyk. 1981. The crest of the East Pacific Rise. *Scientific American* **244**(5):100–116.

Macdonald, K. C., and P. J. Fox. 1990. The mid-ocean ridge. *Scientific American* **262**(6):72–79.

Mace, G. M., et al. 1983. Vertebrate home-range size and energetic requirements. In I. R. Swingland and D. J. Greenwood, eds., *The Ecology of Animal Movement*, Clarendon Press, Oxford, pp. 33–53.

Mach, E. 1896. *Die Prinzipien der Wärmelehre historisch-kritisch entwickelt.* J. A. Barth, Leipzig.

Macriss, R. A. 1983. Space heating by gas and oil. *Annual Review of Energy* **8**:247–267.

Malmberg, T. 1980. *Human Territoriality.* Mouton, The Hague.

Mandelbrot, B. 1977. *Fractals.* W. H. Freeman, San Francisco.

Manibog, F. R. 1984. Improved cooking stoves in developing countries. *Annual Review of Energy* **9**:199–227.

Mann, K. H. 1984. Fish production in open ocean ecosystems. In M. J. R. Fasham, ed., *Flows of Energy and Materials in Marine Ecosystems*, Plenum, New York, pp. 435–458.

Marchetti, C., and N. Nakicenovic. 1979. *The Dynamics of Energy Systems and the Logistic Substitution Model.* International Institute for Applied Systems Analysis, Laxenburg.

Martinez-Alier, J. 1987. *Ecological Economics.* Basil Blackwell, New York.

Maxwell, J. C. 1872. *Matter and Motion.* D. van Nostrand, New York.

Mayer, J. R. 1851. *Bemerkungen über das mechanische Aequivalent der*

Wärme. J. V. Landherr, Heilbronn.

McCullough, D. R. 1973. Secondary production of birds and mammals. In D. E. Reichle, ed., *Analysis of Temperate Forest Ecosystems*, Springer-Verlag, New York, pp. 107–130.

McGilvery, R. W. 1975. The use of fuels for muscular work. In H. Howard and J. R. Portmans, eds., *Metabolic Adaptation to Prolonged Physical Exercise.* Birkhäuser Verlag, Basel, pp. 12–30.

McInnes, W., et al., eds. 1913. *The Coal Resources of the World.* Morang & Company, Toronto.

McLaren, D. J., and B. J. Skinner, eds. 1987. *Resources and World Development.* Wiley, Chichester.

McMahon, T. 1973. Size and shape in biology. *Science* **179**:1201–1204.

McMahon, T., and J. T. Bonner. 1983. *On Size and Life.* Scientific American Library, New York.

Meadows, D. H., et al. 1972. *The Limits to Growth.* Universe Books, New York.

Meentemeyer, V., et al. 1982. World patterns and amounts of terrestrial plant litter production. *BioScience* **32**:125–128.

Mendelssohn, K. 1974. *The Riddle of the Pyramids.* Thames and Hudson, London.

Mermel, T. W. 1988. Major dams of the world—1988. *International Water Power & Dam Construction Handbook 1988*, pp. 47–65.

Merriam, M. F. 1977. Wind energy for human needs. *Technology Review* **79**(3):29–39.

Merriam, M. F. 1978. Wind, waves, and tides. *Annual Review of Energy* **3**:29–56.

Mesarovic, M. D. 1961. *Systems Research and Design.* Wiley, New York.

Meyers, R. A., ed. 1981. *Coal Handbook.* Marcel Dekker, New York.

Mielke, J. E. 1977. Environmental trade-offs of energy supply options. In *Project Interdependence*, Congressional Research Service, Washington, D.C., pp. 488–555.

Miller, A. I. 1981. *Albert Einstein's Special Theory of Relativity.* Addison-Wesley, Reading, Mass.

Miller, D. H. 1981. *Energy at the Surface of the Earth.* Academic Press, New York.

Minchinton, W., and P. Meigs. 1980. Power from the sea. *History Today* **30**(3):42–46.

Minge-Klewana, W. 1980. Does labor time decrease with industrialization? A survey of time-allocation studies. *Current Anthropology* **21**:278–298.

Mirowski, P. 1988. *More Heat than Light.* Cambridge University Press, New York.

Mitchell, B. R. 1975. *European Historical Statistics 1750–1970.* Columbia

University Press, New York.

Molenaar, A. 1956. *Water Lifting Devices for Irrigation.* FAO, Rome.

Monod, T., ed. 1975. *Pastoralism in Tropical Africa.* Oxford University Press, Oxford.

Monteith, J. L. 1978. Reassessment of maximum growth rates for C_3 and C_4 crops. *Experimental Agriculture* **14**:1–5.

Moritz, L. A. 1958. *Grain-Mills and Flour in Classical Antiquity.* Oxford University Press, Oxford.

Morone, J. G., and E. J. Woodhouse. 1986. *Averting Catastrophe: Strategies for Regulating Risky Technologies.* University of California Press, Berkeley.

Morowitz, H. J. 1968. *Energy Flow in Biology.* Academic Press, New York.

Morowitz, H. J. 1978. *Foundations of Bioenergetics.* Academic Press, New York.

Morris, I., ed. 1980. *The Physiological Ecology of Phytoplankton.* University of California Press, Berkeley.

Motor Vehicle Manufacturers Association. 1988. *Facts & Figures '88.* MVMA, Detroit.

Mudahar, M. S., and T. P. Hignett. 1981. *Energy and Fertilizer.* IFDC, Muscle Shoals, Ala.

Murdock, G. P. 1967. Ethnographic atlas. *Ethnology* **6**:109–236.

Nadel, E. R. 1985. Physiological adaptations to aerobic training. *American Scientist* **73**:334–343.

Nader, L., and S. Beckerman. 1978. Energy as it relates to the quality and style of life. *Annual Review of Energy* **3**:1–28.

National Academy of Sciences. 1975. *Productivity of World Ecosystems.* NAS, Washington, D.C.

National Academy of Sciences. 1977. *Energy and Climate.* NAS, Washington, D.C.

National Academy of Sciences. 1980a. *Energy and the Fate of Ecosystems.* NAS, Washington, D.C.

National Academy of Sciences. 1980b. *Firewood Crops.* NAS, Washington, D.C.

National Academy of Sciences. 1980c. *Materials Aspects of World Energy Needs.* NAS, Washington, D.C.

National Academy of Sciences. 1983. *Acid Deposition: Atmospheric Processes in Eastern North America.* NAS, Washington, D.C.

National Academy of Sciences. 1986. *Acid Deposition: Long-Term Trends.* NAS, Washington, D.C.

National Research Council. 1984. *Nutrient Requirements of Beef Cattle.* NAS, Washington, D.C.

National Research Council. 1986. *Electricity in Economic Growth.* NAS,

Washington, D.C.

Needham, J. 1954–. *Science and Civilisation in China* (VI vols.). Cambridge University Press, Cambridge.

Needham, J. 1965. *Science and Civilisation in China. Volume 4, Part II. Physics and Physical Technology.* Cambridge University Press, Cambridge.

Nef, J. U. 1932. *The Rise of the British Coal Industry.* Routledge, London.

Nef, J. U. 1957. Coal mining and utilization. In C. Singer et al., eds., *A History of Technology,* Volume 3. Clarendon Press, Oxford, pp. 73–83.

Nehring, R. 1978. *Giant Oil Fields and World Oil Resources.* Rand Corporation, Santa Monica, Calif.

Norgan, N. G., et al. 1974. The energy and nutrient intake and the energy expenditure of 204 New Guinean adults. *Philosophical Transactions of the Royal Society* **B268**:309–348.

O'Brien, W. J., et al. 1990. Search strategies of foraging animals. *American Scientist* 78:152–160.

Odell, P. R., and K. E. Rosing. 1983. *The Future of Oil.* Kogan Page, London.

Odum, H. T. 1971. *Environment, Power, and Society.* Wiley, New York.

Odum, H. T. 1975. Energy analysis and net energy. In *Report of the NSF–Stanford Workshop on Net Energy Analysis.* Institute for Energy Studies, Stanford, Calif., pp. 90–115.

Odum, H. T. 1984. Energy analysis of the environmental role in agriculture. In G. Stanhill, ed., *Energy and Agriculture,* Springer-Verlag, Berlin, pp. 24–51.

Odum, H. T. 1988. Self-organization, transformity, and information. *Science* **242**:1132–1139.

Office of Technology Assessment. 1987. *Starpower: The US and the International Quest for Fusion Energy.* OTA, Washington, D.C.

Okamoto, S. 1984. *Introduction to Earthquake Engineering.* University of Tokyo Press, Tokyo.

Okigbo, B. N. 1984. Improved production systems as an alternative to shifting intermittent cultivation. In *Improved Production Systems as an Alternative to Shifting Cultivation,* FAO, Rome, pp. 1–100.

Oktay, S., et al. 1986. High heat from a small package. *Mechanical Engineering* **108**(3):36–42.

Olson, J. S. 1963. Energy storage and the balance of producers and decomposers in ecological systems. *Ecology* **44**:322–331.

Olson, J. S. 1982. Earth's vegetation and atmospheric carbon dioxide. In W. C. Clark, ed., *Carbon Dioxide Review: 1982.* Oxford University Press, London, pp. 388–398.

Oltenacu, P. A., and M. S. Allen. 1980. Resource-cultural energy requirements of the dairy production system. In Pimentel, D., ed. *op.cit.*, pp. 363–378.

O'Neill, R. V., and D. L. De Angelis. 1981. Comparative productivity and biomass relations of forest ecosystems. In D. E. Reichle, ed., *Dynamic Properties of Forest Ecosystems*, Cambridge University Press, Cambridge, pp. 411–450.

Openshaw, K. 1978. Woodfuel—A time for re-assessment. *Natural Resources Forum* **3**:35–51.

Organization for Economic Cooperation and Development. 1977. *The OECD Programme on Long Range Transport of Air Pollutants.* OECD, Paris.

Organization for Economic Cooperation and Development. 1982a. *The Energy Problem of the Agro-food Sector.* OECD, Paris.

Organization for Economic Cooperation and Development. 1982b. *Eutrophication of Waters: Monitoring, Assessment and Control.* OECD, Paris.

Ostrander, C. E. 1980. Energy use in agriculture poultry. In Pimentel, D., ed., *op.cit.*, pp. 379–392.

Ostwald, W. 1909. *Energetische Grundlagen der Kulturwissenschaften.* Alfred Kröner, Leipzig.

Parsons, T. R., et al. 1984. *Biological Oceanographic Processes.* Pergamon Press, Oxford.

Partl, R. 1977. *Power from Glaciers: The Hydropower Potential of Greenland's Glacial Waters.* International Institute for Applied Systems Analysis, Laxenburg.

Pasqualetti, M. J., and B. A. Miller. 1984. Land requirements for the solar and coal options. *Geographical Journal* **150**:192–212.

Paul, C. H. 1983. *Irrigation.* The Irrigation Association, Arlington, Va.

Pedley, T. J., ed. 1977. *Scale Effects in Animal Locomotion.* Academic Press, New York.

Peng, S. S., and H. S. Chiang. 1984. *Longwall Mining.* Wiley, New York.

Perkins, D. H. 1969. *Agricultural Development in China, 1368–1968.* Aldine, Chicago.

Perrodon, A. 1985. *Histoire des Grandes Decouvertes Petrolieres.* Elf Aquitaine, Paris.

Perry, H., and S. H. Streiter. 1977. *Multiple Paths for Energy Policy.* National Economic Research Associates, Washington, D.C.

Persson, T., ed. 1980. *Structure and Function of Northern Coniferous Forests.* Ecological Bulletins, Stockholm.

Peters, R. H. 1983. *The Ecological Implications of Body Size.* Cambridge University Press, Cambridge.

Petzet, G. A. 1988. Operators make wider use of horizontal drilling technology. *Oil & Gas Journal* **86**(15):15–17.

Phillips, R. E., et al. 1980. No-tillage agriculture. *Science* **208**:1108–1113.

Pierce, G. J., and J. G. Ollason. 1987. Eight reasons why optimal foraging theory is a complete waste of time. *Oikos* **49**:111–117.

Pimentel, D., ed. 1980. *Handbook of Energy Utilization in Agriculture.* CRC Press, Boca Raton, Fla.

Pimentel, D., et al. 1973. Food production and energy crisis. *Science* **182**:443–449.

Pochin, E. 1984. *Nuclear Radiation Risks and Benefits.* Oxford University Press, Oxford.

Prentice, A. M. 1984. Adaptations to long-term low energy intake. In E. Pollitt and P. Amante, eds., *Energy Intake and Activity.* Alan R. Liss, New York, pp. 3–31.

Price, T. D., and J. A. Brown, eds. 1985. *Prehistoric Hunter-Gatherers.* Academic Press, Orlando, Fla.

Prigogine, I. 1947. *Etude thermodynamique des phenomenes irreversibles.* Dunod, Paris.

Pritchard, S. Z. 1987. *Oil Pollution Control.* Croom Helm, London.

Protzen, J.-P. 1986. Inca stonemasonry. *Scientific American* **254**(2):94–105.

Pryde, P. R. 1985. Land requirements for solar electricity alternatives. In F. J. Calzonetti and B. D. Solomon, eds., *Geographical Dimensions of Energy.* D. Reidel, Dordrecht, pp. 255–275.

Pryor, F. L. 1983. Causal theories about the origin of agriculture. *Research in Economic History* **8**:93–124.

Putnam, P. C. 1954. *Energy in the Future.* Van Nostrand, New York.

Ramsey, J. B. 1981. *The Economics of Exploration for Energy Resources.* JAI Press, Greenwich, Conn.

Rand, P. J. 1982. *Land and Water Issues Related to Energy Development.* Ann Arbor Science Publishers, Ann Arbor, Mich.

Rapp, A. 1960. Recent developments of mountain slopes in Karkevagge and surroundings, northern Scandinavia. *Geografiska Annaler* **42**:71–200.

Rappaport, R. A. 1968. *Pigs for the Ancestors.* Yale University Press, New Haven.

Ratcliffe, K. 1985. *Liquid Gold Ships: History of the Tanker (1859–1984).* Lloyds, London.

Rawitscher, M., and J. Mayer. 1977. National outputs and energy inputs in seafood. *Science* **198**:261–264.

Rayner, J. M. V. 1981. Flight adaptations in vertebrates. *Symposia of the Zoological Society of London* **48**:137–172.

Reaven, S. J. 1984. The concept of net energy. *Explorations in Knowledge* **1**:191–231.

Reed, C. A., ed. 1977. *Originals of Agriculture.* Mouton, the Hague.

Reichle, D. E., ed. 1981. *Dynamic Properties of Forest Ecosystems.* Cambridge University Press, Cambridge.

Reid, J. T., et al. 1980. Cultural energy, land, and labor requirements of swine production systems in the U.S. In Pimentel, D., ed., *op.cit.*, pp. 393–403.

Reitz, J. R. 1985. Potential for energy savings in old and new auto engines. In D. Hafemeister et al., eds., *Energy Sources: Conservation and Renewables*. American Institute of Physics, New York, pp. 368–381.

Reynolds, J. 1970. *Windmills and Watermills*. Hugh Evelyn. London.

Ricci, P. F., and L. S. Molton. 1986. Health risk assessment: Science, economics, and law. *Annual Review of Energy* **11**:77–94.

Richter, C. F. 1935. An instrumental earthquake magnitude scale. *Bulletin Seismological Society of America* **25**:1–32.

Ridley, B. K. 1979. *The Physical Environment*. Ellis Horwood, West Sussex.

Riva, J. P. 1983. *World Petroleum Resources and Reserves*. Westview Press, Boulder, Colo.

Robbins, C. T. 1983. *Wildlife Feeding and Nutrition*. Academic Press, New York.

Roberts, P. C. 1981. *Energy and Value*. Institute for Energy Analysis, Oak Ridge. Tenn.

Rochereau, S. P. 1980. The energy requirements for inshore and offshore fishing crafts—the case of the Northeast fishery. In Pimentel, D., ed., *op.cit.*, pp. 441–446.

Rogin, L. 1931. *The Introduction of Farm Machinery*. University of California Press, Berkeley.

Rose, D. J. 1986. *Learning about Energy*. Plenum, New York.

Rose, D. J. 1974. Nuclear eclectic power, *Science* **184**:351–359.

Rosenberg, N. J., et al. 1983. *Microclimate*. Wiley, New York.

Rosenfeld, A. H., and D. Hafemeister. 1988. Energy-efficient buildings. *Scientific American* **258**(4):78–85.

Ross, D. 1979. *Energy from Waves*. Pergamon Press, Oxford.

Rouse, J. E. 1970. *World Cattle*. University of Oklahoma Press, Norman.

Royal Dutch-Shell. 1983. *Petroleum Handbook*. Elsevier, Amsterdam.

Rubin, E. S. 1989. Implications of future environmental regulation of coal-based electric power. *Annual Review of Energy* **14**:19–45.

Rubner, M. 1902. *Nahrungsmittel- und Ernährungskunde*. E. H. Moritz, Stuttgart.

Rühlmann, G. 1962. *Kleine Geschichte der Pyramiden*. Verlag der Kunst, Dresden.

Ryder, H. W., et al. 1976. Future performance in footracing. *Scientific American* **234**(6):109–119.

Sacher, E. 1881. *Grundzüge einer Mechanik der Gesselschaft*. Gustav

Fischer, Jena.

Sagan, L. A. 1987. What is hormesis and why haven't we heard about it before? *Health Physics* **52**:521–525.

Salzman, P. C., ed. 1981. *Contemporary Nomadic and Pastoral Peoples: Asia and the North.* College of William and Mary, Williamsburg, Va.

Sampl, F. R., and M. E. Shank. 1985. Aircraft turbofans: New economic and environmental benefits. *Mechanical Engineering* **107**(9):47–53.

Schaeffer, J. T., et al. 1980. Tornado track characteristics and hazard probabilities. In J. E. Cermak, ed., *Wind Engineering*, Pergamon Press, Oxford, pp. 95–109.

Schaller, G. B. 1972. *The Serengeti Lion.* University of Chicago Press, Chicago.

Scheidegger, A. E. 1982. *Principles of Geodynamics.* Springer-Verlag, Berlin.

Schipper, L., et al. 1985. Explaining residential energy use by international bottom-up comparisons. *Annual Review of Energy* **10**:341–405.

Schmidt-Nielsen, K. 1979. *Desert Animals.* Dover, New York.

Schmidt-Nielsen, K. 1984. *Scaling: Why Is Animal Size So Important?* Cambridge University Press, Cambridge.

Schneider, S. H. 1989. *Global Warming.* Sierra Club, San Francisco.

Schnell, R. C., ed. 1986. *Geophysical Monitoring for Climatic Change.* U.S. Department of Commerce, Boulder, Colo.

Schrödinger, E. 1944. *What is Life?* Cambridge University Press, Cambridge.

Schofield, W. N., et al. 1985. Basal metabolic rate in man: Survey of the literature and computation of equations for prediction. *Human Nutrition Clinical Nutrition* **39**(Supplement 1):1–96.

Scholander, P. F., et al. 1958. Cold adaptation in Australian Aborigines. *Journal of Applied Physiology* **14**:605–615.

Schroeder, H. 1919. Die jährliche Gesamtproduktion der grünen Pflanzendecke der Erde. *Naturwissenschaften* **7**:8–12.

Schurr, S. H. 1984. Energy use, technological change, and productive efficiency: An economic-historical interpretation. *Annual Review of Energy* **9**:409–425

Schurr, S. H., and B. C. Netschert. 1960. *Energy in the American Economy 1850–1975.* Johns Hopkins University Press, Baltimore.

Sclater, J. G., et al. 1980. The heat flow through oceanic and continental crust and the heat loss of the Earth. *Reviews of Geophysics and Space Physics* **18**:269–311.

Scott, W. 1972. HV and EHV transmission planning. *Energy International* **9**(7):27–30.

Seavoy, R. E. 1986. *Famine in Peasant Societies.* Greenwood Press, New York.

Selby, M. J. 1982. *Hillslope Materials and Processes*. Oxford University Press, Oxford.

Selby, M. J. 1985. *Earth's Changing Surface*. Clarendon Press, Oxford.

Service, E. R. 1979. *The Hunters*. Prentice-Hall, London.

Shannon, C. E., and W. Weaver, 1949. *The Mathematical Theory of Information*. University of Illinois Press, Urbana, Ill.

Shea, J. H., ed. 1985. *Plate Tectonics*. Van Nostrand Reinhold, New York.

Sheehan, G. W. 1985. Whaling as an organizing focus in Northwestern Eskimo society. In T. D. Price and J. A. Brown, eds., *Prehistoric Hunter–Gatherers*. Academic Press, Orlando, Fla., pp. 123–154.

Shephard, R. J. 1978. *Human Physiological Work Capacity*. Cambridge University Press, Cambridge.

Shurcliff, W. A. 1986. Superinsulated houses. *Annual Review of Energy* **11**:1–24.

Silberbauer, G. B. 1981. *Hunter and Habitat in the Central Kalahari Desert*. Cambridge University Press, Cambridge.

Singer, C., et al., eds. 1954–1984. *A History of Technology* (VIII vols.). Clarendon Press, Oxford.

Singh, R. P., ed. 1986. *Energy in Food Processing*. Elsevier, Amsterdam.

Sleep, N. H., and R. T. Langan. 1981. Thermal evolution of the Earth: Some recent developments. *Advances in Geophysics* **23**:1–37.

Slesser, M. 1978. *Energy in the Economy*. Macmillan, London.

Slicher van Bath, B. H. 1963. *The Agrarian History of Western Europe, AD 500–1850*. Arnold, London.

Slingo, J. 1982. A study of the earth's radiation budget using a general circulation model. *Quarterly Journal of the Royal Meteorological Society* **108**:379–405.

Smil, V. 1983. *Biomass Energies*. Plenum, New York.

Smil, V. et al. 1983. *Energy Analysis and Agriculture*. Westview Press, Boulder, Colo.

Smil, V. 1984. *The Bad Earth*. M. E. Sharpe, Armonk, N.Y.

Smil, V. 1985. *Carbon Nitrogen Sulfur Human Interferences in Grand Biospheric Cycles*. Plenum, New York.

Smil, V. 1986. Eating better: farming reforms and food in China. *Current History* **84**:248–251, 273–274.

Smil, V. 1987. *Energy Food Environment: Realities Myths Options*. Oxford University Press, Oxford.

Smil, V. 1987. Fossil-fuelled civilization and the atmosphere: How much should we worry? In D. J. McLaren and B. J. Skinner, eds., *Resources and World Development*. Wiley, Chichester, pp. 363–375.

Smil, V. 1988. *Energy in China's Modernization*. M. E. Sharpe, Armonk, N.Y.

Smith, D. R. 1987. The wind farms of the Altamont Pass area. *Annual Review of Energy* **12**:145–183.

Smith, E. 1857. Inquiries into the quantity of air inspired throughout the day and night under the influence of exercise, food, medicine, temperature, etc. *Philosophical Magazine* **14**:546–572.

Smith, N. 1980. The origins of the water turbine. *Scientific American* **242**(1):138–148.

Socolow, R. H. 1977. The coming age of conservation. *Annual Review of Energy* **2**:239–289.

Socolow, R. H. 1985. The physicist's role in using energy efficiently. In D. Hafemeister et al., eds., *Energy Sources: Conservation and Renewables*, American Institute of Physics, New York, pp. 15–32.

Soddy, F. 1912. *Matter and Energy*. Henry Holt, New York.

Soddy, F. 1926. *Wealth, Virtual Wealth and Debt: The Solution of the Economic Paradox*. Dutton, New York.

Sonenblum, S. 1978. *The Energy Connections between Energy and the Economy*. Ballinger, Cambridge, Mass.

Spady, D. W., et al. 1976. Energy balance during recovery from malnutrition. *American Journal of Clinical Nutrition* **29**:1073–1088.

Spencer, J. E. 1966. *Shifting Cultivation in Southeastern Asia*. University of California Press, Berkeley.

Sperling, D. 1988. *New Transportation Fuels: A Strategic Approach to Technological Change*. University of California Press, Berkeley, Calif.

Splinter, W. E. 1976. Center-pivot irrigation. *Scientific American* **234**(6):90–99.

Spotila, J. R., and D. M. Gates. 1975. Body size, insulation and optimum body temperatures of homeotherms. In D. M. Gates and R. B. Schmerl, eds., *Perspectives of Biophysical Ecology*, Springer-Verlag, New York, pp. 291–301.

Spreng, D. T. 1978. *On Time, Information, and Energy Conservation*. Institute for Energy Analysis, Oak Ridge, Tenn.

Stallo, J. B. 1900. *Concepts and Theories of Modern Physics*. D. Appleton & Co., New York.

Stanford Research Institute. 1971. *Patterns of Energy Consumption in the United States*. SRI, Menlo Park, Calif.

Stanhill, G., ed. 1984. *Energy and Agriculture*. Springer-Verlag, Berlin.

Stanley, S. M. 1987. *Extinction*. Scientific American Library, New York.

Starr, C., et al. 1976. Philosophical basis for risk analysis. *Annual Review of Energy* **1**:629–662.

Stephens, D. W., and J. R. Krebs. 1987. *Foraging Theory*. Princeton University Press, Princeton, N.J.

Stern, A. C., ed. 1976–1986. *Air Pollution*. 8 vols. Academic Press, New York.

Stevenson, R. D. 1985. Body size and limits to the daily range of body temperature in terrestrial ectotherms. *American Naturalist* **125**:102–117.

Store, L. G., and J. A. Teeri. 1978. The geographical distribution of C_4 species of the Dicotyledonae in relation to climate. *American Naturalist* **112**:609–623.

Stout, B. A. 1979. *Energy for World Agriculture.* FAO, Rome.

Stumm, W., ed. 1977. *Global Chemical Cycles and Their Alterations by Man.* Dahlem Konferenzen, Berlin.

Subcommittee on Horse Nutrition. 1978. *Nutrient Requirements of Horses.* NAS, Washington, D.C.

Summers, R., and A. Heston. 1984. Improved international comparisons of real product and its composition: 1950–1980. *Review of Income and Wealth* **30**:207–262.

Swedish Ministry of Agriculture. 1982. *Acidification Today and Tomorrow.* Swedish Ministry of Agriculture, Stockholm.

Tanaka, J. 1980. *The San Hunter-Gatherers of the Kalahari.* University of Tokyo Press, Tokyo.

Tansley, A. G. 1935. The use and abuse of vegetational concepts and terms. *Ecology* **16**:284–307.

Taylor, J. W. R., ed. 1986. *Jane's All the World's Aircraft.* Jane's Publications, London.

Teal, J. M., and R. W. Howarth. 1984. Oil spill studies: A review of ecological effects. *Environmental Management* **8**:27–44.

Technocracy, Inc. 1937. *The Energy Certificate.* Technocracy, Inc., New York.

Testart, A. 1982. The significance of food storage among hunter-gatherers: Residence patterns, population densities, and social inequalities. *Current Anthropology* **23**:523–537.

Thekaekara, M. P. 1977. Solar irradiance, total and spectral. In A. A. M. Sayigh, ed., *Solar Energy Engineering.* Academic Press, New York, pp. 37–59.

Thibau, C. E. 1978. *Industrial Charcoal from Mixed Tropical Forests.* Paper presented at the 8th World Forestry Congress, Jakarta.

Thirring, H. 1958. *Energy for Man.* Indiana University Press, Bloomington.

Thomas, J., ed. 1979. *Energy Analysis.* Westview Press, Boulder, Colo.

Thomson, W. 1853. On the dynamical theory of heat. *Royal Society of Edinburgh Transactions* **20**:261–298, 475–482.

Tillman, D. A. 1978. *Wood as an Energy Resource.* Academic Press, New York.

Tilton, J. E., and B. J. Skinner, 1987. The meaning of resources. In D. J. McLaren and B. J. Skinner, eds., *Resources and World Development.* Wiley, Chichester, pp. 13–27.

Tiratsoo, E. N., ed. 1980. *Natural Gas*. Gulf Publishing, Houston.

Titus, J. G. 1986. *Effects of Changes in Stratospheric Ozone and Global Change*. U.S. Environmental Protection Agency, Washington, D.C.

Tompkins, P. 1971. *Secrets of the Great Pyramid*. Harper & Row, New York.

Tont, S. A., and R. A. Delistraty. 1977. Food resources of the ocean. *Environmental Conservation* **4**:243–252.

Torr, G. 1964. *Ancient Ships*. Argonaut Publishers, Chicago.

Torrey, V. 1976. *Wind-Catchers. American Windmills of Yesterday and Tomorrow*. Stephen Greene Press, Brattleboro, Vt.

Tracey, M. V. 1977. Human nutrition. In R. Duncan and M. Weston-Smith, eds., *Encyclopaedia of Ignorance*. Pergamon Press, Oxford, pp. 355–360.

Travis, C., and E. L. Etnier, eds. 1982. *Health Risks of Energy Technologies*. Westview Press, Boulder, Colo.

Trenkle, A., and R. L. Willham. 1977. Beef production efficiency. *Science* **198**:1009–1015.

Trowell, H. C., et al. 1985. *Dietary Fibre, Fibre-Depleted Foods and Disease*. Academic Press, London.

Tsai, S. C. 1982. *Fundamentals of Coal Beneficiation and Utilization*. Elsevier, Amsterdam.

Tucker, V. A. 1973. Bird metabolism during flight: Evaluation of a theory. *Journal of Experimental Biology* **58**:689–709.

Ubbelohde, A. R. 1954. *Man and Energy*. Hutchinson, London.

Uman, M. A. 1982. *Lightning*. Dover, New York.

UNESCO. 1978. *Tropical Forest Ecosystems*. UNESCO, Paris.

Unger, R. 1984. Energy sources for the Dutch Golden Age. *Research in Economic History* **9**:221–253.

Unger, P. W., and T. M. McCalla. 1980. Conservation tillage systems. *Advances in Agronomy* **33**:1–58.

United Nations Organization. 1956. World energy requirements in 1975 and 2000. In *Proceedings of the International Conference on the Peaceful Uses of Atomic Energy*, Volume 1. UNO, New York, pp. 3–33.

United Nations Organization. 1976. *World Energy Supplies 1950–1974*. UNO, New York.

United Nations Organization. 1980–. *Yearbook of World Energy Statistics*. UNO, New York.

United Nations Scientific Committee on the Effects of Atomic Radiation. 1982. *Ionizing Radiation: Sources and Biological Effects*. United Nations, New York.

United States Department of Agriculture. 1980. *Energy and U.S. Agriculture: 1974 and 1978*. USDA, Washington, D.C.

U.S. Bureau of the Census. 1975. *Historical Statistics of the United States.* U.S. Department of Commerce, Washington, D.C.

Valiela, I. 1984. *Marine Ecological Processes.* Springer-Verlag, New York.

Van der Post, L., and J. Taylor. 1984. *Testament to the Bushmen.* Viking, New York.

Van Hylckama, T. E. A. 1979. Water, something peculiar. *Hydrological Sciences* **24**(4):499–507.

Van Winkle, T. L., et al. 1978. Cotton versus polyester. *American Scientist* **66**:280–290.

Van Valen, L. 1973. Body size and numbers of plants and animals. *Evolution* **27**:27–35.

Verbraeck, A., ed. 1976. *The Energy Accounting of Materials, Products, Processes and Services.* TNO (Netherlands Institute for Applied Scientific Research), Rotterdam.

Verhoogen, J. 1980. *Energetics of the Earth.* NAS, Washington, D.C.

Vernadsky, V. I. 1929. *La Biosphère.* Alcan, Paris.

Veziroglu, T. N., and N. Getoff, eds. 1986 *Hydrogen Energy Progress, Volume VI.* Pergamon Press, New York.

von Braun, W., and F. I. Ordway. 1975. *History of Rocketry and Space Travel.* Thomas Y. Crowell, New York.

von Neumann, J. 1955. Can we survive technology? *Fortune* **51**(6):106–107, 151–152.

von Tunzelmann, G. N. 1978. *Steam Power and British Industrialization to 1860.* Clarendon Press, Oxford.

Wadsworth, G. 1984. *The Diet and Health of Isolated Populations.* CRC Press, Boca Raton, Fla.

Ward, C. R., ed. 1984. *Coal Geology and Coal Technology.* Blackwell Scientific, Melbourne.

Waterbury, J. 1979. *Hydropolitics of the Nile Valley.* Syracuse University Press, Syracuse, N.Y.

Watt, B. K., and A. L. Merrill. 1975. *Handbook of the Nutritional Contents of Foods.* Dover Publications, New York.

Watters, R. F. 1971. *Shifting Cultivation in Latin America.* FAO, Rome.

Weast, R. C., ed. 1988. *CRC Handbook of Chemistry and Physics.* CRC Press, Boca Raton, Fla.

Weber, B. H., et al., eds. 1988. *Entropy, Information, and Evolution.* The MIT Press, Cambridge, Mass.

Webster, P. J. 1981. Monsoons. *Scientific American* **245**(2):108–118.

Weinberg, A. M., and R. P. Hammond. 1970. Limits to the use of energy. *American Scientist* **58**:412–418.

Westoby, M. 1984. The self-thinning rule. *Advances in Ecological Research* **14**:167–225.

White, D. A. 1987. Conventional oil and gas resources. In D. J. McLaren and B. J. Skinner, eds., *Resources and World Development*. Wiley, Chichester, pp. 113–128.

White, K. D. 1970. *Roman Farming*. Thames & Hudson, London.

White, L. 1978. *Medieval Religion and Technology*. University of California Press, Berkeley.

White-Stevens, R. 1977. Perspectives on fertilizer use, residue utilization and food production. In *Food, Fertilizer and Agricultural Residues*, R. C. Loehr, ed., Ann Arbor Science Publishers, Ann Arbor, Mich. pp. 5–26.

Whitt, F. R., and D. G. Wilson. 1982. *Bicycling Science*. MIT Press, Cambridge, Mass.

Whittaker, R. H., and G. E. Likens. 1975. The biosphere and man. In H. Lieth and R. H. Whittaker, eds., *Primary Productivity of the Biosphere*. Springer-Verlag, New York, pp. 305–328.

Widdowson, E. M. 1983. How much food does man require? In J. Mauron, ed., *Nutritional Adequacy, Nutrient Availability and Needs*. Birkhäuser Verlag, Basel, pp. 11–25.

Williams, E. R. 1988. The electrification of thunderstorms. *Scientific American* 259(5):88–99.

Williams, Q., et al. 1987. The melting curve of iron to 250 gigapascals: A constraint on the temperature of Earth's center. *Science* 236:181–182.

Williams, R. H. 1978. Industrial cogeneration. *Annual Review of Energy* 3:313–356.

Wilshire, H., and D. Prose. 1987. Wind energy development in California, USA. *Environmental Management* 11:13–20.

Wilson, D. C. 1979. Energy conservation through recycling, *Energy Research* 3:307–323.

Wilson, P. N. 1956. *Watermills: An Introduction*. Times Printing, Mexborough.

Winterhalder, B., and E. A. Smith, eds. 1981. *Hunter-Gatherer Foraging Strategies*. University of Chicago Press, Chicago.

Wischmeier, W. H., and D. D. Smith. 1978. *Predicting Rainfall Erosion Losses*. U.S. Department of Agriculture, Washington, D.C.

Wolff, A. R. 1900. *The Windmill as Prime Mover*. Wiley, New York.

World Bank. 1988. *World Development Report 1988*. World Bank, Washington, D.C.

World Energy Conference. 1974. *Survey of Energy Resources*. World Energy Conference, London.

World Energy Conference. 1986. *Survey of Energy Resources*. World Energy Conference, London.

Wyndham, C. H. 1969. Adaptation to heat and cold. *Environmental Research* 2:442–469.

Yesner, D. R. 1980. Maritime hunter-gatherers: Ecology and prehistory. *Current Anthropology* **21**:727–750.

Yorks, T. P., et al. 1980. Energy minimised vs. cost minimised alternatives for the US beef production system. *Agricultural Systems* **6**:121–129.

Zachar, D. 1982. *Soil Erosion*. Elsevier, Amsterdam.

Zvelebil, M. 1986. Postglacial foraging in the forests of Europe. *Scientific American* **254**(5):104–115.

Zweibel, K., and P. Hersch. 1984. *Basic Photovoltaic Principles and Methods*. Van Nostrand Reinhold, New York.

INDEX